THE TWENTY-SECOND CHESAPEAKE SAILING YACHT SYMPOSIUM

THE INTERNATIONAL COMMUNITY FOR MARITIME AND OCEAN PROFESSIONALS

I0468775

CSYS

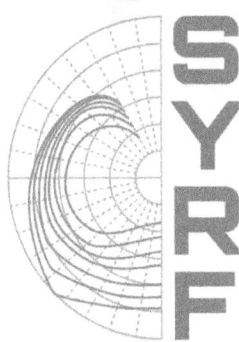

March 18 –19, 2016
Annapolis, Maryland, USA

Presented by:

The Society of Naval Architects and Marine Engineers

Sailing Yacht Research Foundation

THE 22nd CHESAPEAKE SAILING YACHT SYMPOSIUM
ANNAPOLIS, MARYLAND, MARCH 18 – 19th, 2016

TABLE OF CONTENTS

Papers Presented on Friday March 18th, 2016

Pressure Measurements on Yacht Sails: Development of a new system for wind tunnel and full scale testing

Fabio Fossati, Ilmas Bayati, Sara Muggiasca, Ambra Vandone, Department of Mechanical Engineering, Politecnico di Milano, Milano, Italy.
Gabriele Campanardi, Department of Aerospace Science and Technology, Politecnico di Milano, Milano, Italy
Thomas Burch, CSEM, Centre Suisse d'Electronique et de Microtechnique SA, Alpnach Dorf, Switzerland
Michele Malandra, North Sails Group, Carasco, Italy

Modal Analysis of Pressures on a Full Scale Spinnaker

Julien Deparday, Naval Academy Research Institute, Brest, France
Patrick Bot, Naval Academy Research Institute, Brest, France
Fréderic Hauville, Naval Academy Research Institute, Brest, France
Benoit Augier, Naval Academy Research Institute, Brest, France
Marc Rabaud, Laboratoire FAST, Univ.Paris-Sud, CNRS, Université Paris-Saclay, F-91405, Orsay,France
Dario Motta, University of Auckland, New-Zealand
David Le Pelley, University of Auckland, New-Zealand

Effect of Dynamic Trimming on Upwind Sail Aerodynamics in a Wind Tunnel

Aubin N., Naval academy research Institut - IRENAV, France 1
Augier B., Naval academy research Institut - IRENAV, France
Bot P., Naval academy research Institut - IRENAV, France
Hauville F., Naval academy research Institut - IRENAV, France
Sacher M., Naval academy research Institut - IRENAV, France
Flay R. G. J., Yacht Research Unit, Department of Mechanical Engineering, The University of Auckland, New Zealand

THE 22nd CHESAPEAKE SAILING YACHT SYMPOSIUM
ANNAPOLIS, MARYLAND, MARCH 18 – 19th, 2016

TABLE OF CONTENTS

Papers Presented on Saturday March 19th, 2016

THE 22ⁿᵈ CHESAPEAKE SAILING YACHT SYMPOSIUM
ANNAPOLIS, MARYLAND, MARCH 18 – 19th, 2016

Steering Committee:

Event Chair:
Jaye Falls BSc, MSc, PhD
United States Naval Academy

Naval Academy Liaison:
Paul Miller, D.Eng., P.E.
United States Naval Academy

Papers Committee Chair:
Britton Ward, BSc, MSc
Farr Yacht Design, Ltd.

Event Coordination:
Alana Anderson
Society of Naval Architects and Marine
Engineers

Promotion/Arrangements:
Dobbs Davis, BSc, MSc, PhD
Sailing Yacht Research Foundation

Website/IT Support:
Mat Bird, BSc
DRS Advanced Marine Technology Center

Technical Review Committee:

Britton Ward BSc, MSc
Farr Yacht Design, Ltd.
britton@farrdesign.com

Jaye Falls, BSc, MSc, PhD
United States Naval Academy

William Lasher BSE, MSE, PhD
Penn State Behrend, Erie, PA
lasher@psu.edu

Gregory Buley BSc, MSc
CDI Marine, Band Lavis Division
Greg.Buley@cdicorp.com

Frank W. DeBord Jr. PE, BE, ME
USCG Surface Forces Logistics Center

Jim Schmicker BSc, MSc
Farr Yacht Design, Ltd.
jim@farrdesign.com

Chris Cochran BSc, MSc
Farr Yacht Design, Ltd.
chris@farrdesign.com

Myles Cornwell, BSc, MSc
Persak & Wurmfeld

David Kring, BSc, PhD
Navatek Ltd.

Jonathan Binns, BE (hons), MSc, PhD
Australian Maritime College, Uni. of Tasmania
jrbinns@amc.edu.au

David Helgerson, BSc
David Helgerson & Associates, LLC
dh@dhelgerson.com

Patrick Bot, PhD
Ecole Navale

David Le Pelley, ME
Doyle Sails

Richard Royce BSIM, BSE, MS, MSE, PhD
Webb Institute

George Hazen, BSc, MSc
DRS Advanced Marine Technology Center

Robert Ranzenbach, BSc, PhD
Quantum Sails
robert@quantumsails.com

Giorgio Provinciali, BSc, ME
Hydros Innovations SA

THE 22nd CHESAPEAKE SAILING YACHT SYMPOSIUM

ANNAPOLIS, MARYLAND, MARCH 18 - 19, 2016

Sponsors

THE INTERNATIONAL COMMUNITY FOR MARITIME AND OCEAN PROFESSIONALS

The Society of Naval Architects and Marine Engineers
99 Canal Center Plaza, Suite 310
Alexandria, VA 22314
www.sname.org

Sailing Yacht Research Foundation
1643 Warwick Avenue, Box 300
Warwick, Rhode Island 02889
www.sailyachtresearch.org

The Twenty-Second CSYS was held on March 18-19, 2016
The papers were presented in Rickover Hall
Located on the campus of the United States Naval Academy
Annapolis, Maryland, USA.

ABSTRACTS IN ORDER OF PRESENTATION

Prediction and Optimization of Aerodynamic Forces and Boat Speed of Foiling Catamaran with a Rig of a Rigid Wing and a Jib

K. Graf, H. Renzsch, J. Meyer - University of Applied Sciences Kiel, Germany

This paper describes a method to calculate the aerodynamic forces generated by a rigid two-element wing together with a jib. Additionally, investigations of hydrodynamic flow forces generated by water-piercing L-shaped foils are introduced. The aerodynamic and hydrodynamic flow force prediction methods are combined in a velocity prediction program featuring a constraint optimization method in order to predict boat speed and wing and foil trimming parameters for its maximization.

A velocity polar calculated by applying this method to a 50-foot catamaran is shown and the result of some studies are presented, varying design parameters of the catamaran.

A comparison of a RANS based VPP to on the water Sailing Performance

T. Doyle, B. Knight, D. Swain - Doyle CFD

This paper compares performance predictions from a Reynolds Averaged Navier Stokes (RANS) based Velocity Prediction Program (VPP) to on the water testing of a J70. The J70 has been outfitted with a system to determine sail flying shapes, apparent wind conditions and performance data. The on the water testing is conducted in both racing and controlled sailing conditions. Data taken during racing conditions is analyzed to determine optimal performance envelopes while data taken in controlled conditions is used to match exact sailing and VPP states. The data acquisition system combines a number of standard marine sensors including a sonic anemometer, a GPS, a digital compass, an accelerometer and a gyroscope with custom sensors that measure rudder and boom angles as well as a custom sail shape acquisition system. The RANS based VPP developed by Doyle CFD has three main components; an aerodynamic force model, a hydrodynamic force model and an algorithm to balance the forces. The force balance routine uses four degrees of freedom; boat speed, yaw, heel and rudder angle to balance the aerodynamic and hydrodynamic forces for a given true wind speed and angle. The force models are derived from RANS CFD data calculated using OpenFOAM. The aerodynamic forces are calculated using steady state RANS as a function of apparent wind angle, apparent wind speed and sail flying shape. The VPP force model is derived by fitting response surfaces to this data. The aerodynamic CFD is run with sail flying shapes recorded from on the water testing. Using accurate flying shapes is critical for picking out slight aerodynamic differences in sail and rig setup. The hydrodynamic CFD data points are calculated using RANS Volume of Fluid CFD (VOF) as a function of boat speed, rudder angle, yaw angle, heel angle and displacement. Response surfaces are generated from a 64 data point array of RANS VOF simulations.

Experimental and numerical trimming optimizations for a mainsail in upwind conditions

M. Sacher, F. Hauville, R. Duvigneau, O. Le Maitre, N. Aubin, M. Durand - IRENAV France

This paper investigates the use of meta-models for optimizing sails trimming. A Gaussian process is used to robustly approximate the dependence of the performance with the trimming parameters to be optimized. The Gaussian process construction uses a limited number of performance observations at carefully selected trimming points, potentially enabling the optimization of complex sail systems with multiple trimming parameters. We test the optimization procedure on the (two parameters) trimming of a scaled IMOCA mainsail in upwind conditions. To assess the robustness of the Gaussian process approach, in particular its sensitivity to error and noise in the performance estimation, we contrast the direct optimization of the

physical system with the optimization of its numerical model. For the physical system, the optimization procedure was fed with wind tunnel measurements, while the numerical modeling relied on a fully non-linear Fluid-Structure Interaction solver. The results show a correct agreement of the optimized trimming parameters for the physical and numerical models, despite the inherent errors in the numerical model and the measurement uncertainties. In addition, the number of performance estimations was found to be affordable and comparable in the two cases, demonstrating the effectiveness of the approach.

Fully Integrated fluid-structural analysis for the design and performance optimization of fiber reinforced sails
S. Malpede, D. MacVicar, F. Nasato, P. Semeraro - SMAR Azure, UK

This paper presents an advanced and accurate integrated system for the design and performance optimization of fiber reinforced sails -commonly named string sails- developed by SMAR Azure Ltd. This integrated design system allows sail designers not only to design sail-shapes and the reinforcing fiber paths, but also to validate the performance of the flying sail-shape and have accurate production details including the overall sail weight, material used, which means costs, and length of the fiber paths, which means production time.

The SMAR Azure design and analysis method includes a validated and computationally efficient structural analysis method coupled with a modified vortex lattice method, with wake relaxation, to enable a proper aero-elastic simulation of sails in upwind conditions. The structural analysis method takes into account the geometric non-linearity and wrinkling behavior of membrane structures –such as sails-, the fiber layout, the influence of battens, trimming loads and interaction with rigging elements, e.g. luff sag calculation on a headstay, in a timely manner. This method has been extensively validated and used to optimize several racing and super-yachts sailing plans. Specifically, this paper presents a validated optimization of a real fiber reinforce membrane sail plan of 140', 240 ton aluminium Super Yacht, carried out in collaboration with Paolo Semeraro (from Banks Sails Europe), who designs and produces the MEMBRANE™ and BFAST™ string sails, the latter with Marco Semeraro. Both BFAST and Bank Sails have been using the SMAR Azure technology for almost a decade and notwithstanding the long experience of Mr. Semeraro in using the technology, given the sailplan-size and detailed customer requirements, among which improved durability, strength and reliability and smooth use of in-boom furling, this project was carried out in-cooperation with the SMAR Azure technical team. A total of 1000 sqm of upwind sailing area was analyzed and optimized. A combination of Dyneema TM Sk 90 and black Twaron 2200 was chosen for the fibers and a triple step lamination under hi-pressure plus laminated patches utilizing the same fibers where added to prevent local deformation of the corners. A long term vacuumed post-curing period sealed the production phases. The final sail plan is -as anticipated by the analysis results- holds the desired shape and is stronger. The final fiber layout shows a reduction in maximum stress by 22% compared to the initial design; this was achieved with only 11% (4kg) gain in fiber weight.

Unsteady Sail Dynamics due to Bodyweight motions
R.Schutt, C.H.K. Williamson - Cornell University

In small sailboats, the bodyweight of the sailor is proportionately large enough to induce significant unsteady dynamics of the boat and sail. Sailors use a variety of techniques to create sail dynamics which can provide an increment in thrust, increasing the boatspeed. In this study, we experimentally investigate the unsteady aerodynamics associated with two such techniques, "upwind leech flicking" and "downwind S-turns". We employ a two-part approach.

First, on-the-water experiments are carried out using a Laser class sailboat sailed by Olympic and world championship level sailors. Data collected from an on-board GPS, IMU, anemometer, and camera array is used to generate characteristic motions of the boat and sail relative to the apparent wind.

Second, laboratory experiments using the characteristic motion of the sail are run in a computer-controlled 3 degree-of-freedom (X, Y, and θ) towing tank. We use water as the working fluid. Rather than directly experiment with three-dimensional sail shapes, we represent the primary effects of the sail dynamics using rapidly prototyped two-dimensional flexible sail geometries. Shapes are based on extruded draft stripes from the upper third of the sail. The laboratory experiments approximately match the key non-dimensional parameters of the on-the-water sailing conditions, including the reduced frequency and heave-to-chord ratio. Particle Image Velocimetry and force measurements are used to analyze the flow field and thrust generated by the model sail during the dynamic motions.

On-the-water testing shows that the characteristic sail motion in leech flicking is a combination of periodic heave caused by the actions of the sailor and a passive twisting of the sail due to rig flexibility. The heaving sail motions are due to rotation (roll) of the rig around the longitudinal axis of the hull. This is at an angle to the apparent wind, resulting in heave that has components both perpendicular and parallel to the oncoming wind flow. This is distinct from classical aerodynamic studies with heave purely perpendicular to the incoming flow.

In laboratory experiments, the characteristic flicking motion is applied to a NACA 0012 airfoil and a 2D sail, both angled at 15 deg to the flow. Lift increases and drag decreases, leading to an overall increase in resultant driving force of the boat. The beneficial effect of this dynamic motion becomes greater as the apparent wind angle increases. In the case of leech flicking, the experiments show that the formation of vortex pairs is fundamental to the augmented thrust due to heaving.

The presence of S-turns, whereby the sailor changes the boats direction simultaneous with rolling the boat, generally in the downwind direction, is also associated with vortex formation and pairing, which will be described at the conference. During downwind S-turns, large amplitude heaving motions are paired with substantial rotations of the sail caused by both adjustments of the main sheet and changes in heading. Increased velocity made good downwind is measured from the on-the-water experiments, and is associated with an increase of thrust during characteristic dynamics of the airfoil or sail shape in the laboratory.

A Comparison of RANS and LES for Upwind Sailing Aerodynamics
S. Nava, S. Norris, J. Cater – University of Auckland

Yacht sails experience complex aerodynamics that are challenging to reproduce with numerical methods. The experiments of Fluck on an idealized upwind sail plan [1] [2] showed areas of flow separation and vortex structures that subsequent Reynolds Averaged Navier-Stokes (RANS) calculations have struggled to correctly simulate [3] [4]. Therefore it is of interest to see if other methods, such as Large Eddy Simulation (LES), are able to more accurately predict this flow. To this end the experiments of Fluck have been reproduced in Fluent, using RANS and LES. Both RANS and LES accurately model the attached flow on the lower region of both sails, but RANS fails in capturing the separation bubble occurring at the top section of the mainsail, predicting it as either too short, non-existent, or extended for the entire chord length. However, LES correctly models the leading edge separation bubble with its reattachment point located approximately halfway along the sail chord. The LES calculations are sensitive to the free stream turbulence intensity with an increase in the turbulence intensity from 3% to 12% shortening the separation bubble by half. Accurately reproducing the experimental geometry also improves the numerical results, with modelling the computational domain as a simple box or as the real wind tunnel results in different flow, with the angle of attack varying by 3°. In conclusion, LES, even if it is very sensitive to the mesh and the choice of the solver, is more accurate than a steady RANS calculation, although the computational time increases by a factor of 50.

Pressure Measurements on Yacht Sails: Development of a new system for wind tunnel and full scale testing

F. Fossati, I. Bayati, S. Muggiasca, A. Vandone, G Campanardi, T Burch, M. Malandra -Politecnico di Milano, Italy

The paper presents an overview of a joint project developed among Politecnico di Milano, CSEM and North Sails, aiming at developing a new sail pressure measurement system based on MEMS sensors (an excellent compromise between size, performance, costs and operational conditions) and pressure strips and pads technology. These devices were designed and produced to give differential measurement between the leeward and windward side of the sails. The project has been developed within the Lecco Innovation Hub Sailing Yacht Lab, a 10 m length sailing dynamometer which intend to be the reference contemporary full scale measurement device in the sailing yacht engineering research field, to enhance the insight of sail steady and unsteady aerodynamics [1].

The pressure system is described in details as well as the data acquisition process and system metrological validation is provided; furthermore, some results obtained during a wind tunnel campaign carried out at Politecnico di Milano Wind Tunnel, as a benchmark of the whole measuring system for future full scale application, are reported and discussed in details.

Moreover, the system configuration for full scale testing, which is still under development, is also described.

Modal Analysis of Pressures on a full scale spinnaker

J. Deparday, P. Bot, F. Hauville, B. Augier, M. Rabaud, D. Motta, D. Le Pelley - Naval Academy Research Institute, France

While sailing offwind, the trimmer typically adjusts the downwind sail "on the verge of luffing", letting occasionally the luff of the sail flapping. Due to the unsteadiness of the spinnaker itself, maintaining the luff on the verge of luffing needs continual adjustments. The propulsive force generated by the offwind sail depends on this trimming and is highly fluctuating. During a flapping sequence, the aerodynamic load can fluctuate by 50% of the average load. On a J/80 class yacht, we simultaneously measured time resolved pressures on the spinnaker, aerodynamic loads, boat and wind data. Significant spatio-temporal patterns are detected in the pressure distribution. In this paper we present averages and main fluctuations of pressure distributions and of load coefficients for different apparent wind angles as well as a refined analysis of pressure fluctuations, using the Proper Orthogonal Decomposition (POD) method. POD shows that pressure fluctuations due to luffing of the spinnaker can be well represented by only one proper mode related to a unique spatial pressure pattern and a dynamic behavior evolving with the Apparent Wind Angles. The time evolution of this proper mode is highly correlated with load fluctuations. Moreover, POD can be employed to filter the measured pressures more efficiently than basic filters. The reconstruction using the first few modes allows to restrict to the most energetic part of the signal and remove insignificant variations and noises. This might be helpful for comparison with other measurements and numerical simulations.

Effect of dynamic trimming on upwind sail aerodynamics in a wind tunnel

N. Aubin, B. Augier, P. Bot, F. Hauville, M.Sacher, R. G. J. Flay - IRENAV France, University of Auckland

An experiment was developed at the Yacht research Unit's Twisted Flow Wind Tunnel (University of Auckland) to test the effect of dynamic trimming on three 60 IMOCA inspired main sails models in upwind configuration. This study presents dynamic fluid structure interaction results in well controlled conditions

(wind, sheet length) with a dynamic trimming system. First the optimum optimization target CF_{obj} coefficient with a steady trim for AWA = 60 degrees using the car traveler position and main sail sheet length is located. Oscillation are then done around this optimum value using the main sheet length L_{sheet} oscillation. Different oscillation amplitudes and frequencies of trimming are investigated. Measurements are done with a 6 components force balance and a load sensor giving access to the unsteady main sail sheet load. The driving CF_x and optimization target CF_{obj} coefficient first decrease at low reduced frequency f_r for quasi-steady state then increase, becoming higher than the steady state situation. The driving force CF_x and the optimization target coefficient CF_{obj} show an optimum for the three different design sail shapes located at $f_r = 0:255$. This optimum is linked to the power transmitted to the rig and sail system by the trimming device. The effect of the camber of the design shape is investigated too. The flat mainsail design benefits more than the other mainsail designs from the dynamic trimming compared to their respective steady situation. This study presents dynamic results that cannot be accurately predicted with a steady approach. These results are therefore valuable for future FSI numerical tools validation in unsteady conditions.

Towards a new mathematical model for investigating course stability and maneuvering motions of sailing yachts
E. Angelou, K. Spyrou - National Technical University, Athens, Greece

In order to create capability for analyzing course instabilities of sailing yachts in waves, the authors are at an advanced stage of development of a mathematical model comprised of two major components: an aerodynamic, focused on the calculation of the forces on the sails, taking into account the variation of their shape under wind flow; and a hydrodynamic one, handling the motion of the hull with its appendages in water.

Regarding the first part, sails provide the aerodynamic force necessary for propulsion. But being very thin, they have their shape adapted according to the locally developing pressures. Thus, the flying shape of a sail in real sailing conditions differs from its design shape and it is basically unknown. The authors have tackled the fluid-structure interaction problem of the sails using a 3d approach where the aerodynamic component of the model involves the application of the steady form of the Lifting Surface Theory, in order to obtain the force and moment coefficients, while the deformed shape of each sail is obtained using a relatively simple Shell Finite Element formulation. The hydrodynamic part consists of modeling hull reaction, hydrostatic and wave forces.

A Potential Flow Boundary Element Method is used to calculate the Side Forces and Added Mass of the hull and its appendages. The Side Forces are then incorporated into an approximation method to calculate Hull Reaction terms. The calculation of resistance is performed using a formulation available in the literature. The wave excitation is limited to the calculation of Froude - Krylov forces.

The SYRF Wide Light Project
M. Prince – Sailing Yacht Research Foundation/Wolfson Unit

Modern racing yacht semi-planing hull forms provide a number of complex challenges for designers and a minefield for those involved in yacht rating.

The SYRF Wide Light Project was initiated as a means of providing data with which to assess a range of alternative computation methodologies to analyze sailing yacht hydrodynamic forces and moments, making this data available to the entire sailing yacht research community and demonstrating how this type of study can be used to inform the rating process.

This paper presents a comprehensive set of tank test results in both canoe body only and appended configurations to be used as a benchmark for a defined geometry of a modern semi-planing hull.
Five different CFD stakeholders carried out 'blind' CFD analysis on the same test matrix using a range of different computational codes and approaches. The results are presented here along with feedback detailing the software, methods and resources used to generate the results.

This project offers a comprehensive set of public domain data which researchers may use to validate and develop their numerical tools and highlights how successfully commercial CFD codes may be used to confidently predict the variation of the forces on a sailing yacht hull as speed, heel and leeway change.
Finally, discussion will be made on how this first phase of the project may be used to inform handicap rule makers.

Numerical Simulations of a Surface Piercing A-Class Catamaran Hydrofoil and Comparison against model tests
T. Keller, J. Hendrichs, K. Hochkirch, A. C. Hochbaum - TU Berlin, DNV GL

Hydrofoil supported sailing vessels gained more and more importance with in the last years. Due to new processes of manufacturing it is possible to build slender section foils with low drag coefficients and heave stable hydrofoil geometries are becoming possible to construct. These surface piercing foils often tend to ventilate and cavitate at high speeds. The aim of this work is to define a setup to calculate the hydrodynamic forces on such foils with RANSE CFD and to investigate whether the onset of ventilation and cavitation can be predicted.

Therefore a surface piercing hydrofoil of an A-Class catamaran is simulated by using the RANSE software FineMarine with its volume of fluid method. The C-shaped hydrofoil is analysed for one speed at Froude Number 7.9 and various angles of attack (AoA). The rake was defined and a leeway angle was applied to simulate realistic set ups. It is presented how the rake and drift angle influence the lift to drag ratio. Over a wide range of AoA there is no ventilation predicted but cavitation may occur from AoA > 10°. Due to the very small aspect ratio (Λ=2.64), the maximum AoA before stall is increased. The simulations have been verified by extensive analyses, including domain size verification for unrestricted water, mesh refinement and y+ verification. The influence of the K27 (cavitation tunnel of the Technical University of Berlin) on the flow around the hydrofoil and the wave system is presented. It is shown how test section of the K27 influences the flow around the foil, the forces and the wave elevation. Finally the CFD results are compared against the experiments conducted in the K27.

Advanced CFD Simulations of free-surface flows around modern sailing yachts using a newly developed openFOAM solver
J. Meyer, H, Renzsch, K. Graf, T. Slawig - University of Applied Sciences Kiel, Germany

While plain vanilla OpenFOAM has strong capabilities with regards to quite a few typical CFD-tasks, some problems actually require additional bespoke solvers and numerics for efficient computation of high-quality results. One of the fields requiring these additions is the computation of large-scale free-surface flows as found e.g. in naval architecture. This holds especially for the flow around typical modern yacht hulls, often planing, sometimes with surface-piercing appendages. Particular challenges include, but are not limited to, breaking waves, sharpness of interface, numerical ventilation (aka streaking) and a wide range of flow phenomenon scales. A new OF-based application including newly implemented discretization schemes, gradient computation and rigid body motion computation is described. In the following the new code will be validated against published experimental data; the effect on accuracy, computational time and solver stability will be shown by comparison to standard OF-solvers (interFoam / interDyMFoam) and Star-

CCM+. The code's capabilities to simulate complex "real-world" flows are shown on a well-known racing yacht design.

Insights from the Load Monitoring Program for the 2014-2015 Volvo Ocean Race
S. Russell, G. Vanhollebeke, P. Manganelli – Gurit Composite Components

This paper describes insights into keel and rigging loads obtained through a data acquisition system fitted on the fleet of Volvo 65 yachts during the 2014-2015 Volvo Ocean Race. In the first part keel fin stress spectra are derived from traces of canting keel ram pressures and keel angle; these are reviewed and compared against equivalent spectra obtained by applying methods proposed by Det Norske Veritas - Germanischer Lloyd ("DNVGL") guidelines and the ISO 12215 standard. The differences between stress spectra and their validity are discussed, considering two types of keel: milled from a monolithic cast of steel, and fabricated from welded metal sheets. The second part discusses predicted and actual rigging working loads for the Volvo 65 yachts, considering how safety factors vary between design loads proposed by DNVGL and actual recorded loads.

Influence of sailor position and motion on the performance prediction of racing dinghies
J. Taylor, J. Banks, M. Toward, D. Taunton, S. Turnock - University of Southampton, UK

The time-varying influence of a sailor's position is typically neglected in dinghy velocity prediction programs. When applied to the assessment of dinghy race performance the position and motions of the crew become significant but are practically hard to measure as they interact with the motions of the sailboat. As the initial stage in developing a time accurate dinghy velocity prediction program this work develops an on-water system capably of measuring the applied hiking moment due to the sailor's pose and compares this with the resultant dinghy motion. The sailor's kinematics are captured using a network of inertial motion sensors (IMS) synchronized to a video camera and dinghy motion sensor. The hiking moment is evaluated using a 'stick man' body representation with the mass and inertial terms associated with the main body segments appropriately scaled for the representative sailor. The accuracy of the pose capture is validated using laboratory based pose measurements. The completed work will provide a platform to model how sailor generated forces interact with the sailboat to affect boat speed. This will be used alongside realistic modelling of the wind and wave loadings to extend an existing time-domain dynamic velocity prediction program (DVPP). The results are demonstrated using a single handed Laser and demonstrate an acceptable level of accuracy.

Development of a routing software for Inshore Match Races
F. Tagliaferri, I. M. Viola - Newcastle University UK

Yacht races are won by good sailors racing fast boats. A good skipper takes decisions at key moments of the race based on the anticipated wind behavior and on his position on the racing area and with respect to the competitors. His aim is generally to complete the race before all his opponents, or, when this is not possible, to perform better than some of them. In the past two decades some methods have been proposed to compute optimal strategies for a yacht race. Those strategies are aimed at minimizing the expected time needed to complete the race and are based on the assumption that the faster a yacht, the higher the number of races that it will win (and opponents that it will defeat). In a match race, however, only two yachts are competing. A skipper's aim is therefore to complete the race before his opponent rather than completing the race in the shortest possible time. This means that being on average faster may not necessarily mean winning the majority of races. This papers present the development of software to compute a sailing strategy for a match race that can defeat an opponent who is following a fixed strategy that minimizes the

expected time of completion of the race. The proposed method includes two novel aspects in the strategy computation:

- A short-term wind forecast, based on an Artificial Neural Network (ANN) model, is performed in real time during the race using the wind measurements collected on board.
- Depending on the relative position with respect to the opponent, decisions with different levels of risk aversion are computed. The risk attitude is modeled using Coherent Risk Measures.

The software is tested in a number of simulated races. The results confirm that maximizing the probability of winning a match race does not necessarily correspond to minimizing the expected time needed to complete the race.

Teamwork as Joint Activity in Sailing

F. Forsman, C. Finnsgard - Chalmers University of Technology, Sweden

Sailing is a sport and activity that takes a long time both to learn and to master, as much of its competence-based knowledge is acquired through experience. Experience based learning is very important time-intensive, and the factors for success are often tacit and hidden. Should these success factors become explicit and salient, learning would occur faster and produce obvious competitive advantages.

This research was conducted by embedding on-going research results into two competitive sailing teams racing in different classes, one offshore keelboat racing with a crew of 8, and a one-design Star-class racing yacht with a crew of two. The data collection consisted of observations, interviews, and video recordings. The results were also verified with the crews to catch biases in the analysis process. A jibe, a specific but common maneuver was analyzed from the perspective of Common Ground within Joint Activity.

Maneuvering a competitive offshore sail racer or a previously Olympic Star-class yacht are tasks that fulfill the requirements for Joint Activity. A high level of Common Ground is required for the effective coordination needed in order to perform at a high level and maintain the safety of the crew and equipment.

Breakdowns in the coordination of maneuvers were observed, although they must be recorded on video for higher analysis reliability. To achieve greater validity, more and different maneuvers should be considered within the analysis.

By better understanding the factors for success, sail racing teams can more quickly gain competence and thus competitive advantages.

The research analyzes the teamwork found in sailing from the perspective of Joint Activity and Common Ground and provides insight into how to achieve performance improvements more efficiently.

Prediction and optimization of aerodynamic and hydrodynamic forces and boat speed of foiling catamarans with a wing sail and a jib

Kai Graf, University of Applied Sciences Kiel, Germany
Hannes Renzsch, Fluid Engineering Solutions GmbH, Schleswig, Germany
Janek Meyer, Yacht Research Unit Kiel, Germany

ABSTRACT

This paper describes a method to calculate the aerodynamic forces generated by a rigid two-element wing together with a jib. Additionally, investigations of hydrodynamic flow forces generated by water-piercing L-shaped foils are introduced. The aerodynamic and hydrodynamic flow force prediction methods are combined in a velocity prediction program featuring a constraint optimization method in order to predict boat speed and wing and foil trimming parameters for its maximization.

A velocity polar calculated by applying this method to a 50-foot catamaran is shown and the result of some studies are presented, varying design parameters of the catamaran.

NOTATION

CSYS	Chesapeake Sailing Yacht Symposium
$\mathbf{A}$	Start point of vortex filament, $\mathbf{A}=\{x_A, y_A, z_A\}^T$
AoA	Angle of Attack (-)
A_{RW}	Area of rudder wing (m²)
AWS	Apparent wind speed (m/s)
AWA	Apparent wind angle (-)
$\beta, beta$	Flap angle (-)
c	Profile Length (m)
c_D	Drag coefficient (-)
c_L	Lift coefficient (-)
$Dind$	Induced drag (N)
D_{AX}, D_{AY}	Drag area in x- and y-direction (m²)
ds	Integrator (m)
F_{AX}, F_{AY}	Windage force in x- and y-direction (N)
$\mathbf{F}$	$=\{F_X, F_Y, F_Z\}^T$, global force (N)
$\mathbf{L}$	Lift (N)
L	$|\mathbf{L}|$ (N)
L_{VRW}	Vertical lift of rudder wing (N)
$\mathbf{l}, \mathbf{t}$	profile leading and trailing edge, $\mathbf{l}=\{x_l, y_l, z_l\}^T$
$\mathbf{M}$	$=\{M_X, M_Y, M_Z\}^T$, global moment (Nm)

$\mathbf{P}$	Collocation point, $\mathbf{P}=\{x_P, y_P, z_P\}^T$
R_I	Induced resistance (N)
R_T	Total resistance of hull (N)
R_{RW}	Resistance of rudder wing (N)
$\mathbf{r}$	Vector from point to integrator $d\mathbf{s}$ (m)
r	$|\mathbf{r}|$ (m)
$\mathbf{s}$	vector along filament (m)
S_{RW}	Span of rudder wing (m)
s_{Eff}	Effective span (m)
T_{Eff}	Effective draft or span (m)
t/c	Jib profile depth over chord length (-)
$\mathbf{u}$	Flow Velocity vector (m/s)
u_B	Boat speed
$\mathbf{v}$	Induced Velocity vector (m/s)
v	Induced wind speed (m/s)
v_i	Induced wind speed at profile i (m/s)
v_j^*	Induced wind speed at panel j (m/s)
w_{FVF}	Lower free vortex filament weighting (-)
x,y,z	Coordinates in space (m)
y^+	Dimensionless wall distance (-)
z	Vertical span-wise coordinate (m)
z_{fvf}	Vertical coordinate of free vortex filament (m)
z_C	Vertical coordinate of the panel center (m)
z_1, z_2	Coordinates of lower and upper bound of wake sheet (m)
Δ	Displacement (kg)
λ	Aspect ratio of wing (-)
λ_{RW}	Aspect ratio of rudder wing (-)
Γ	Vorticity (m²/s)
$\Delta\Gamma_{bound}$	Stepwise change of bound vorticity (m²/s)
Γ_{bound}	Vorticity of bound vortex filament (m²/s)
Γ_{fvf}	Vorticity of free vortex filament (m²/s)
Γ_1, Γ_2	Vorticity at lower and upper bounds of wake sheet (m²/s)
ρ, ρ_H	Density of air, density of water (kg/m³)
ω	Relaxation factor (-)
SI units only	

INTRODUCTION

In a recent paper Graf et.al, 2014 published a method for the calculation of aerodynamic forces of rigid wings as they are used on modern racing catamarans, see Figure 1. The method is based on 2D-RANS investigations for the flow around wing profiles and a nonlinear lifting line method to account for three-dimensional flow phenomena. It has been combined with a constraint optimization method to find the trim of the wing for maximum driving forces and given heeling moment. This method successfully has been successfully applied to the AC 72 as well as the C-class catamaran wings.

This paper describes an enhancement of that method. It employs a lifting line method with multiple rather than single bound vortex filaments, thus allowing to take into account an additional jib, which is fully integrated into the trim optimization process. In addition the optimization procedure now is based on a maximization of boat speed rather than wing driving force. In order to do so, a hydrodynamic model for foiling catamarans has been added, taking into account an L-shaped, water surface piercing foils, which has been investigated using 3D-RANS free surface flow investigations. Rudders and hull windage is also taken into account.

The method is used to predict flow forces, boat speed and flying behavior of a 50 feet catamaran, similar to the yachts intended to be sailed in the next Americas Cup. It is used to predict a velocity polar for a range of wind speeds and true wind angles. Some parameter studies predict the impact of wing trim variations on the performance of the yacht.

Figure 1: Oracle Team USA in the 2013 America's Cup, Foto: Donan.Raven, License: CreativeCommons by-sa-3.0-de

AERODYNAMIC MODEL

Aerodynamic forces generated by the wing and the jib are calculated using a lifting line method. This method relies on lift and drag coefficients of the wing profiles and the jib, which are calculated by planar flow RANSE simulations.

Lifting Line Method

The lifting line method is based on the following principal: a bound vortex filament of piecewise linear, optionally discontinuous, vorticity distribution along span is running from root to tip of the wing. An additional bound vortex filament is located along the span of the jib, longitudinally in a point 1/4 of the profile length aft of the luff. Individual free vortex filaments and/or wake sheets (in any combination) are shedding from bound vortex filaments into infinity in the direction of incident flow, see Figure 2: Bound and free vortex filaments on a wing. Discrete vorticities of the free vortex filaments and the distributed vorticity per length of the wake sheet are calculated conforming to *Thompsons* rule, see Truckenbrodt, 1964. Induced wind is calculated from the free vortex filaments and the wake sheet using *Biot Savart's* law. Lift and induced drag is calculated from *Kutta's* law taking into account the sum of undisturbed flow velocity and induced wind. The discrete form of this method calculates the total drag as integration over span of the induced drag per span plus the viscous profile drag. In a similar manner, the profile lift is integrated over span, taking into account the local effective incident flow.

To take into account nonlinear lift coefficients with respect to angle of attack, in particular lift coefficient changes due to flow separation, an iterative approach is employed here. It is based on an iterative correction of induced wind from the spanwise lift distribution, which takes into account the induced wind as a correction of angle of attack.

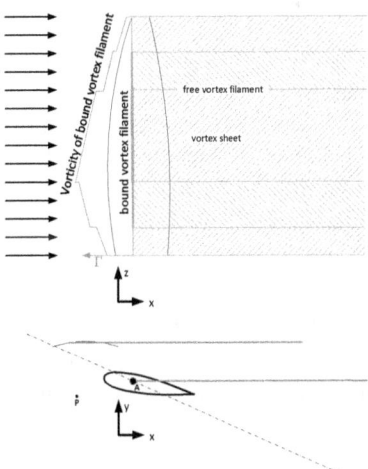

Figure 2: Bound and free vortex filaments on a wing

Theorems of *Kutta*, *Biot-Savart* and *Thompson* are used as follows:

Kutta's Law is used to calculate the lift generated by a vortex filament:

$$\mathbf{L} = \int_{span} \rho\, \mathbf{u}\ \Gamma \times d\mathbf{s} \tag{1}$$

where $\mathbf{L}$ is the generated lift, ρ the density of flow, $\mathbf{u}$ the incident flow, Γ the vorticity of filament and $d\mathbf{s}$ an integrator along the filament.

Biot-Savart's Law is used to calculate the induced wind generated by a vortex filament:

$$\mathbf{v} = \frac{-1}{4\pi} \int_s \Gamma \frac{\mathbf{r} \times d\mathbf{s}}{|\mathbf{r}|^3} \qquad (2)$$

where $\mathbf{v}$ is the induced velocity generated at $\mathbf{P} = \{x_P, y_P, z_P\}^T$ by the vortex filament of vorticity Γ. $\mathbf{r}$ is a vector from $\mathbf{P}$ to $\{x, y, z\}^T$, the current location of the integrator $d\mathbf{s}$ along the filament.

We assume that the span of the wing is oriented in z-direction of the coordinate system used and the incident velocity is directed parallel to the x-axis. A change of the angle of attack due to induced wind is then generated only by the y-component of the induced wind. This yield:

$$\mathbf{r} \times d\mathbf{s} = \begin{Bmatrix} x - x_P \\ y - y_P \\ z - z_P \end{Bmatrix} \times \begin{Bmatrix} dx \\ 0 \\ 0 \end{Bmatrix} = \begin{Bmatrix} 0 \\ (z - z_P)dx \\ -(y - y_P)dx \end{Bmatrix} \qquad (3)$$

and $|\mathbf{r}|^3 = ((x - x_P)^2 + (y - y_P)^2 + (z - z_P)^2)^{3/2}$ (4)

For the y-component of the induced velocity v we then get

$$v = \frac{-\Gamma}{4\pi} \int_s \frac{(z - z_P)dx}{\left((x - x_P)^2 + (y - y_P)^2 + (z - z_P)^2\right)^{3/2}} \qquad (5)$$

Free vortex filaments are calculated by *Thompsons* rule, saying that a vortex filament may only end at a fixed wall or at infinity. *Prandtl's* concept of horseshoe vortices is used, saying that any change of the bound vortex filament results in a free vortex filament. This results in:

$$d\Gamma_{fvf} = \frac{\partial \Gamma_{bound}(z)}{\partial z} dz \qquad (6)$$

For a stepwise change of the vorticity of the bound vortex filament $\Delta\Gamma_{bound}$, the vorticity of the free vortex filament at the position of the stepwise change calculates from:

$$\Gamma_{fvf} = \Delta\Gamma_{bound} \qquad (7)$$

Generated from a stepwise change of Γ_{bound} at location $\mathbf{A} = \{x_A, y_A, z_A\}^T$ a free vortex filament of constant vorticity Γ_{fvf} starts at $\mathbf{A}$ and is running at constant y, z to $\{\infty, y_A, z_A\}^T$. It generates induced wind in $\mathbf{P} = \{x_P, y_P, z_P\}^T$ in the direction of the y-axis by (see Figure 3):

$$v = \frac{-\Delta\Gamma_{bound}}{4\pi}(z_A - z_P) \int_{x_A}^{\infty} \frac{dx}{\left((x - x_P)^2 + (y_A - y_P)^2 + (z_A - z_P)^2\right)^{3/2}} \qquad \mathbf{(8)}$$

Integration gives:

$$v = \frac{-\Delta\Gamma_{bound}}{4\pi} \frac{\Delta z}{\Delta y^2 + \Delta z^2} \left(1 - \frac{\Delta x}{\left(\Delta x^2 + \Delta y^2 + \Delta z^2\right)^{1/2}}\right) \qquad (9)$$

with $\Delta x = x_A - x_P, \Delta y = y_A - y_P, \Delta z = z_A - z_P$

(9) holds for an infinite small vorticity $\partial\Gamma$ starting at $\{x_A, y_A, z\}^T$ to $\{\infty, y_A, z\}^T$, yielding:

$$v = \frac{-\partial\Gamma}{4\pi} \frac{z - z_P}{\Delta y^2 + (z - z_P)^2} \left(1 - \frac{\Delta x}{\left(\Delta x^2 + \Delta y^2 + (z - z_P)^2\right)^{1/2}}\right) \qquad (10)$$

For a vortex sheet we assume a span-wise change of $\Gamma_{bound}(z)$, varying linearly between $\mathbf{A}_1 = \{x_A, y_A, z_1\}^T$ and $\mathbf{A}_2 = \{x_A, y_A, z_2\}^T$ from Γ_1 to Γ_2:

$$\Gamma_{bound}(z) = \Gamma_1 + \frac{\Gamma_2 - \Gamma_1}{z_2 - z_1}(z - z_1) \qquad (11)$$

Introducing (11) into (6) yields: $\partial\Gamma_{fvf} = \frac{\Gamma_2 - \Gamma_1}{z_2 - z_1} dz$.

Introducing this into (10) and integrating from z_1 to z_2 yields:

$$v = -\frac{\Gamma_2 - \Gamma_1}{4\pi(z_2 - z_1)} \ln\left(\frac{\sqrt{(z_2 - z_P)^2 + \Delta y^2 + \Delta x^2} + \Delta x}{\sqrt{(z_1 - z_P)^2 + \Delta y^2 + \Delta x^2} + \Delta x}\right) \qquad (12)$$

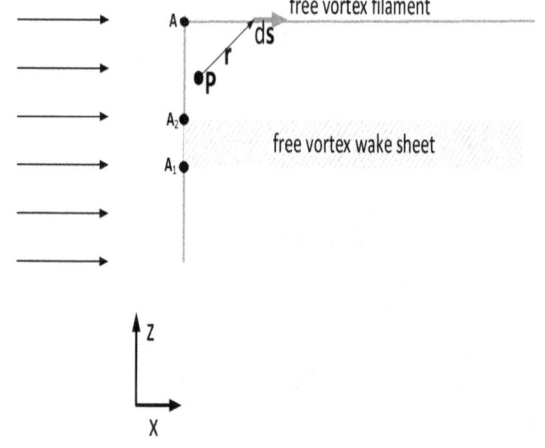

Figure 3: Free vortex filament and wake sheet, generating induced velocity at arbitrary point P

Profile lift per span length $L = d|\mathbf{L}|/dz$ is calculated from a lift coefficient c_L, the dynamic pressure $0.5\rho AWS^2$ and the profile chord length c. c_L depends nonlinear on angle of attack AoA of incident flow. For a profile of a wing of finite span the angle of attack is AoA reduced by induced wind:

$$L = 0.5\rho AWS^2 c \ c_L(AoA - v / AWS) \qquad (13)$$

The induced drag then calculates from

$$D_{ind} = Lv / AWS \qquad (14)$$

The bound vorticity of the profile is calculated from the lift per span length due to (1)

$$\Gamma_{bound} = \frac{L}{\rho \, AWS} \qquad (15)$$

Since vorticity of free vortex filament depends on vorticity of bound vortex filament, which in turn depends on lift, itself depending on free vortex filament vorticity due to induced wind, this procedure is iterative by nature.

For a wing of finite span with varying lift distribution over span the iterative procedure is:

(1) Assume $v=0$
(2) Calculate profile lift per span from (13) for a given geometric angle of attack, flow speed and chord length. Additional parameters like a flag angle can be taken into account here.
(3) Calculate Γ_{bound} over span from profile lift from (15)
(4) Calculate v from (12) and (9)
(5) redo 2.-5. until v converges
(6) calculate induced drag
(7) calculate total drag by adding parasitic profile drag to induced drag

Discretization

Generally this method allows for an arbitrary number of wings and sails in any combination as long as it is taken care that geometry and free vortex filaments do not intersect. Here we focus on the combination of a wing with a hinged flap and a jib as they are used for the AC class catamarans..

The envelope of a wing or a sail is discretized with an arbitrary number of horizontal profiles, numbered $i=1,2,...,N$. Each profile is described by its z-coordinate z_i along with leading and trailing edge x- and y-coordinates:

$$\mathbf{l}_i = \{x_{li}, y_{li}, z_i\} \text{ and } \mathbf{t}_i = \{x_{ti}, y_{ti}, z_i\} \qquad (16)$$

In addition, for each profile an individual incident flow speed AWS_i, an incident (geometric) angle of attack AoA_i, a lift coefficient c_{Li} and a parasitic profile drag coefficient c_{DPPi} has to be known. No profile geometry is needed since the property of the profile is entirely described by the lift coefficient.

Flow force coefficients c_L, c_{DPP} and can be provided by tabulated data for given AoA and additional parameters like a flap angle β_t. This approach allows taking into account profile properties from any source, being it linear or nonlinear, from inviscid or viscous calculation methods.

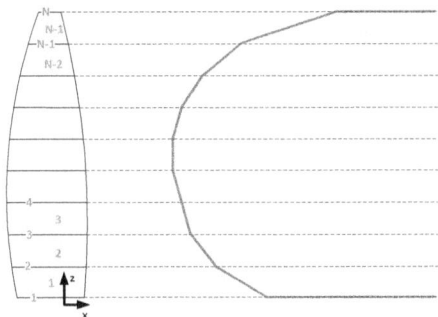

Figure 4: Discretization of wing planform

A bound vortex filament aligned with or approximately parallel to the z-axis for incident flow aligned with the x-axis is assumed. The vorticity Γ of the bound vortex filament segment between two profiles changes linearly, generating a free vortex wake sheet. At bottom profile (root) and top profile (tip) discrete free vortex filaments are generated in order to satisfy zero bound vorticity for $z<z_1$ and $z>z_N$. Consequently $\Gamma_{bound}(z)$ and $\Gamma_{fvf\,root}$ as well as $\Gamma_{fvf\,tip}$ can be calculated from profile definition information: for any profile i. $1 <= i <= N$:

$$\Gamma_i = 0.5 \, AWS_i \, c_i \, c_{Li}(AoA_i - \frac{v_i}{AWS_i}) \qquad (17)$$

where the chord length is given by the distance of leading and trailing edge, $c_i = |\mathbf{t}_i - \mathbf{l}_i|$. Piecewise linear distribution of circulation and discrete circulation at the extends of the wing are defined by: for $z_i < z < z_{i+1}$

$$\Gamma_{bound}(z) = \Gamma_i + \frac{\Gamma_{i+1} - \Gamma_i}{z_{i+1} - z_i}(z - z_i) \qquad (18)$$

$$\Gamma_{fvf\,root} = -\Gamma_1 \, w_{FVF} \qquad (19)$$

$$\Gamma_{fvf\,tip} = \Gamma_N \qquad (20)$$

Here w_{FVF} is a factor taking into account to which degree the root free vortex filament is suppressed by a wall. If the root of the wing is fixed to a wall without a gap, $w_{FVF}=0$. If no wall is present at all, $w_{FVF}=1$. w_{fwf} can be calculated from a known result at a particular AoA.

The end points of a bound vortex filament section within a panel is given by the quarter-points of the profiles at the lower and upper end of the panel. The quarter-point is a point on the chord of the profile 25% of the chord length from leading to trailing edge. The collocation point $\mathbf{P}_j$- where the induced velocity is calculated in panel j - is located in the middle of the bound vortex filament section, see

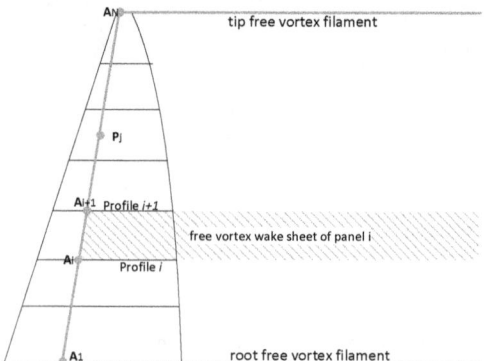

Figure 5: Bound and free vortex filaments and free vortex wake sheet due to linear change of vorticity of bound vortex filament i

For the panel i $\mathbf{A}_i = \{x_{Ai}, y_{Ai}, z_{Ai}\}^T = 0.75\mathbf{l}_i + 0.25\mathbf{t}_i$ and $\mathbf{A}_{i+1} = \{x_{Ai+1}, y_{Ai+1}, z_{Ai+1}\}^T$ are the quarter points of the lower and upper profile. $\mathbf{P}_j = \{x_{Pj}, y_{Pj}, z_{Pj}\}^T = 0.5(\mathbf{A}_j + \mathbf{A}_{j+1})$ is the collocation point for the panel j to calculate induced velocity for the bound vortex filament segment j.

The total induced velocity in point $\mathbf{P}_j$ is calculated by summing up the induced wind generated by any free vortex wake sheet and the discrete free vortex filaments at root and tip.

$$v_j^* = -\frac{\Gamma_1}{4\pi}\frac{\Delta z_{root}}{\Delta y_{root}^2 + \Delta z_{root}^2}\left(1 - \frac{\Delta x_{root}}{\left(\Delta x_{root}^2 + \Delta y_{root}^2 + \Delta z_{root}^2\right)^{1/2}}\right)$$
$$+\frac{\Gamma_N}{4\pi}\frac{\Delta z_{tip}}{\Delta y_{tip}^2 + \Delta z_{tip}^2}\left(1 - \frac{\Delta x_{tip}}{\left(\Delta x_{tip}^2 + \Delta y_{tip}^2 + \Delta z_{tip}^2\right)^{1/2}}\right) \quad (21)$$
$$-\sum_{i=1}^{N-1}\frac{\Gamma_{i+1} - \Gamma_i}{4\pi(z_{i+1} - z_i)}\ln\left(\frac{\sqrt{(z_{i+1} - z_{Pj})^2 + \Delta y^2 + \Delta x^2} + \Delta x}{\sqrt{(z_i - z_{Pj})^2 + \Delta y^2 + \Delta x^2} + \Delta x}\right)$$

where

$$\Delta x_{root} = x_{A1} - x_{Pj}; \quad \Delta y_{root} = y_{A1} - y_{Pj}; \quad \Delta z_{root} = z_{A1} - z_{Pj}$$
$$\Delta x_{tip} = x_{AN} - x_{Pj}; \quad \Delta y_{tip} = y_{AN} - y_{Pj}; \quad \Delta z_{tip} = z_{AN} - z_{Pj} \quad (22)$$
$$\Delta x = x_{Pi} - x_{Pj}; \quad \Delta y = y_{Pi} - y_{Pj}$$

(22) assumes $y_{Ai} = y_{Ai+1} = y_{Pi}$ and $x_{Ai} = x_{Ai+1} = x_{Pi}$, the latter expression only being true for a bound vortex filament parallel to the z-axis of the coordinate system. For wings of only small rake (22) is a reasonable approximation.

After calculating induced velocity in each panel, in the middle of each bound vortex filament segment, linear interpolation is used to calculate induced wind v_i at profile vertical location z_i:

$$v_i = 0.5(v_{i-1}^* + v_i^*) \qquad \text{for } 2<=i<=N-1 \quad (23)$$

At root and tip induced wind is calculated using linear extrapolation:

$$v_1 = 2v_1^* - v_2^* \quad (24)$$

$$v_N = 2v_{N-1}^* - v_{N-2}^* \quad (25)$$

An iterative procedure has to be used in order to calculate induced wind. We assume that for any profile i, the local height z_i, the profile length c_i, the local geometric angle of attack AoA_i and the local wind speed AWS_i is given. We also assume that lift coefficient can be calculated using the above values. Induced wind then is calculated iteratively starting with a zero guess:

(1) Set $v_i = 0$ for any profile i=1,2,...,N
(2) Predict profile lift coefficient $c_{Li}(AoA_i - v_i / AWS_i)$ using tabular data from 2D RANS profile solutions
(3) Calculate Γ_i for any profile using (17)
(4) Calculate induced wind in panel center v_C from (21)
(5) Calculate induced wind at profile height from (23), (24) and (25)
(6) Repeat (2) to (5) until convergence, if convergence achieved continue
(7) Calculate lift per span for any profile using (13)
(8) Calculate induced drag per span for any profile using (14) and add parasitic profile drag per span, taken from tabulated RANSE results
(9) calculate driving and side force per span from trigonometric relationship
(10) Integrate over span by trapezoidal integration

It has to be taken into account, that direct integration of lift and drag over span is not feasible since their direction changes with height due to a twisted incident wind over wing span. Hence lift and drag is transformed to driving and side forces which allow integration over span independent of local incident wind.

To achieve convergence, the iterative procedure needs under-relaxation. If k denotes the current iteration step, the induced wind is calculated as a weighted average of the result of the current and last iteration step:

$$v_i^{*k} = v_i^* \omega + v_i^{*k-1}(1-\omega) \quad (26)$$

where v_i^* is the induced wind as calculated from (21). Some attention has to be paid to the angle of attack AoA, usually defined by the angle between incident wind and a reference line of the profile (for the profile of a symmetric main element this is its center line). The angle of incidence AoA is calculated from the apparent wind angle AWA and the local rotation of the profile, given by wing rotation and wing geometric twist. This allows taking into account the sheeting of the wing and jib, their geometric twist a twist of the incident wind.

PROFILE LIFT AND DRAG COEFFICIENTS

The lifting line algorithm as outlined above relies on profile lift and drag coefficients for the wing profile as well as for the jib profile. For the wing profile they are predicted using planar RANSE flow investigations, while for the jib profiles standard lift and drag coefficients from literature are used.

Wing profile

The geometry of the wing is given by the AC Class Rules, see Americas Cup Race Management, 2015. The rules come with a surface description of the wing (IGES-file). For the profile investigations some simplifications are applied: a single profile at 50% of the span of the wing has been investigated at a single Reynolds number of $Rn=10^7$, corresponding to an apparent wind speed of $AWS \approx 24\text{m}/\text{s}$. It is generally assumed that a change of the Reynolds number by a factor of less than 10 has a negligible impact on the profile lift and drag coefficients. Figure 6 shows the profile.

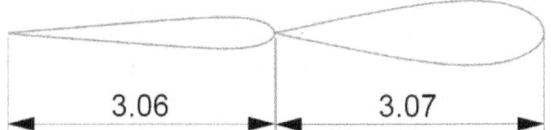

Figure 6: AC Class wing profile at 50% of wing span

Profile lift and drag coefficients are predicted using planar flow RANSE investigations. A computational grid of approximately $3*10^5$ grid cells has been used, see Figure 7. Flow simulations have been carried out using the OpenFOAM framework, employing a solver for turbulent transient flow (pisofoam) and the SST turbulence model, see *Menter, 1994.* Intensive grid sensitive studies and validation investigation have been carried out in *Graf et.al, 2014,* hence only the main results of the wing profile study are shown here.

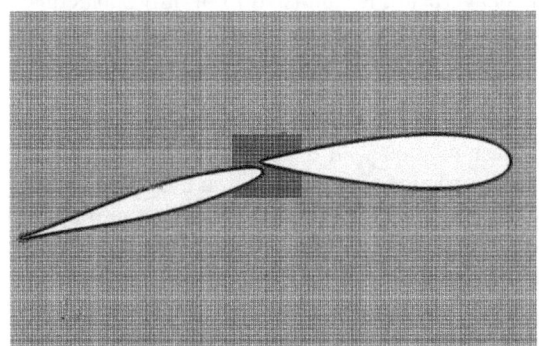

Figure 7: Computational Grid for planar flow around wing profile

A test matrix of combinations of angle of attack AoA and flap angle β has been investigated:
- 14 AoA with dense distribution of angles close to maximum lift
- 4 flap angles ranging from 0° to 30°

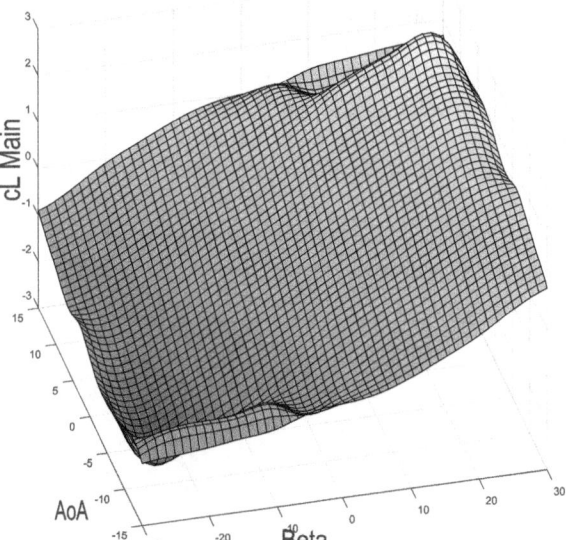

Figure 8: Wing profile lift coefficient

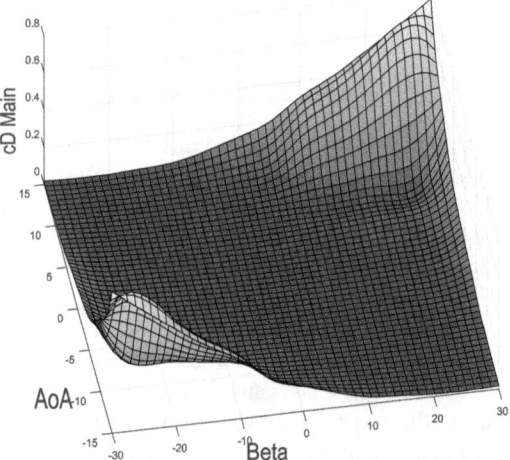

Figure 9: Wing profile drag over lift coefficient

Jib Profile

For the study presented here only minor attention has been paid to proper prediction of the forces generated by a jib. It is generally known that the jib only marginally contributes to the driving force of a catamaran equipped with a small jib and a large wing. The main reason for taking a jib into account was proper prediction of its impact on the incident wind and optimal trim settings of the wing. Hence only little effort has been used for proper prediction of the jib profile lift and drag coefficients.

The jib profile lift coefficient has been predicted for a rigid profile based on potential theory, see Truckenbrodt, 1964:

$$c_L = 2(\pi\beta + t/c) \tag{27}$$

where t/c is the relative profile depth, which has been fixed to the value $t/c=0.1$. In order to take softness of the jib fabric and flow separation into account, the lift coefficients of the rigid jib profile in ideal flow has been manipulated: the ideal flow lift coefficient is used for angles of attack larger than the profile leading edge entrance angle and may never exceed values of $c_L=1.6$. For smaller angles of incidence the lift coefficient is linearly reduced to the value of 0 at an entrance angle of 0. The drag coefficient is predicted from ITTC 57 friction line and a drag factor of $c_{DPP}/c_L^2 = 0.1$. Figure 10 shows jib profile lift and drag over angle of attack.

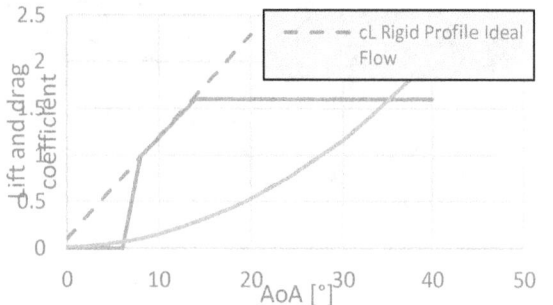

Figure 10: Coefficients of lift and parasitic profile drag of jib profile

L-FOILS AT FREE SURFACE

Geometry

L-Foils are used on sailing catamarans as daggerboard to produce side forces as well as vertical forces allowing the catamaran to fly (the hulls lift out of the water). The design of the L-foil is regarded to play a key role for the performance of any flying catamaran and intensive research is carried out to develop an optimized design. The design used for the present study is pretty much a standard design. The authors do not claim that it is an elaborated one.

The rules for the AC Class catamaran describe only little constraints on the foils: the maximum dimension is given (3.5m), the deepest point of the foil may not exceed a draft of 2.3m with respect to a measurement water plane, and there are some constraints on the maximum rotation angle around the longitudinal axis (cant axis) and the transverse axis (rake axis). Based on these constrains a simplified L-shape has been designed. It uses a NACA 64$_2$-010 profile, extruded along an L-shaped curved, see Figure 11.

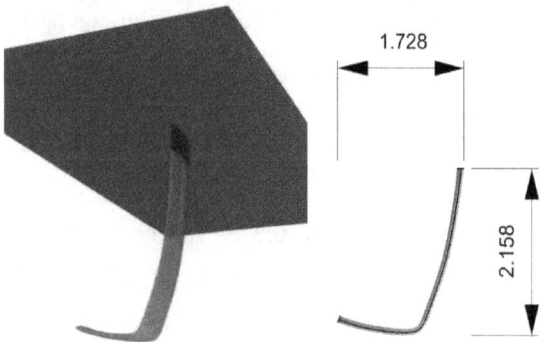

Figure 11: Water surface piercing L-shaped daggerboard

RANSE free surface flow solver

The hydrodynamic properties of the daggerboard have been investigated using a proprietary free surface RANSE code. This code has been developed based on the OpenFOAM field operation toolbox. Free surface flow simulation is based on the principle of the well-known Volume-of-Fluid-method, involving the solution of an additional conservation equation for the fluid fraction.

OpenFOAM provides solvers for the simulation of free surface flow. However these standard solvers are hardly capable to solve real world free surface flow problems with multi-million grid cells, since they lack computational efficiency and robustness. Hence a proprietary solver for free surface simulations has been developed.

Compared to the standard solvers the new solver allows large time steps, while remaining stable, thus making it possible to achieve a steady-state solution even for large computational grids within reasonable computational run time. To this end the BRICS scheme has been implemented for the discretization of the convective term in the fluid fraction conservation equation. Additional modifications of the method implement special treatment of the pressure gradient in the vicinity of the free surface.

This new OpenFOAM flow solver is presented at this conference, for details of its principles and implementation please consult Meyer et.al, 2016. An intensive validation study is also given there.

Test matrix and setup

The daggerboard acts as a water-piercing foil even when it is fully submerged and the catamaran hull starts to float. When the catamaran hull's hydrostatic lift exceed a given threshold (20% of weight) the daggerboard is assumed to act fully submerged under a flat plate. In this case the flow around the daggerboard is simulated for a fully submerged daggerboard with a rigid water surface, not taking account of any wave generation. Linear blending is used between foil at free surface and foil under flat plate for hull buoyancy between 0 and 0.6 m³. This assumption is used to avoid free surface flow simulation with daggerboard and hull, which are too time-consuming to be done within this study. While this approach will model the onset of transition into flying state reasonable accurate, the process of transition will not be modelled correctly.

As a test matrix a full permutation of the following trim settings of the daggerboard has been investigated:

- Speed 6 m/s, 12 m/s and 18 m/s
- Rake angle -2°, 0° and 2°
- Leeway angle 0° and 2°
- Cant angle 0° and 15°
- Immerse 0.4, 0.7 and 1.0

This results in 108 individual daggerboard trim parameter combinations, which have investigated using the RANSE-solver. Here *Immerse* is a parameter defining the immersion of the foil at rest. *Immerse=1* is the daggerboard at maximum draft while *immerse=0* describes a daggerboard just touching the water surface. Immerse approximately but not exactly scales with the vertical draft of the daggerboard.

The ACC rules only allow either fully down or fully retracted daggerboard positions. Hence the *immerse* factor is used to take into account lifting of the entire catamaran out of the water due to the vertical forces generated by the foils.

Flow simulations are carried out using a finite volume discretization approach for the solution of the discrete, time averaged Navier Stokes equation. The Shear Stress Transport turbulence model (Menter, 1994) is used for turbulence modelling. The solution procedure solves the transient form of the equation since the Volume-of-Fluid method as described above is inherently transient, however a steady state solution is achieved by continuing the simulation until the unsteady disturbance of the flow field has faded away.

Computational grid

Computational grids are generated for the daggerboard configurations of two cant angles and the three immerse states. Figure 12 shows the grid for a daggerboard cant angle of 15° and an immersion of 0.7. The computational grid consists of about 3.5 million grid cells. Note the grid refinement in the vicinity of the water plane, on the surface of the foil and in the wake field.

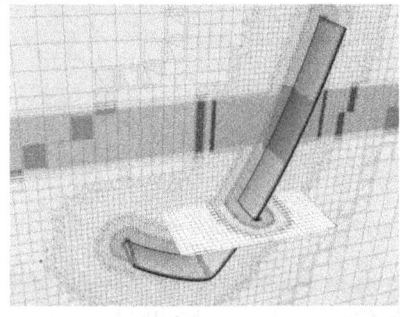

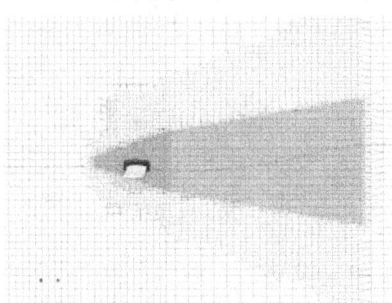

Figure 12: Computational Grid of L-foil daggerboard

ADDITIONAL RESISTANCE ELEMENTS

Additional resistance elements are hydrodynamic hull resistance, windage and rudder resistance. Only rough estimates are used for these resistance elements and they are subject to future research activities. For the method presented here, targeting wing trim optimization, the estimates are assumed to be sufficiently accurate.

Hydrodynamic Hull Resistance

Hydrodynamic hull resistance is calculated from two parameters only: the velocity and the buoyancy. No attempt has been made to take into account the impact of leeway, heel and pitch on hull resistance. The bare hull resistance is derived from an available AC 72 hull resistance curve, scaled down to the length of $L=14.65$ m and a displacement of $\Delta=3$t, the latter assumed to be the displacement in sailing trim, see Figure 13. The resistance at displacements smaller than 3t R_T is calculated using:

$$R_T(\Delta) = R_T(\Delta = 3t)\left(\frac{\Delta}{3t}\right)^{2/3} \qquad (28)$$

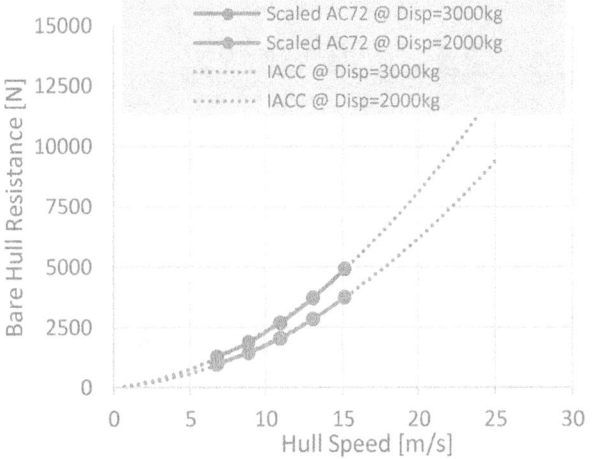

Figure 13: IACC Bare Hull Resistance

The floating displacement is calculated from the displacement at rest $\Delta=3$t of the catamaran and the vertical lift generated by the daggerboard L_H:

$$\Delta = 3t - L_H / g \qquad (29)$$

Obviously some means has to be used to avoid negative displacement.

Windage

A very simple approach is used to estimate wind resistance of the yacht's platform. Resistance in longitudinal and transverse direction are calculated from a drag area in the respective direction and the longitudinal and transverse wind speed. The drag areas are:

$D_{A\,X} = 3.5$ m²
$D_{A\,Y} = 14.5$ m²

Forces in longitudinal and transverse direction are then calculated using:

$$F_{AX} = 0.5\rho_A(AWS_{z=1m}\cos(AWA_{z=1m}))^2 D_{AX} \qquad (30)$$

8

$$F_{AY} = 0.5\rho_A(AWS_{z=1m}\sin(AWA_{z=1m}))^2\, D_{AY} \qquad (31)$$

Rudder

Rudders and rudder wings are taken into account as flat plate lifting foils, their hydrodynamic properties estimated by simple formulas derived from potential theory.

AC class rules provide values for rudder wing area and span. For the rudder itself area and span is estimated. The following values are used:

Rudder wing area: A_{RW}=0.18 m²
Rudder wing effective span: S_{RW}=1.25 m
Rudder immersed area: A_R=0.28 m²
Rudder effective span: S_R=1.4 m

The vertical lift of the rudder wing is calculated from

$$L_{V\,RW} = 0.5\rho_H u_B^2\, A_{RW}\, 2\pi\frac{\lambda_{RW}(\lambda_{RW}+1)}{(\lambda_{RW}+2)^2}\delta_{RW} \qquad (32)$$

where the rudder wing aspect ratio is:

$$\lambda_{RW} = S_{RW}^2 / A_{RW} \qquad (33)$$

The drag of the rudder wing is estimated from a viscous contribution, calculated from the ITTC-57 friction line and a form factor $1+k$=1.3, and from a induced drag contribution, calculated with the effective span S_{RW}:

$$R_{RW} = 0.5\rho_H u_B^2\, 2A_{RW}c_{F\,ITTC}(1+k)+$$
$$\frac{L_V^2}{S_{RW}^2}\frac{1}{0.5\rho_H u_B^2\,\pi} \qquad (34)$$

The rudder drag is calculated from a similar expression. No rudder lift is calculated since yawing moments are not taken into account in this study.

VPP-IMPLEMENTATION

The method for the calculation of boat speed is based on constrained optimization. Boat speed is maximized for the following constraints:
- Equilibrium of longitudinal aerodynamic and hydrodynamic forces and respective side forces and vertical forces.
- Heeling moment and pitching moment may not exceed given thresholds
- In addition, upper and lower bounds are given for most of the free variables of the optimization.

The free variables of the optimization are:
- Wing main element angle of attack α_M
- Wing main element twist: τ_M
- Flap angle, given at some control points over height $\beta_i = f(z_i)$
- Jib angle of attack and twist: α_J and τ_J
- Boat velocity u_B
- Leeway angle ψ
- Daggerboard immerse factor im
- Daggerboard pitch angle φ
- Daggerboard cant angle ξ
- Rudder wing angle δ_{RW}

$$\max(u_B) \qquad (35)$$

subject to the following nonlinear equality and inequality constraints:

$$\sum F_X = 0 \qquad (36)$$

$$\sum F_Y = 0 \qquad (37)$$

$$\sum F_Z = 0 \qquad (38)$$

$$\sum M_X \le 135000\,\text{Nm} \qquad (39)$$

$$\sum M_Y \le 100000\,\text{Nm} \qquad (40)$$

and the following bounds:

$$\begin{aligned}
-10 &\le \alpha_M \le 20\\
0 &\le \tau_M \le 5\\
-30 &\le \beta_i \le 30\\
5 &\le \alpha_J \le 20\\
0 &\le \tau_J \le 15\\
0.4 &\le im \le 2\\
-2 &\le \varphi \le 2\\
0 &\le \xi \le 15\\
-3 &\le \delta_{RW} \le 3
\end{aligned} \qquad (41)$$

As to (39) the maximum heeling moment at the break angle of the catamaran has been set to 135 kNm, which has been estimated from the yacht's main dimensions and the crew weight.

To take into account flying and non-flying states of the leeward hull, the following expression is used to calculate the total vertical force:

$$\begin{aligned}
F_Z &= F_{Z\,Daggerboard}\,\min(im,1)\\
&+ \max(im-1,0)\Delta_R g
\end{aligned} \qquad (42)$$

This expression says that for $0<im<1$ the leeward hull is flying and the vertical force is generated entirely by the daggerboard, while for $im>1$ the hull generates hydrostatic lift. im=1 indicates transition from non-flying to flying state or vice-versa while im=2 indicates that hydrostatic lift according to buoyancy at rest is generated.

The VPP method is implemented using the *MatLab* © programming language. For the constraint optimization a derivative-free interior point method is chosen. This method is used pretty much as a black box, so no theory is given here.

RESULTS

Comparison of wing with and without a jib

The predecessor of this publication, Graf et.al. 2014, investigated aerodynamic properties of a wing without a jib. One of the main motivations of the investigation shown here was to include a jib as it is used on an AC catamaran. So a comparison of a wing system with and without jib is of high interest.

Base of comparison has to be chosen thoroughly. Obviously a jib generates some extra driving force, but heeling moment as well and it can be assumed that it reduces the effective span of the entire system. Consequently the only reasonable comparison of wings with and without a jib is the driving force of the entire system with an optimized trim and a given heeling moment constraint. Free optimization parameters are sheeting angles and twist of the wing and the jib as well as the angle of the flap depending on height, given by four control points, distributed evenly over the wing span.

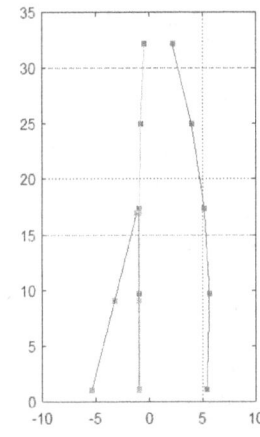

A wing similar to the one given by the new AC class rules (ACRM 2015) has been used, see right.

The span of the wing is 31.1m, the span of the jib approximately half of it.

The maximum heeling moment was set to $M_{X\,Max}$=135 kN.

The study has been carried out at a true wind speed of TWS=8m/s, 9m/s and 10m/s and the boat speed has been set to wind speed. True wind angle was set to TWA=45°.

Two test cases have been studied: wing without and with a jib. For the latter case the jib sheeting has been constrained to a value where it does not curl. Figure 14 left depicts the optimized trim of the wing with the jib at TWS=8m/s. The jib is sheeted to an angle of approximately 10° while the main element angle is relatively open (26.7°). Without a jib, Figure 14 right, the main element is sheeted at an angle of 22.6°. The trim of the flap is almost the same for the two cases with and without a jib.

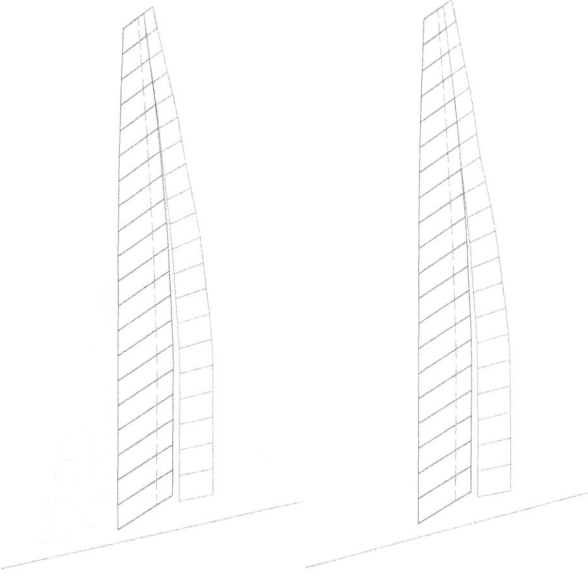

Figure 14: Trim of wing, left with jib, right no jib

Figure 15 and Figure 16 show lift coefficient of the wing and the jib over height for both test cases. The jib sheeting is set to a value with almost constant $c_{L\,Jib}\approx1$ over its entire span, while the lift of the wing changes significantly over height, showing a maximum at approximately 1/4 of the wing height and being negative at the top, see Figure 15. When no jib is present, maximum lift of the wing is a bit higher and negative lift angles at the top are more pronounced, see Figure 16.

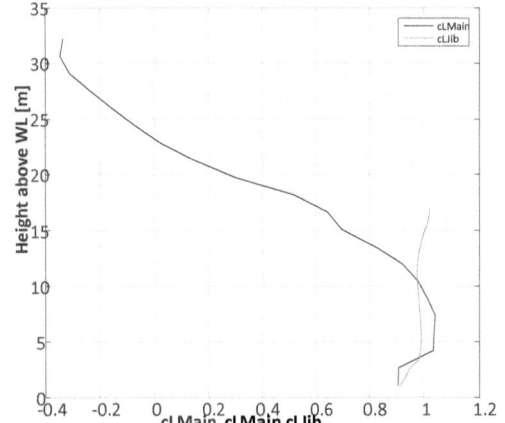

Figure 15: Lift coefficient over height for wing and close hauled jib

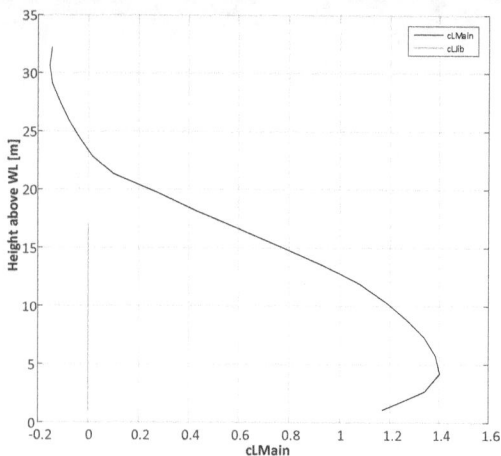

Figure 16: Lift coefficient over height for wing, no jib

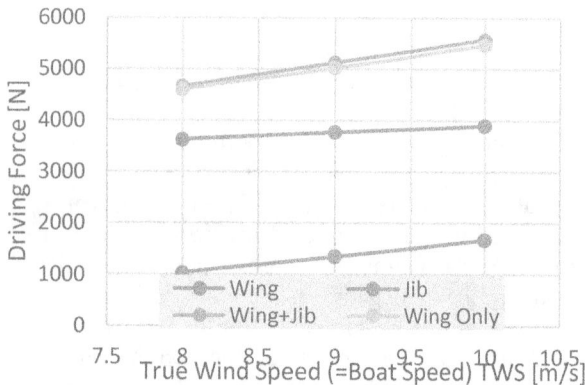

Figure 17: Driving force of wing and jib, optimized trim for heeling moment constraint

Figure 17 shows driving force of the aggregate wing with a jib and for the wing alone for optimized trim and a heeling moment constraint of $M_{X\,Max}$=135kN. While driving forces generally increase slightly with increasing wind speed, it can be seen that the jib contributes approximately 25 % to the driving force of the aggregate. However for the same heeling moment the wing without a jib can be sheeted closer, giving a driving force only slightly lower than for the aggregate. Here the difference is about 1 to 2 %. In the light of expected inaccuracy of the jib profile polar this can almost be neglected.

Free Surface Flow Simulations of Daggerboard

A single computational run took about 3 h on a single compute node with dual CPU and 24 cores. The entire test matrix has been executed in about 3 days on four nodes running parallel.

Figure 18 shows contour plots of the wave pattern for the daggerboard canted 15° at an immerse factor of 0.4 at the highest speed of 18 m/s. The wave pattern looks very different to well know wave pattern of slender hulls sailing at large Froude numbers. With a Froude number of the test case, calculated with the profile length of the daggerboard being Fn=9, it can be expected that the conventional

interpretation of Froude number as a speed related to the speed of a wave as long as the flow body (Fn=0.4) is no longer helpful.

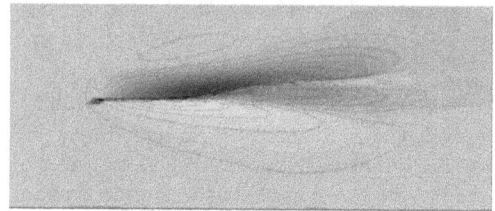

Figure 18: Wave pattern of water piercing L-foil daggerboard

Figure 19 shows total drag over total lift squared. Here total lift L is calculated from horizontal and vertical lift L_V and L_H as follows:

$$L^2 = L_H^2 + L_V^2 \qquad (43)$$

The diagram shows that the simulated results can well be approximated by a straight line. It can concluded from this, that horizontal as well as vertical lift contribute to induced drag with the same effective span. This is somewhat unexpected since the geometric span in horizontal and vertical direction differ significantly, see Figure 11.

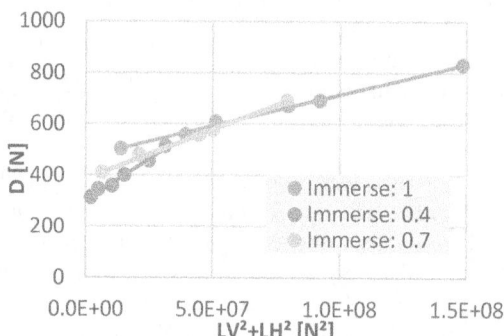

Figure 19: Drag over lift squared (cant=0°, velocity=6m/s)

Figure 20 shows horizontal lift coefficient of the daggerboard at cant angle 0° (as to Figure 11). The diagram depicts that leeway and pitch both generate horizontal lift in the same order of magnitude.

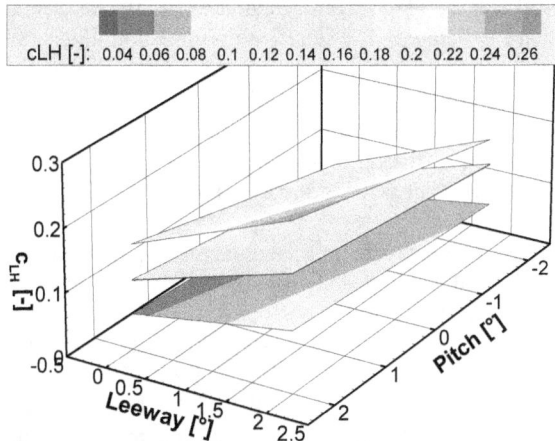

Figure 20: Daggerboard horizontal lift coefficient at cant angle 0°, plotted over pitch (rake) and leeway angle

Figure 21 shows the effective draft of the daggerboard, which is calculated from induced resistance using:

$$R_I = L^2 / (T_{Eff}^2 \, 0.5 \rho u^2 \pi) \tag{44}$$

Two surfaces are shown in Figure 21: effective draft for cant angle 0° and 15° over immerse and velocity. These two surfaces are equidistant with a small delta between them. A clear dependency of effective draft on the immersion of the foil can be observed as expected. A slight dependency of effective draft on velocity is owed to the wave system, which also can be expected. However, the effective draft is only slightly dependent on cant angle which does not match expected behavior. It may be a consequence of the property of the foil to generate induced drag regardless whether lift is generated in horizontal or vertical direction.

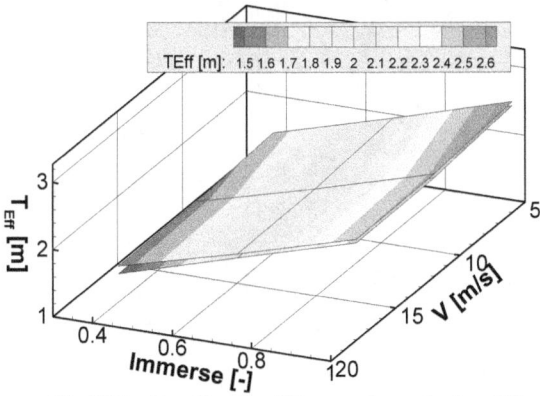

Figure 21: Effective Draft of Daggerboard at u=12m/s, two surfaces for cant angle of 0° and 15°

The hydrodynamic forces from the L-foil investigation are integrated into the VPP by a multidimensional spline-interpolated surface of the native variables (forces rather than coefficients).

Velocity Polar of AC-Catamaran

For the test case as above a velocity polar, giving boat speed and other state variables depending on true wind speed TWS and true wind angle TWA. The following test matrix has been executed:

6 m/s ≤ TWS ≤ 11 m/s
35° ≤ TWA ≤ 140°

The result of the velocity prediction is shown in Figure 22. No validation data for comparison is available and the only way for assessment of the result is a plausibility check. Generally the polar diagram shows the expected pattern. True wind angles for maximum VMG can clearly be detected and are reasonable. However the total speeds on upwind courses TWA<50° appear to be too large. For TWA=35° no reasonable results can be achieved at all. There are various possible reasons for the overestimation of speed: windage is known to have a strong impact on performance of these yachts and it is estimated only in this study. The hydrodynamic properties of the foils are predicted only up to a speed of 18m/s and are extrapolated for higher speeds. This may lead to erroneous low resistance at higher speeds.

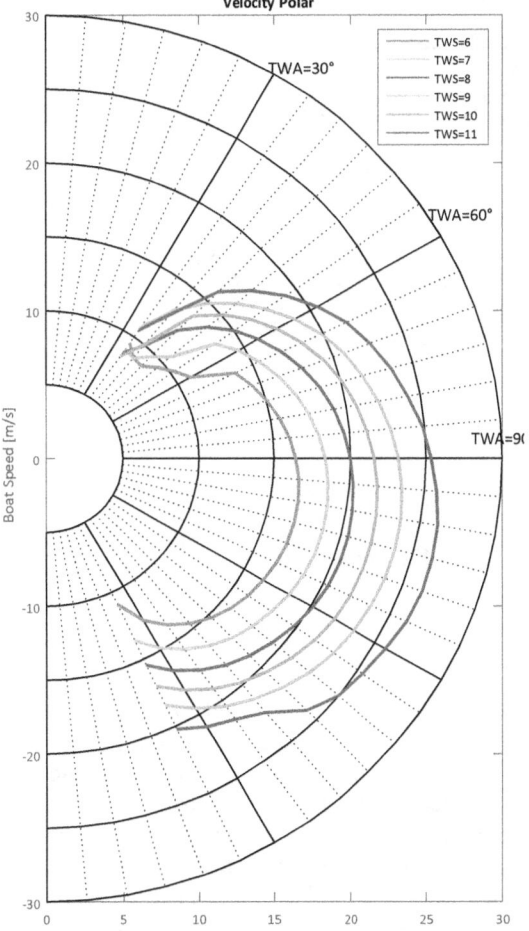

Figure 22: AC Class velocity polar

Figure 23 shows the immerse factor of the catamaran. Immerse factors below 1 indicate a flying state, while immerse factor above 1 indicate the fraction of hydrostatic lift on the entire vertical forces. The diagram depicts that for TWS=6 m/s the catamaran starts to fly at TWA≥65°, while for TWS=7m/s this value is TWA≥55°. For higher wind speed the catamaran flies even on the smallest true wind angles. This seems to be very optimistic and a consequence of the high velocities achieved.

for wind speed TWS≥10m/s and TWA≥70° no useful results are generated. The iterative optimization algorithm does not converge for these wind conditions. Since the optimization solver does not completely diverge either, it is assumed that some oscillation occurs in the results, which may be caused by flat optima, however an in-depth study of the convergence behavior of the solver is beyond the scope of this study.

Figure 25 shows the immerse factor for the case of increased wind resistance. It can be clearly detected that flying of the leeward hull is postponed to higher true wind angles. At TWS=6m/s, the yacht starts to fly at TWA≥70°, at TWS=7m/s this value falls to TWA=57° and at TWS=8m/s flying starts at TWA=52°. However for TWS=9m/s the yacht flies for any investigated true wind angle, which still appears to be a bit too optimistic.

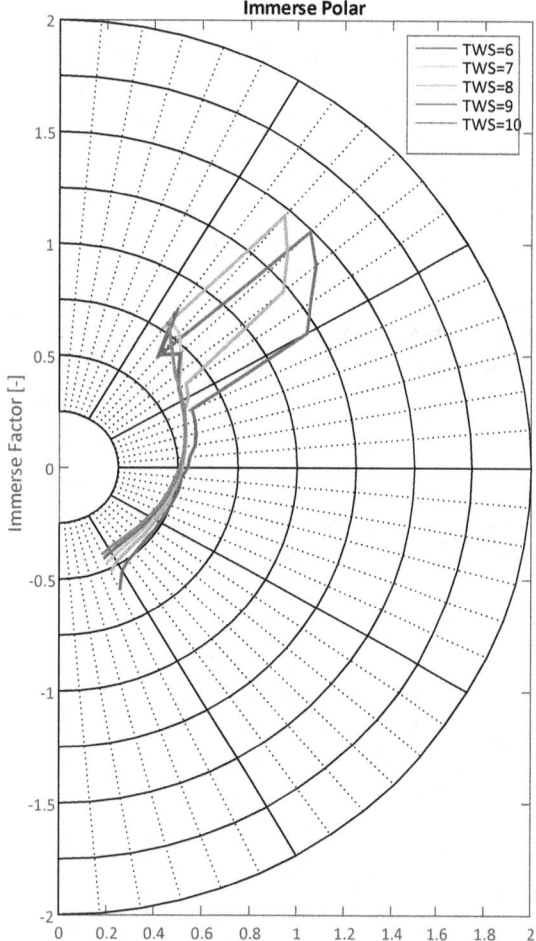

Figure 23: Polar of the immerse factor

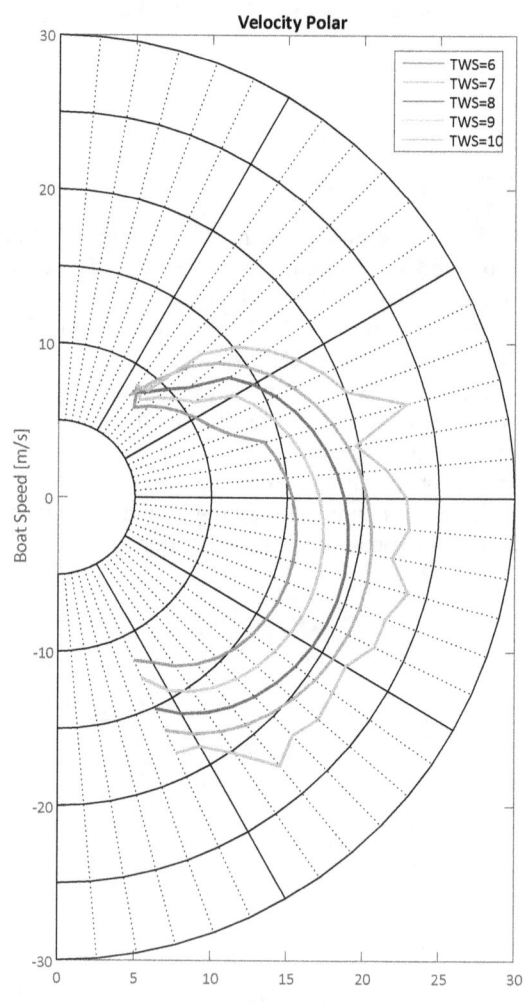

Figure 24: Velocity polar with increased wind resistance

Variation of Wind Resistance

The polars above have been generated for a drag area D_X=3.5 m², which is a quite low value. To assess the impact of wind resistance an additional polar has been generated with a drag area twice as high, D_X=7m². Taken the beam of the platform being about 7m this is still a reasonable value.

Figure 24 shows the velocity polar for increased wind resistance. A significantly lower boat speed can be observed in particular for wind speeds TWS≤8m/s. Again for TWA=35° no reasonable results are generated. In addition

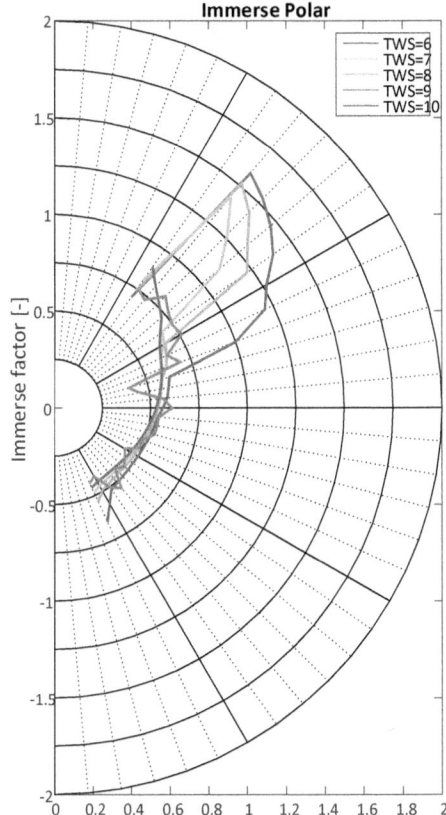

Figure 25; Polar of immerse factor, increased wind resistance

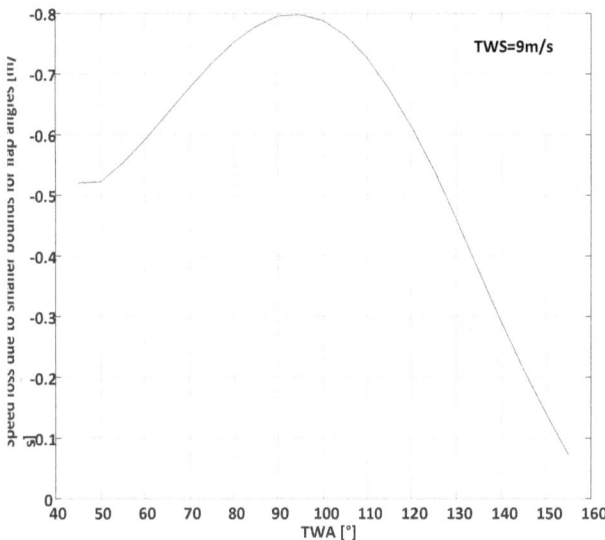

Figure 26: Speed losses due to smaller bounds for flap trimming

Impact of wing trim constraints

The results achieved so far have be calculated with a constraint on the wing flap restricting it to angles $-5 \leq \beta \leq 30°$. This results in a strong twist of the wing, with negative flap angles as well as negative lift at the top of the wing, see Figure 15. This is due to the heeling moment constraint, letting the optimizer search for a trim with a low vertical center of effort. Achieving such a large range of flap angles is not easily done in reality and needs some extra development effort of the wing builders. Consequently it is of high interest, if such high twist angles for the flap are really necessary.

The following calculations generate an excerpt of the polars, comparing the results achieved so far with a yacht with a constraint of the flap angle, restricting it to $5 \leq \beta \leq 30°$. This means the flap was not able to be opened as much as it has been done so far.

Figure 26 shows the result of a comparison of the speed achieved with wide bounds for the flap trim and the speed achieved with the restricted bounds for the trim of the flap. It can be seen that allowing an inversed trim of the flap gives significantly faster boat speed than a trim with no inversed flap trim. Restricting the trim of the flap to an angle larger than 5° results in a speed loss of up to 0.8m/s at TWA=95°. On an upwind tack the speed loss is about 0.5m/s, still a very large number. It can be concluded that even small changes in the handling of the yacht can result in quite significant speed changes.

Conclusion

This paper presents a method for the prediction of wing generated forces on a catamaran equipped with a jib. It is combined with a hydrodynamic model for a water surface piercing daggerboard to predict the optimized trim of the wing and jib and the catamarans speed achieved with the optimized wing trim. It is demonstrated that the general approach works reasonably well and can be used to predict the optimized trim of the wing with jib and the speed of the yacht for various conditions. However much room is left for further research. The hydrodynamic model is rudimentary and needs refinement. The aerodynamic profile properties of a jib are predicted with a simplified model and the windage drag is only estimated. In addition pitching moment and heeling moment should be considered as equality rather than inequality constraint.

An additional topic of further research will be validation of the achieved results. This remains to be done after the next Americas Cup when respective data is getting available to the public.

REFERENCES

Graf, K., van Hoeve, A. and Watin, S., "Comparison of full 3D-RANS simulations with 2D-RANS/lifting line method calculations for the flow analysis of rigid wings for high performance multihulls", J Ocean Engineering,

2014.

Menter, F.R., "Two-equation eddy-viscosity turbulence models for engineering applications", AIAA journal, 1994.

Meyer, J., Renzsch, H., Graf, K. and Slawig, T., "Advanced CFD-Simulations of free-surface flows around modern sailing yachts using a newly developed OpenFOAM solver", Society of Naval and Marine Engineers, The Twenty-Second Chesapeake Sailing Yacht Symposium, Annapolis, Maryland/USA, March 2016.

Truckenbrodt, E., Schlichting, H., „Aerodynamik des Flugzeuges: Grundlagen der Strömungsmechanik Aerodynamik des Tragflügels" Teil 1 und 2, 2nd ed. Springer, Berlin, 1967.

Anderson, "Fundamentals of aerodynamics", 5th ed. McGraw-Hill, Singapore, 2011

America's Cup Race Management, America's Cup Class Rule Version 1.2, https://drive.google.com/file/d/0B57wxypH8i5sWm1iTDNnZVU3bjA/view?pli=1, 23rd Oct. 2015

A Comparison of a RANS Based VPP to on the Water Sailing Performance

Tyler Doyle, Doyle CFD, Salem, MA
Bradford Knight, Doyle CFD, Salem, MA
Duncan Swain, Doyle CFD, Salem, MA

ABSTRACT

This paper compares performance predictions from a Reynolds Averaged Navier Stokes (RANS) based Velocity Prediction Program (VPP) to on the water testing of a J70. The J70 has been outfitted with a system to determine sail flying shapes, apparent wind conditions and performance data. The on the water testing is conducted in both racing and controlled sailing conditions. Data taken during racing conditions is analyzed to determine optimal performance envelopes while data taken in controlled conditions is used to match exact sailing and VPP states. The data acquisition system combines a number of standard marine sensors including a sonic anemometer, a GPS, a digital compass, an accelerometer and a gyroscope with custom sensors that measure rudder and boom angles as well as a custom sail shape acquisition system. The RANS based VPP developed by Doyle CFD has three main components; an aerodynamic force model, a hydrodynamic force model and an algorithm to balance the forces. The force balance routine uses four degrees of freedom; boat speed, yaw, heel and rudder angle to balance the aerodynamic and hydrodynamic forces for a given true wind speed and angle. The force models are derived from RANS CFD data calculated using OpenFOAM. The aerodynamic forces are calculated using steady state RANS as a function of apparent wind angle, apparent wind speed and sail flying shape. The VPP force model is derived by fitting response surfaces to this data. The aerodynamic CFD is run with sail flying shapes recorded from on the water testing. Using accurate flying shapes is critical for picking out slight aerodynamic differences in sail and rig setup. The hydrodynamic CFD data points are calculated using RANS Volume of Fluid CFD (VOF) as a function of boat speed, rudder angle, yaw angle, heel angle and displacement. Response surfaces are generated from a 128 data point array of RANS VOF simulations.

NOTATION

CSYS	Chesapeake Sailing Yacht Symposium
γ	Yaw Angle
θ	Rudder Angle
ϕ	Heel Angle
ρ_{air}	Air density
a	Hellmann Exponent
AW	Apparent Wind Vector
AWA	Apparent Wind Angle

AWS	Apparent Wind Speed
BS	Boat Speed Vector
C	Crew Position
F_{air}	Aerodynamic Forces and Moments Vector
F_{hydro}	Hydrodynamic Forces and Moments Vector
FX_{air}	Aerodynamic Driving Force
FY_{air}	Aerodynamic Side Force
FX_{hydro}	Hydrodynamic Resistance Force
FY_{hydro}	Hydrodynamic Side Force
LWL	Length of Waterline
MH	Mast Height
MX_{air}	Aerodynamic Heeling Moment
MZ_{air}	Aerodynamic Yaw Moment
MX_{hydro}	Hydrodynamic Heeling Moment
MZ_{hydro}	Hydrodynamic Yaw Moment
S	Sail Shape
SA	Total Sail Area
TW	True Wind Vector
TWA	True Wind Angle
TWS	True Wind Speed
v_{10}	Velocity 10 meters above waterline
h	Height above waterline
h_{10}	Height at ten meters= 10 meters

INTRODUCTION

This paper compares performance predictions from a RANS CFD based VPP program to sailing data logged on an instrumented J70. High performance computing is now fast and cheap enough to make running RANS based VPP's a reality for many types of projects. The authors have developed a RANS based VPP system that is used for performance, balance and optimization studies on boats ranging from mega yachts to J70's. For studies where sailing data has been available we have attempted to make comparisons with the VPP predictions. The comparisons have been encouraging, however it has been difficult to obtain enough high quality data to make exact comparisons using typical onboard instrument systems. To validate, tune and continue to develop our RANS based VPP system a custom data logging system for a J70 was designed and built. Sail flying shape, AWS, AWA, boat speed, heel angle, and rudder angle are recorded to enable a direct comparison. The J70 is selected as a test platform because one of the authors owns and actively races one providing for ample opportunities to collect data and experiment with various sensors systems. This paper describes our first attempt at comparing upwind J70 performance to VPP predictions. The on the water testing program also included taking data on downwind performance that will be compared to predictions in a later study.

RANS based VPP's have been used by researchers mainly in high end race programs such as the America's cup and other grand prix racing for twenty years. RANS CFD models the entire flow field around the boat from vortices coming off the tips of the sails to waves being generated by the bow. RANS CFD takes into account turbulent flow phenomena important to sailing such as viscous drag and flow separation. Most other numerical force prediction methods that have been used for VPP force analysis such as panel codes only model the surface pressures. In addition to providing force data to the VPP program, the RANS solutions can be post-processed to look in detail at flow features of interest. The large amount of flow information comes at a high computational cost. Until recently the cost of computing needed for RANS based VPP's limited their use to high budget projects. Today high performance computing is an affordable commodity that allows the use of RANS based VPP's in many type of projects.

Data from RANS CFD solutions can be used in one of two ways to build the VPP force models. The data can either be used to create response surfaces from precomputed data sets as is typically done with wind tunnel or tow tank testing, or it can be calculated as needed by integrating the VPP solver directly in one or both of the aerodynamic or hydrodynamic RANS solvers. Korpus gives a good description of the response surface implementation while Bohm gives a good description of a RANS based VPP that only calculates needed data points (Korpus, 2007), (Bohm, 2014). A combination of both implementation methods can also be used where the hydrodynamic force models are precomputed and the aerodynamic simulations are run as needed or vice versa. This method may be especially useful in doing shape optimization studies because as the shape changes only local data needs to be computed. The response surface method of force model implementation is used in this study. For each comparison point an appropriate sail flying shape is selected from recorded data and the aerodynamic solver sweeps through a small range of AWA's and heel angles around the sailing point of interest. A response surface of force coefficients as a function of AWA and heel is generated. The hydrodynamic response surfaces are calculated once and used for all analysis point as functions of boat speed, heel, yaw and rudder angle

No matter how the VPP is implemented many RANS simulations will be needed, requiring robust CFD modeling and efficient case management. Open source CFD software OpenFOAM is used to meet the computation requirements with high efficiency, robustness and reduced cost.

INSTRUMENTATION

Purpose of Instruments

High performance racing yachts, such as those in the America's Cup and other competitive racing circuits have many systems that constantly monitor the boat performance with a system of instruments on the hull and rig.

The goal of designing and creating a data collection and monitoring system for the J70 was to produce a system with capabilities similar to those that are in use on larger vessels, while retaining the ability to access the collected data in an open source environment. The data collection system was also intended to be a low impact device enabling the yacht to be sailed during competitions without the crew having to change their sailing style. Data on sailing performance was able to be obtained in racing conditions when the boat was sailing near peak performance, as well as in non-racing conditions.

The data collection system operated with four hardwired sensors and incorporated three remote cameras for capturing images of the sails. The hardwired sensors were composed of two solid state anemometers, one masthead mounted, the other on the bowsprit, a combined GPS/IMU unit and a rudder angle sensor. The system was connected as indicated in Figure 1. The bowsprit anemometer was only deployed during non-racing conditions, since the location of the sensor made usage impractical in conjunction with the spinnaker.

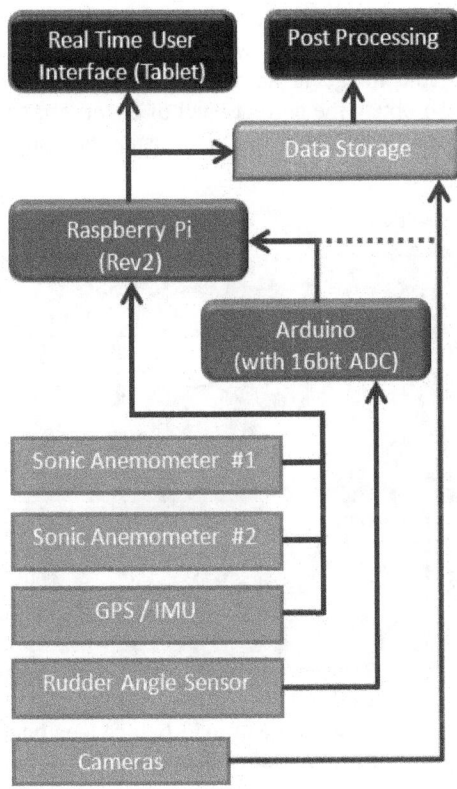

Figure 1 – System Schematic: Overview of data collection system implemented for testing on the J-70

Sensors

A Raspberry Pi microcomputer functioned as the brains of the system, running a Linux based operating system with custom scripts for processing and recording the data from the sensors. The anemometers and GPS/IMU were standard marine grade electronic sensors that used the NMEA0183 communication protocol. NMEA sentences were obtained from these sensors via RS232 serial converters connected to the Raspberry Pi. The Raspberry Pi also received a running average of the rudder angle from an Arduino microcontroller that was connected to the rudder angle sensor via an external 16bit ADC.

All the data from these sensors was collected, displayed and recorded at a rate of 1Hz. The sampling rate was limited to 1Hz by the stock marine electronics since the NMEA communication protocol baud rate did not allow for high rates of data output. The 1 Hz data sampling rate was adequate for the testing needs of this project, since the primary interest was how the yacht preformed in steady state situations. In future versions, the sensors could be exchanged to allow higher data collection rates enabling more sensitive and accurate measurement of the yacht's dynamic motion.

When running the system as seen in Figure 2, data was collected from both anemometers. Collecting data from the two anemometers at different heights on the yacht allowed a basic wind velocity profile to be constructed. The wind profile data was useful to determine the wind shear effect on the J70's sails.

Figure 2 - Data acquisition system in use during sail testing in Marblehead harbor

The GPS/IMU unit on the boat provided the system with data on the boat's SOG, COG, heel, pitch, and corrected magnetic heading. The AWA and AWS obtained from the anemometers enabled TWA and TWS calculations to be performed with correlation to the other data. All of the data was time stamped and logged automatically by the system while testing was occurring. All of the information was also output to the user in real time via a portable tablet and Wi-Fi network broadcast from the data system.

The live data broadcast to the tablet enabled the system to be evaluated during testing sessions. User evaluation ensured that the system was performing as expected and helped to enhance the performance of the yacht by giving the crew extra information to utilize on the water.

Camera System

Cameras were incorporated into the data collection system to generate three dimensional models of the real world flying shapes of the sails. All three cameras were variations of the GoPro Hero3, a rugged mobile camera system with wireless capabilities. Two of the cameras were used to capture images of the mainsail, one for each tack, while the third was used to capture images of the jib on both tacks. The cameras were all deck mounted in the locations indicated in Figure 3. The deck mounted configuration enabled user access and simplicity. Other sail stripe vision systems use masthead mounted cameras to get a better perspective on the bottom of the sail. (LePelley & Modral, 2008) Wide angle lenses were used to capture the bottom sail stripes.

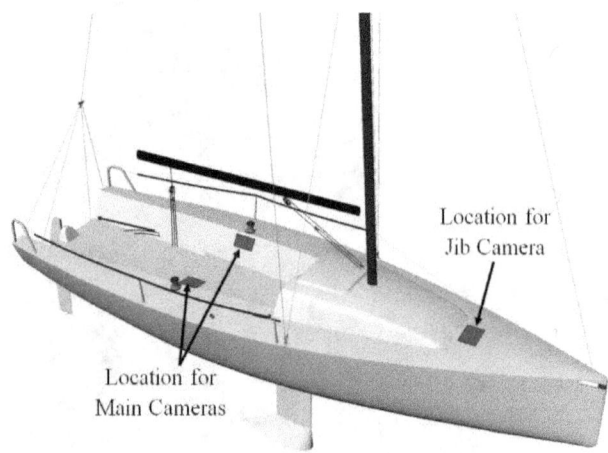

Figure 3 - Placement location of GoPro cameras on the J-70

There were four stripes applied to the jib and five stripes applied to the mainsail as indicated in Figure 2. The number of shape stripes was chosen to accurately capture the shape of the sails, without artificially adding excess structural reinforcement. The stripes were applied parallel to the deck in order to keep the stripes orthogonal to the cameras and minimize the perspective effect.

Lens Distortion Corrections

Two types of lenses are used in the system. Standard GoPro wide angles lenses are used for the mainsail while a fisheye lens is used for the jib to enable full capture of the

bottom shape stripe. The bottom shape stripe of the main is far enough away from the camera that a fisheye lens is not needed. The jib lens is a full-frame fisheye that has the image circle from the lens circumscribed around the sensor area. Both lenses created barrel distortion in the images. The distortion was especially apparent on the images of the jib where an extra wide angle, a 170 degree viewing angle lens, was used to capture the images of all the jib shape stripes.

The images from the wide angle lens as seen in Figure 4 have significant distortion, and need to be corrected before they can be accurately analyzed. To correct the images the image editor GIMP 2.8 was used. In GIMP there is a prebuilt filter that corrects for lens distortion caused by a fisheye lens. (Gimp, 2015) The filter adds a spherical distortion correction to the image to make the image concave, reversing the effects caused by the fisheye lens. The correcting filter followed a cubic interpolation for distorting the image in an attempt to reduce the fisheye effect. To obtain the correct level of distortion, correction images were taken from multiple distances from the grid shown in Figure 4.

Figure 4 - Test image of grid before lens barrel correction was applied

An iterative process was used to obtain the proper correction values for the lens. The process involved making a best guess as to the needed correction values for the image and applying them. The filter was applied iteratively, up to eight times to get the corrected image without large distortion, since the correction algorithm is not a linear operation. The image was then imported into Rhino 3D where the straightness of the lines in the grid was evaluated. If the grid was not straight then the correction values for the filter were modified in an attempt to obtain straighter lines. The process was repeated until the lines were straight and the images looked like that displayed in Figure 5. Some small amount of distortion still remained near the corners of the image where the distortion caused by the lens was the greatest and the corrections were the most extreme. The small remaining distortion at these points was deemed insignificant for the processing of the shape stripes as the stripes were generally centered, and did not extend far into the corners of the images.

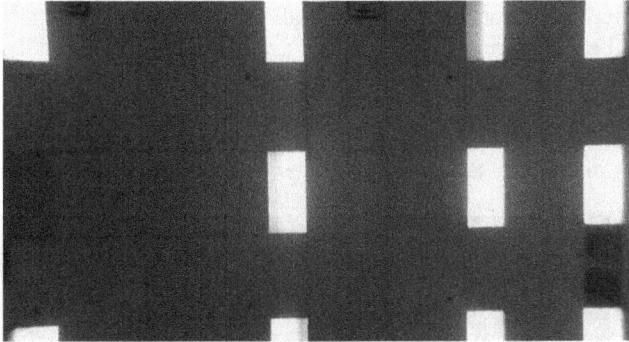

Figure 5 - Test image after lens barrel correction was applied

Figure 6 depicts an image of the jib before the barrel distortion correction is applied and Figure 7 illustrates the image after the correction is made.

Figure 6 - Image of jib before barrel distortion correction is applied

Figure 7 - Image of Jib after barrel distortion correction is applied

FLYING SHAPE GENERATION

Image Stripe Extraction

In this work, three-dimensional sail flying shapes for use in the aerodynamic CFD simulations are generated using an automated procedure driven by python scripts. There are a number of ways to reconstruct flying shape surfaces from images of sails. Some approaches map 2D image coordinates directly to 3D model space. Other approaches, such as the one used in this paper, perform sectional shape measurements in image coordinates and then create 3D points by scaling the sectional parameters and locating them in 3D space. After the image has been corrected for distortions caused by the wide angle lens the stripes are extracted from the images using algorithms from the open-source image processing library OpenCV. OpenCV's image processing algorithms are fast allowing the stripes to be extracted from images in less than half of a second. The extracted stripes are then digitized to extract sectional shape parameters similar to the sectional shape parameters used in sail design programs. The sectional shape data is saved in text files that are used as input for the geometry creation script. A python script generates the geometry by lofting sectional curves into surfaces in the 3D surface modeling program, Rhino 3D. The sail and mast geometry is then exported to the appropriate CFD case directory.

Shape Stripe Extraction and Analysis

The first step in generating a 3D model of the sail flying shape is to extract the sail shape stripes from the images and parameterize them. A summary of the stripe extraction and analysis algorithm is given below:

- Crop and rescale image
- Convert to HSV color space
- Apply HSV threshold
- Calculate image contours
- Filter contours by length and position
- Fit a polynomial to contours that are stripes
- Parameterize stripes by measuring polynomials
- Calculate mast bend or head stay sag

To reduce processing time and filter out unneeded regions in the images the first step is to scale and crop the images. For each camera the system uses separate scale and crop settings. The limiting factor in the scaling operation is the number of pixels needed to capture the width of the top stripes. The shape stripe processing algorithm works best when at least three pixels are captured across the width of the top stripe. For the cameras used in this paper a scale factor of two achieves a good balance between needed resolution and processing speed.

After the pixel count has been reduced the next step is to convert the images from the standard camera pixel color space of red-blue-green (RBG) to the hue-saturation-value (HSV) color space. This transformation makes it easier to extract the stripes according to color because hue values remain steadier through the range of lighting conditions than RGB values do. A number of different stripe colors were tested and it was found that bright orange worked the best because it maintained its hue value range from bright light to shadow conditions. A HSV threshold was applied to the image to create a new binary image with only the orange pixels. Contours, sets of pixels that enclose a contiguous region, are then extracted from the binary threshold image. Along with the shape stripes other unwanted image features are sometimes included in the list of contours. The contours are filtered according to length and position in the image to determine which contours are sail stripes and to order the stripes. This method of extracting the stripes requires that the entire stripe be contained in one contour. This is problematic when the image contains defects such as water droplets on the camera lens. A more robust implementation of this procedure can be developed to handle image defects; however, stripes with defects are not analyzed in this study. During testing it was found that defects such as water droplets, regions of high light intensity and object interference usually were only temporary and were cleared within a few images.

Once the list of shape stripe contours is built a best fit polynomial is fit to the contours which are internally stored as sets of points that define the boundary of the contour. The process creates a centerline curve of the contour. The polynomial fit parameters for each stripe are stored to be used in the shape parameterization. The analysis of the shape stripe curves starts by calculating the section twist defined as the angle of the chord line to the horizontal axis. Next the maximum depth of the section is calculated together with its location on the chord line. The chord line is then broken into a front and back section and the mid points of these sections determined. The front section is split again and the midpoint of this section found. The depth of the shape stripe curve at these four points (mid, front, back and front-front) is then calculated. The maximum depth is divided by the chord length and the depth of the other three points is divided by the maximum depth to create non-dimensional shape parameters. This sectional shape parameterization is sufficient to capture the range of flying shapes of interest including back winding of the luff of the sails. The shape stripe extraction and parameterization steps are shown graphically for the mainsail in Figure 8 and for the jib Figure 9. The polynomial fit centerlines are shown aligned at the top of images in the position where they are measured. The extracted shape stripe contours are shown in the middle of the images.

The final step in the image analysis procedure is to calculate the mast bend or head stay sag using a similar procedure as is used for the stripe analysis. The leading edge points of the stripes are used to fit a polynomial and the maximum depth and location of the maximum depth are calculated. Theoretically the leading edge curve analysis is not as accurate as the sectional analysis because of perspective problems and because the entire length of the head stay or mast is not captured however during

testing it was found that the measured values using this procedure correlated well with more accurate measurements taken from images perpendicular to the mast. In real sailing, mast bend and head stay sag are critical parameters in setting sail flying shape because they have a large influence on the depth and twist of the sails. The accuracy of the sail flying shapes in this system is not as dependent on the leading edge deflection as it is in the real world because the sectional shapes capture the influence of the leading edge deflection directly.

Figure 8 - Overview of extraction of shape stripe geometry from corrected image of main

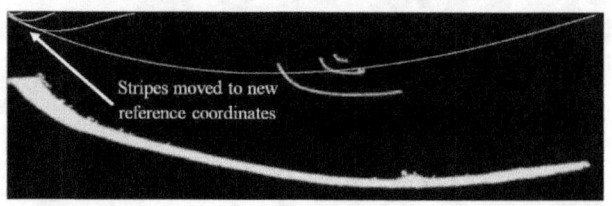

Figure 9 - Shape stripe contours (shown in green) are isolated from other parts the image and then the best fit polynomials are translated to the image origin as indicated for processing

Flying Shape Geometry Generation

A python script is used to generate the sail and mast flying shapes using the surface modeling program Rhino 3D. The sails and mast are generated by lofting sectional curves that are defined by the shape stripe parameters calculated in the stripe analysis. Prior to generating flying shapes a configuration step is necessary to define the end points of the shape stripes along the leech and luff of the sails, the length of stripes and also the shape of the mast bend. The mast bend is defined as the percentage of max bend and the mast sections offset according to bend. The section end points are then located in the 3D model by placing the sails on the mast in the proper location.

The flying shape generation procedure starts by reading in the shape stripe parameter files. The non-dimensional shape stripe parameters are converted to the 3D model coordinates by scaling the parameters according to the

21

shape stripe length calculated in the configuration step. A curve is drawn for each section and then moved to the appropriate base point location. The base points used for the mast, main and jib are shown along with the parameterized section curves in Figure 10. The base points are initially defined along straight lines from the tack to head and then offset in the horizontal plane according to mast rake, mast bend and head stay deflection. The section curve is rotated according to the twist of the section and the sheeting angle of the sail. In addition to the shape stripe sections it's necessary to also define the head and the foot sections of the sails based on the parameters of the closest shape stripe. Simple algebraic expressions are used to link the head and foot curves to their closest shape stripe. The mast sections and main sail sections share the same base point to ensure that the luff of the main and the mast are well aligned. The final step is to loft all of the section curves and convert the mast shell into a closed solid. The sails and mast are then exported as STL surface mesh files to the appropriate CFD case directory.

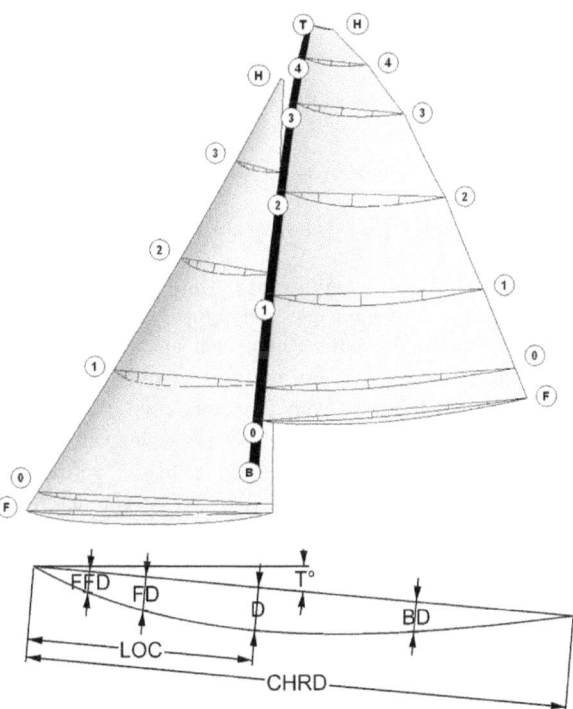

Figure 10 - Sample sail stripe after extraction from image and movement to origin for processing

Generation of Hydrodynamic Model Shape

Before the CFD simulations were run, an accurate model of the J-70 yacht hull had to be created. Since there was no readily available model of the hull a model had to be generated. Section profiles of the hull were obtained by affixing lines to the hull and imaging them from a distance as shown in Figure 11. The images were then traced and the shape of the lines extracted in a similar manner that was used for the sail shape stripes. The lines on the boat were

aligned using a laser level for vertical accuracy and the length of the stripe was measured with a flexible tape. The data from the images was then imported into a modeling program where all the individual hull sections were lofted together to generate the full shape of the hull.

Figure 11 - Image of line on the hull that was used for extraction of a hull shape section

Generating the shape of the appendages on the boat used a combination of the image shape extraction technique and a physical jig. The jig, as pictured in Figure 12, enabled accurate measurement of the foil shapes extending off the main hull.

Figure 12 - Jig measuring shape of keel foil for model generation

The combination of the appendage models and the hull model created an accurate model of the hull that was verified through more physical measurements and overlays of images like Figure 13.

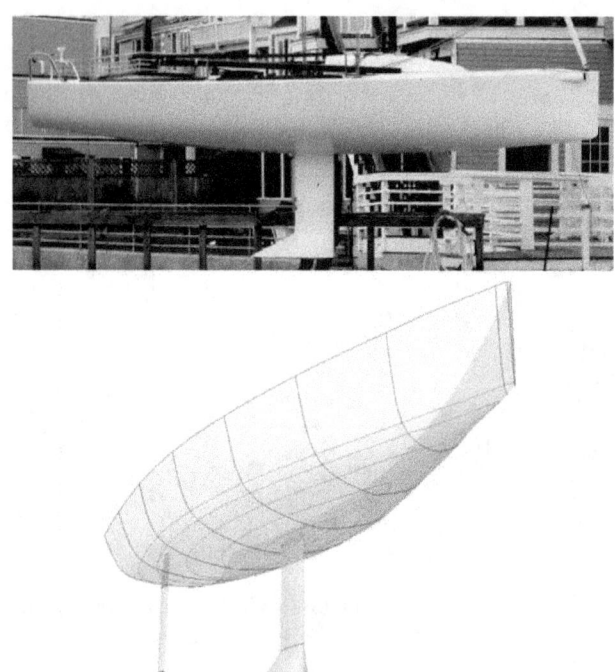

boat speed by assuming no current. Current was not measured but is not large where the data was collected in Salem Sound. Since the masthead anemometer with no heel is at 10m above waterline, Equation 1 uses the true wind velocities calculated from each anemometer, where v_{10} is masthead TW, h is the height above waterline, h_{10} is the masthead height of 10m, and the a is the Hellmann coefficient. The Hellmann exponent is experimentally determined using logarithmic fitting to match the velocities that the upper anemometer calculated (Kaltschmitt, 2007).

$$AW = TW + BS = v_{10} \left(\frac{h}{h_{10}}\right)^a + BS \qquad \text{(Eq 1)}$$

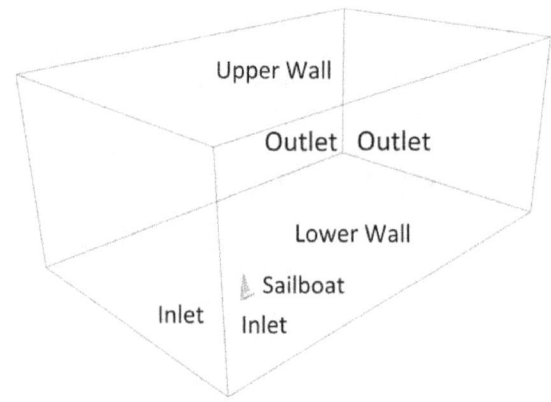

Figure 13 - Image used for hull shape verification shown at the top and on the bottom the final hull sections and waterlines used to ensure the hull is fair

Figure 14 – Aerodynamic Domain

RANS VPP METHOD

The RANS CFD based VPP uses aerodynamic forces generated from steady state RANS CFD, hydrodynamic forces generated from RANS Volume of Fluid (VOF) CFD, and a force balance algorithm. Linux based OpenFOAM version 2.2.1 is used for CFD software. The overall method is the same as Doyle & Knight's 2015 paper, though more efficient. (Doyle & Knight, 2015)

Steady State Aerodynamic CFD Setup

SimpleFOAM, the steady-state incompressible flow solver in OpenFOAM, is used to analyze the aerodynamic forces of the hull, sails, and mast. The kω-SST model turbulence model is used with wall functions.

Figure 14 depicts the boundaries of the domain. The domain is 80 meters tall, 120 meters wide, and 190 meters long. The outlet is specified as an inlet-outlet with 0 inlet velocity, the top and bottom walls are applied as slip walls, and the sailboat is a no-slip wall with wall functions. The inlet is applied as a groovyBC using swak4FOAM so that the wind gradient can be applied. The inlet velocity is specified to match the apparent wind vector (AW) from the sailing data, as a function of the boat speed vector (BS), and the true wind vector (TW). The apparent wind vector applied at the inlet is specified by Equation 1 below. True wind speed varies as a function of height above waterline due to the boundary layer above the water. The apparent wind speed vector is known from the anemometers on the J70. The speed over ground is assumed to be the same as

A 3.8 million cell mesh is generated using OpenFOAM's native mesher, SnappyHexMesh. The hull, mast, jib, and main sail have seven surface refinements applied and prism layers are not used. A refinement box is formed around the J70 from 2 meters forward, 3 meters behind, 1 meter to windward, 4 meters to leeward, and 10 meters above. This refinement zone has 3 refinements applied. Figure 15 illustrates the mesh characteristics. Meshes generated from SnappyHexMesh are hexagonal meshes. The diagonal lines in the image are an artifact of the post processing. The top image depicts a cut plane showing the mesh around the jib, mast, and main. The bottom image shows the mesh representing the ocean surface, the sailboat, and cut-planes behind and to leeward.

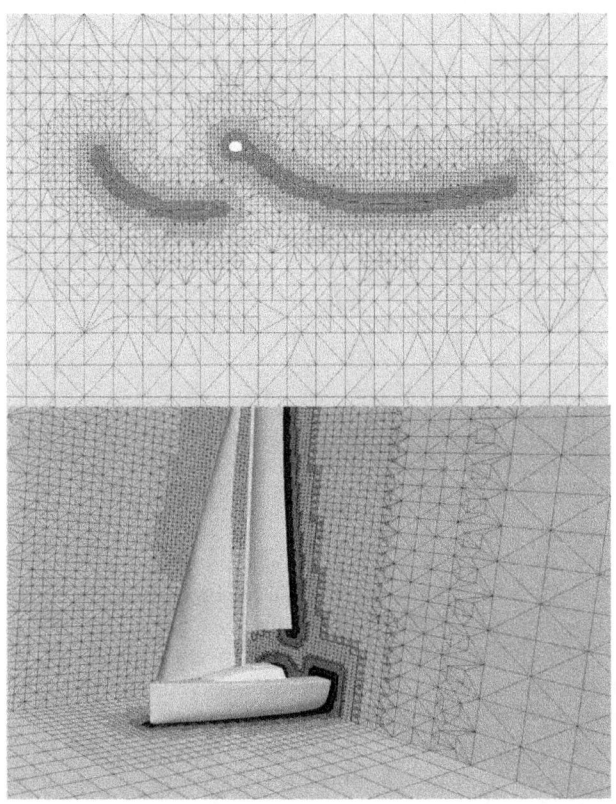

Figure 15 – Aerodynamic CFD Mesh

Prism layers are not used in the mesh since the yplus values are in an acceptable range. Adding extra refinements to the sail surfaces provides a higher quality mesh and less defeaturing. The range of apparent wind speeds investigated was between 10.5 kts and 17.3 kts. The average yplus at 10.5 kts was 61 on the jib, 74 on the main, 66 on the mast, and 75 on the hull. The average yplus for the 17.3 kts case was higher at 86 on the jib, 114 on the main, 80 on the mast, and 128 on the hull. The average yplus is in the logarithmic range, 30-300, required to use wall functions for the full range of apparent wind speed.

VPP aerodynamic data sets are generated for each sail flying shape geometry by varying the apparent wind and heel angles in a range that captures the sailing point of interest. Force and Moment coefficients for the high TWS comparison point geometry are shown in Figure 16. This is a depowered configuration that has 40% less heel force coefficient than the light wind comparison point geometry. The force coefficients are calculated from the forces as follows:

$$CxA = 2*FX_{air}/(\rho_{air}*AWS^2*SA)$$
$$CyA = 2*FY_{air}/(\rho_{air}*AWS^2*SA)$$
$$CmxA = 2*MX_{air}/(\rho_{air}*AWS^2*SA*MH)$$
$$CmzA = 2*MZ_{air}/(\rho_{air}*AWS^2*SA*LWL)$$

The reference sail area used is the three-dimensional surface area. The reference length used for the heel moment coefficient is the mast height and the reference length used for the yaw moment coefficient is the waterline length. The flying shape geometries are imported into the OpenFOAM platform as STL surface mesh files.

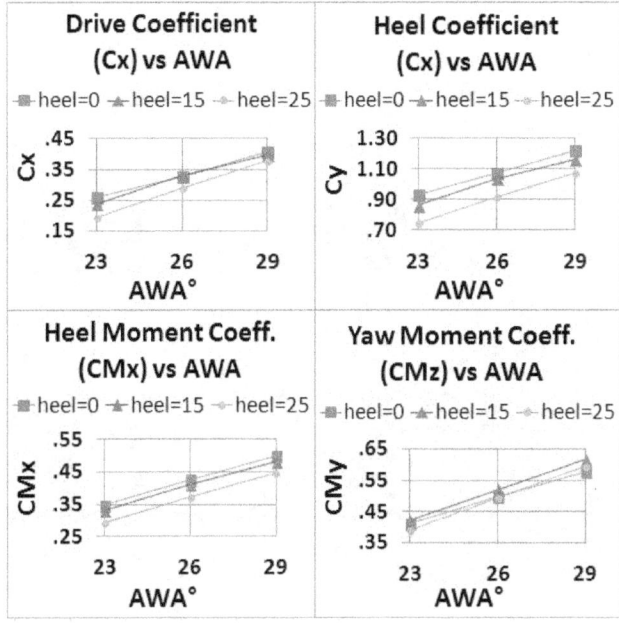

Figure 16 – Aerodynamic force and moment coefficients from the high TWS comparison point geometry are plotted as functions of AWA for three heel angles.

Figure 17 below depicts the RANS CFD representation of the flow for an over eased case represented by 30 degrees AWA on the right and overly trimmed at 20 degrees AWA on the left. The top most images show the location of the cut planes and the following images show velocity magnitudes contours for 8 meters, 5 meters, and 1.5 meters above waterline. For over trimmed case on the left, there is some separation towards the leech of the jib shown by the dark colored contours. At the midsection of the mainsail there is some separation towards windward showing that the main may have been excessively eased even at 20 degrees AWA. On the right separation towards windward on the main is large which shows that the main is too eased for that condition.

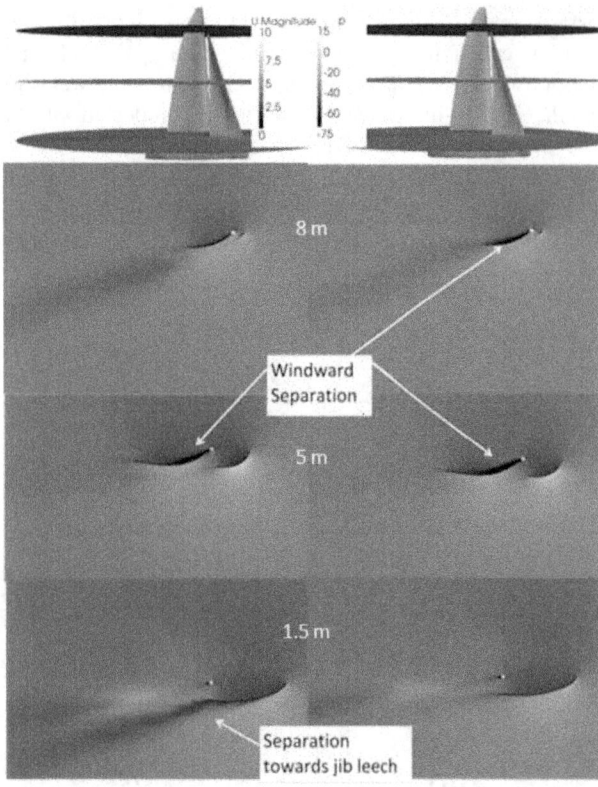

Figure 17 - Velocity magnitude at 1.5, 5, and 8 meters above waterline for over trimmed and under trimmed AWA

Volume of Fluid CFD Setup

OpenFOAM's steady state VOF solver, LTSInterFOAM, is used to determine the hydrodynamic forces at conditions of boat speed, heel angle, yaw angle, and rudder angle. The kω-SST model turbulence model is used with wall functions.

Figure 18 depicts the domain used for the hydrodynamic solutions. The inlet is a velocity inlet, the outlet is specified as an inlet-outlet with 0 inlet velocity, the 'sides' boundary on the bottom of the domain is a symmetry plane, the atmosphere condition at the top is specified as a pressureInletOutletVelocity condition where the atmospheric pressure is specified, and the boat is a no-slip wall with wall functions. The waterline is specified at height equal to zero, with the domain extending 10 meters below and 7.5 meters above. The domain origin is located at the forward most point of the hull at the center and extends 19 meters forward, 50 meters back, and 25 meters on each side. The boat is rotated about the longitudinal axis to vary heel and is translated vertically to evaluate the forces with respect to displacement. Yaw is applied by varying the longitudinal and sideways velocity at the inlet. By applying the yaw at the inlet, the number of meshes necessary to create the solution matrix for the VPP is reduced and the VOF simulations share the same coordinate system as the aerodynamic simulations. Reducing the number of meshes reduces complexity and

troubleshooting time. The other way to apply yaw is to rotate the hull about the vertical axis and apply boat speed only at the forward most boundary of the domain.

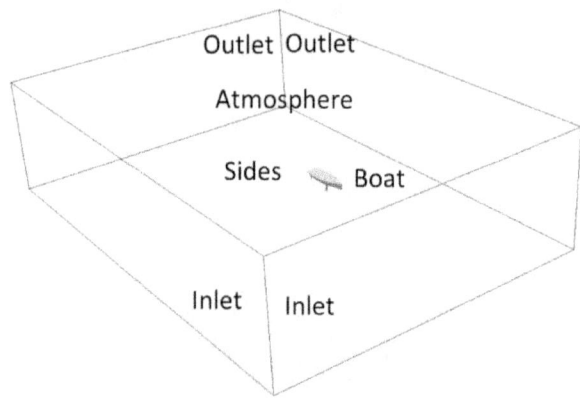

Figure 18 - VOF Domain

SnappyHexMesh is used to generate the meshes. The mesh count must be sufficiently small around the hull, appendages, and the free surface to correctly evaluate the hydrodynamic forces. The OpenFOAM utility blockMesh is used to create an initial biased mesh. The biased blockMesh has a small aspect ratio at the center of the domain, but has a larger aspect ratio towards the lateral and vertical boundaries. By adjusting the blockMesh, the cell count is reduced, while maintaining a high cell count at the waterline and near the boat. Five surface refinements are applied to the appendages, two surface refinements are applied to the bottom and sides of the hull, four are applied to the leading edge of the hull, one to the transom, and none extra to the deck. Prism layers are applied to the surfaces that are wetted, but not to the deck. Five prism layers are applied to the hull and appendages. One prism layer is applied to the transom to improve the mesh quality at the transom edge. Figure 19 depicts the hydrodynamic mesh. The top-most element depicts the surface mesh of the boat, the middle image shows the biased mesh with cut-planes through the middle of the boat and to the side, and the bottom image depicts the mesh close to the keel and bottom of the boat.

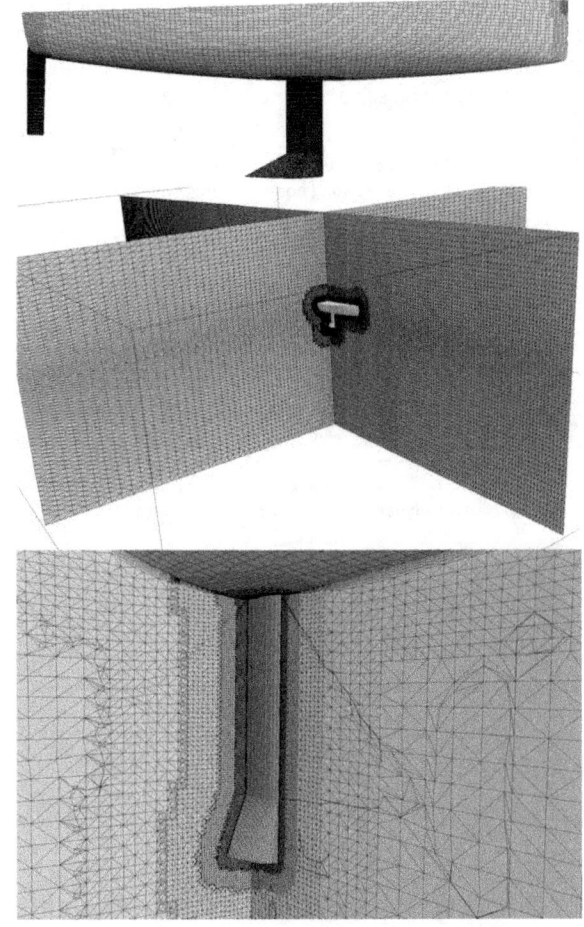

Figure 19 - Hydrodynamic Mesh

The average yplus values for wetted surfaces are within the logarithmic range for all data boat speeds except the hull for Fr= 0.5 and the appendages for Fr=0.2. Fr=0 .5 is for 7.86 kts and Fr= 0.2 is 3.15 kts. The slight yplus flaw should minimally affect the results, but further investigation should be performed. Figure 20 below depicts the average yplus of hull and appendages in the top plot. The two contours of the boat show the yplus of Fr= 0.2 in the middle and Fr= 0.5 on the bottom.

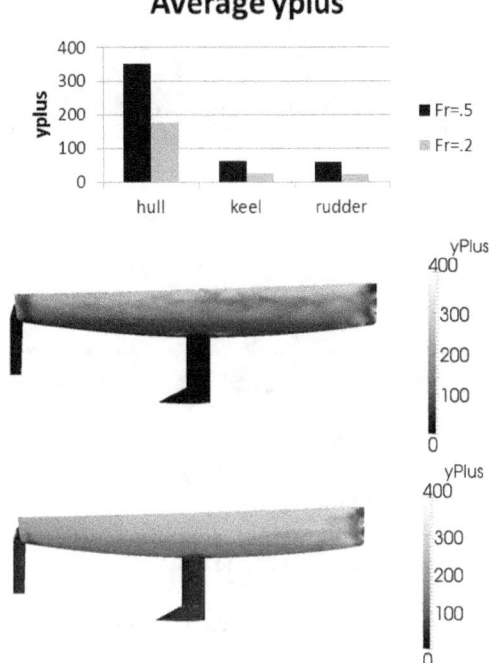

Figure 20 - VOF yplus at Fr= 0.5 and Fr= 0.2

Complete RANS VOF Driven Hydrodynamics

RANS VOF is used to calculate each data point to form a structured input file for the VPP. Four hull speeds are run ranging from Fr= 0.2 to Fr= 0.5 in .1 increments, 3.14 kts, 4.72 kts, 6.30 kts, and 7.87 kts respectively. At each boat speed, the yaw is varied from 0 to 9 degrees in 3 degree increments and for each yaw variation heel is varied from 0 to 30 degrees in 10 degree increments. Each of these cases are simulated at two displacements. The result is a 128 point matrix that is input to the VPP. Figure 21 depicts the wake of two cases, on the top the wake from above is depicted, as a function of wave height, and the bottom shows streamlines around the keel and the waterline at the front of the hull. The image on the left is for the upright boat at 0 yaw at Fr= 0.2, while the image on the right depicts the other maximum extreme of the highest speed, at 9 degrees, yaw and 30 degrees heel.

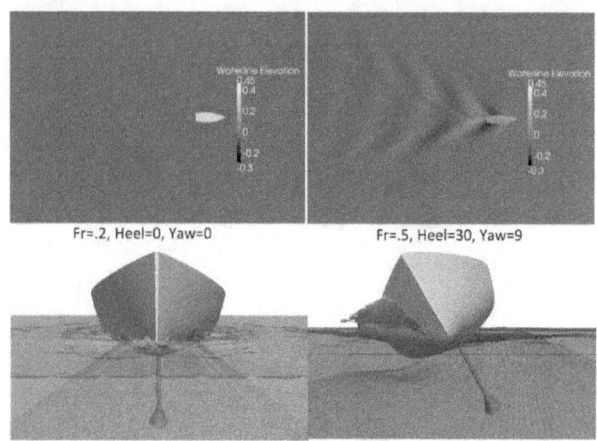

Figure 21 - Wake and flow visualization for Fr= 0.2, heel = 0, yaw = 0 on left & Fr = 0.5, heel = 20, yaw = 0 on right (The minimum speed, heel, and yaw is represented on the left and the maximum speed, heel and yaw on the right)

The effect of the rudder is added as a correction. Eight extra data points were simulated at Fr= 0.4 at 0, 3, 6, and 9 degrees yaw and 0 degree heel at 5 degrees windward helm. The resulting forces are interpolated for each yaw and input into the VPP as a function of rudder angle.

Simulations are performed with fixed sink and pitch. The effect of the pitching moment is ignored herein since time did not allow for more data points. For each data point in the VPP matrix, steady state VOF solutions are analyzed at two displacements and the forces are interpolated between the two displacements to determine the forces acting on the hull and appendages at the correct displacement. In OpenFOAM, the transient moving mesh VOF solver, interDymFOAM could be used to evaluate the forces for each data point with just one solution. InterDymFOAM was not examined in this paper as the steady state solver was deemed sufficient and the added complexity of a moving mesh would reduce the robustness of the solutions. Robustness of solutions is important when generating a large number of solutions since the RANS driven VPP workflow would be significantly hindered if significant troubleshooting of solutions was required. Each solution took approximately 12 hours on a node with 2 Xeon processors and 60 GB of RAM. In general simulations were robust and converged. One data point that had convergence issues was that the upright Fr= 0.2 case began to diverge towards the end of the solution. For this data point forces of the solution before divergence were used in the VPP.

Figure 22 shows the resistance, lift, and yaw moment curves as a function of BS for different yaw angles at 10 degrees heel. As BS increases the resistance increases as a function of wave making forces, skin friction, and induced drag. When analyzed in the boat's reference frame, as yaw increases, the lift increases but the resistance decreases. The angle of the incoming flow is off-center, since the flow direction was altered as a function of yaw instead of rotating the boat itself. The bottom two plots show resistance and lift in the customary reference frame used by tow tanks, with respect to the flow field direction, in which the resistance increases as a function of yaw.

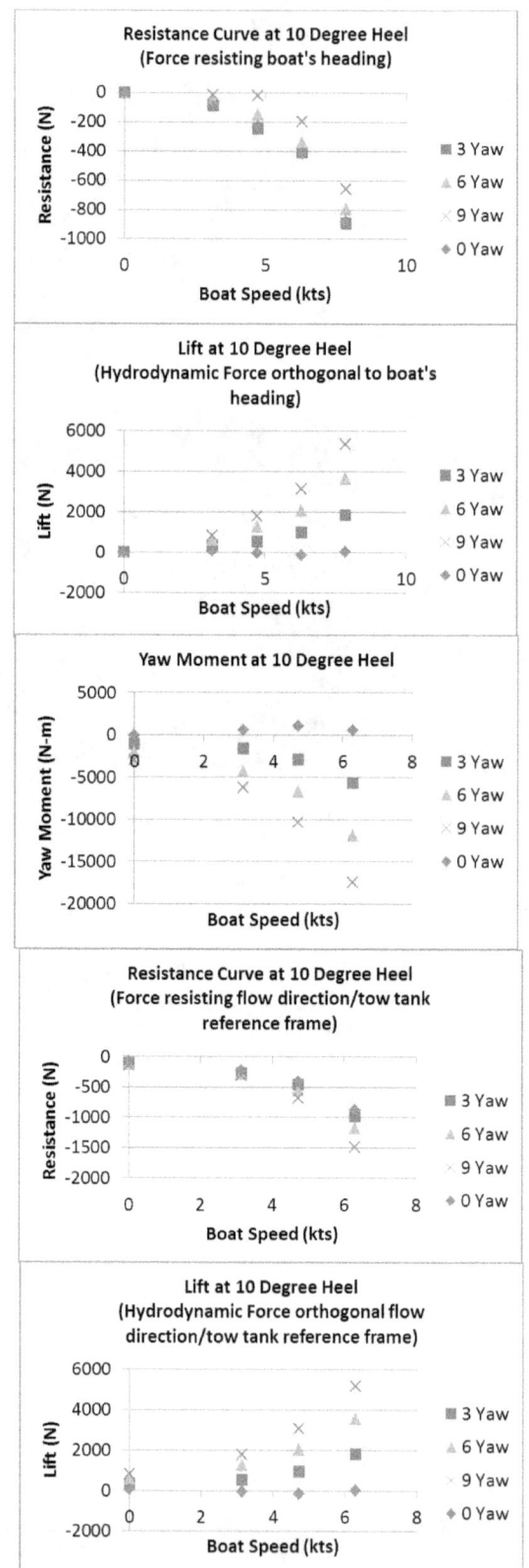

Figure 22 - Resistance and lift curve at 10 degrees heel

VPP Program

Once the aerodynamic and hydrodynamic data sets have been calculated the VPP program is used to balance the forces and predict sailing states. A number of python and shell scripts are used to automate the data set calculations. The VPP program is written in Java. Java was selected so that the program can easily run on most operating systems and high performance plotting and numerical processing libraries are readily available. The program is command line driven. A case configuration file is used to specify run time parameters such as data file names, wind condition range and what type of output is desired. The program is designed to compare multiple data sets at once allowing for efficient parametric studies. The program has both graphical and text based output. Plots can be displayed for any combination of input and output variables. The output data file has a simple structure that is easily read by external data analysis programs. Tables of hydrodynamic forces and aerodynamic force coefficients are input into the program in tabular text file format. The aerodynamic and hydrodynamic force functions are created by directly interpolating the RANS data using cubic splines.

The main flow control of the VPP program consists of a series of three or more nested loops. The outer two loops iterate through the desired true wind angle and wind speed range and the inner loop uses Newton-Rhapson iteration to balance the forces for a given true wind speed and angle. Additional optimization loops can be added to optimize things such as sail trim, sail shape, center of effort location and many others. For this project the VPP program was run with fixed flying shapes over a small range of wind conditions near measured sailing data points that are used for comparison. A separate aerodynamic input file is used for each flying shape and a constant hydrodynamic data file is used for all cases.

The program is designed to solve for four degrees of freedom; boat speed, heel, yaw and rudder angle however rudder angle adjustment can be turned off reducing the degrees of freedom to three. When rudder is turned off the yaw moments are not balanced, however the effect of the rudder at zero degrees is included in the hydrodynamic forces.

VPP Force Balance

The aerodynamic force model consists of four functions that use force coefficients to calculate the aerodynamic driving force, heeling force, heeling moment and yaw moment as functions of apparent wind speed, apparent wind angle, heel angle (ϕ) and sail flying shape (S). CFD data is used to derive the force coefficients. The aerodynamic forces scale with the square of the apparent wind speed so the coefficients are functions only of apparent wind angle, heel angle and sail flying shape.

$$FX_{air} = f1(AWS, AWA, \phi, S) = f1(AWS, c1(AWA, \phi, S))$$
$$FY_{air} = f2(AWS, AWA, \phi, S) = f2(AWS, c2(AWA, \phi, S))$$

$$MX_{air} = f3(AWS, AWA, \phi, S) = f3(AWS, c3(AWA, \phi, S))$$
$$MZ_{air} = f4(AWS, AWA, \phi, S) = f4(AWS, c4(AWA, \phi, S))$$

The hydrodynamic force model uses three functions that are functions of boat speed, yaw angle, heel angle and rudder angle and one function, the righting moment, that is a function of heel angle and crew placement. Both the total weight of the crew and crew placement is taken into account in the righting moment calculation. Crew weight position is accounted for by the variable C, which determines the average crew position from centerline to full-hike. The dynamic effects regarding righting moment were not accounted for in this study.

$$FX_{hydro} = f5(BS, \gamma, \phi, \theta)$$
$$FY_{hydro} = f6(BS, \gamma, \phi, \theta)$$
$$MX_{hydro} = f7(\phi, C)$$
$$MZ_{hydro} = f8(BS, \gamma, \phi, \theta)$$

The aerodynamic forces are defined as functions of AWA, AWS, ϕ and S, however, for each sailing state the program calculates AWA and AWS from TWS, TWA and BS so that both the aerodynamic and hydrodynamic forces are internally defined as functions of BS, γ, ϕ, θ which are considered the sailing state variables. The global coordinate system used by both the aerodynamic and hydrodynamic simulations is shown in Figure 23.

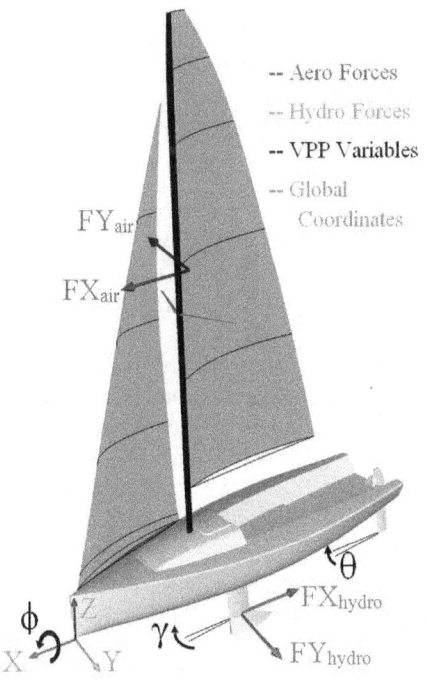

Figure 23 - Global Coordinate System

The balance of forces can be written in matrix form as shown in Equation 2.

$$F_{air} + F_{hydro} = \begin{bmatrix} FX_{air} + FX_{hydro} \\ FY_{air} + FY_{hydro} \\ MX_{air} + MX_{hydro} \\ MZ_{air} + MZ_{hydro} \end{bmatrix} = \begin{bmatrix} FX \\ FY \\ MX \\ MZ \end{bmatrix} = 0 \qquad \text{(Eq 2)}$$

The above set of nonlinear equations is solved for the state variables BS, γ, ϕ, θ using Newton-Raphson iteration. The iteration scheme works by updating the state variable vector with a perturbation vector at each step. The perturbation vector at state n is given by Equation 3. Derivatives are calculated with finite differences. The state vector at state n+1 is then given by Equation 4.

$$\begin{bmatrix} \Delta BS \\ \Delta \gamma \\ \Delta \phi \\ \Delta \theta \end{bmatrix} = - \begin{bmatrix} \partial FX/\partial BS & \partial FX/\partial \gamma & \partial FX/\partial \phi & \partial FX/\partial \theta \\ \partial FY/\partial BS & \partial FY/\partial \gamma & \partial FY/\partial \phi & \partial FY/\partial \theta \\ \partial MX/\partial BS & \partial MX/\partial \gamma & \partial MX/\partial \phi & \partial MX/\partial \theta \\ \partial MZ/\partial BS & \partial MZ/\partial \gamma & \partial MZ/\partial \phi & \partial MZ/\partial \theta \end{bmatrix}^{-1} \begin{bmatrix} Fy \\ Fy \\ Mx \\ Mz \end{bmatrix}_n \qquad \text{(Eq 3)}$$

$$\begin{bmatrix} BS_{n+1} \\ \gamma_{n+1} \\ \phi_{n+1} \\ \theta_{n+1} \end{bmatrix} = \begin{bmatrix} BS_n + \Delta BS \\ \gamma_n + \Delta \gamma \\ \phi_n + \Delta \phi \\ \theta_n + \Delta \theta \end{bmatrix} \qquad \text{(Eq 4)}$$

The iteration continues until the state variables are no longer changing. The program terminates the loop when the norm of the force vector residual drops below the convergence criteria. For well-defined aero and hydro sets the iteration scheme converges rapidly usually within 10 iterations. For each converged solution the program saves all of the state data which can then be displayed in a number of ways.

CASE STUDIES

Sail flying shape and performance data was collected during racing conditions, during two boat testing and while sailing alone. Comparison points were selected from these data sets accordingly to wind range, how steady the data was, what the sail flying shapes looked like and the sea state. The data used in the comparisons captured performance in light to heavy conditions, collected mostly in flat water, but includes data in sea states up to three foot wave conditions. Currently our RANS based VPP only handles flat water, however as the wind increases usually so does the sea state so all of our data in winds above 10 kts contained waves that we are not directly accounting for.

Three comparison points are made for three TWS. Figure 24 below depicts the sailing conditions for each case. The topmost image depicts the hull and sail configuration for each of these speeds with increasing TWS from left to right. The second image displays the drive and heel force coefficients as a function of AWA. The third image portrays C as a function of TWS.

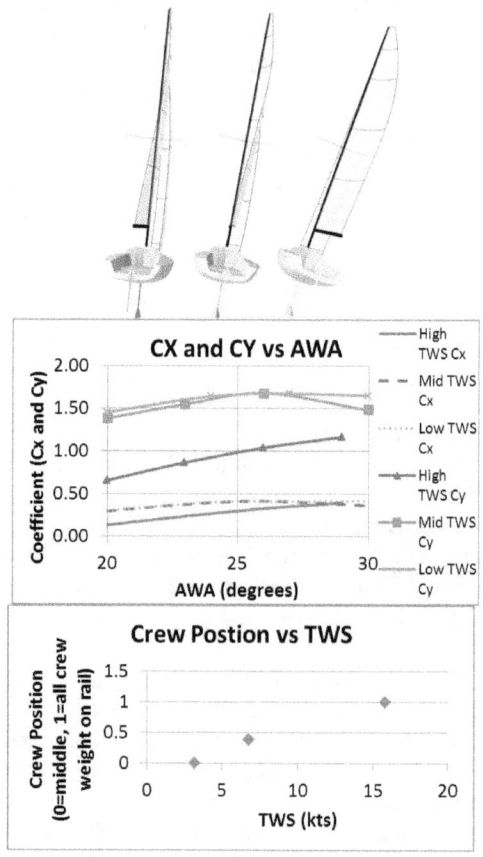

Figure 24 - Range of flying sail shapes and heel angles used for VPP comparison to measured data (TWS increasing from left to right), sail coefficients, and crew position

Results

Three TWS measured data points are compared to VPP output. The VPP is analyzed in .5 kts TWS and 3 degree TWA increments. VPP comparisons are selected by choosing the VPP data point that matches TWS and AWA. The RANS VPP predicts to varying degrees of accuracy depending upon the TWS. VPP predicted boat speed is accurate except at high TWS. The TWS compared in this section are for 3.2 kts, 6.8 kts, and 15.8 kts TWS, discussed as Low, Mid, and High TWS respectively. The measured data is presented by the diamond, the lowest TWS is shown by the square, the highest TWS by the 'x' and the mid speed by the triangle. Comparing the VPP results, the closest VPP prediction for the low, mid, and high TWS cases correlated to 3.5, 7, and 16 kts TWS. The low speed was chosen to be 3.5 kts instead of 3 kts because the AWA was in better agreement with the measured data point. Table 1 presents the mean of the measured data, the standard deviation of the measurements, the corollary VPP results, and the number of standard deviations that separate the measured sailing data from the VPP result.

Table 1: Measured Data vs VPP results

		Measured Sailing Data (mean)					
	TWS (kts)	AWS (kts)	AWA (degrees)	heel (degrees)	BS (kts)	yaw (degrees)	rudder (degrees)
High TWS	15.8	20.40	21.80	21.38	5.27	3.67	1.20
Mid TWS	6.8	10.45	27.48	9.38	4.50	2.80	-0.30
Low TWS	3.2	5.26	32.00	5.60	2.83	2.60	-1.45

		Standard deviation of measured results					
	TWS (kts)	AWS (kts)	AWA (degrees)	heel (degrees)	BS (kts)	yaw (degrees)	rudder (degrees)
High TWS	1.04	1.1	2.87	3.22	0.17	2.75	2.54
Mid TWS	0.38	0.45	3.46	2.11	0.16	1.67	1.16
Low TWS	0.19	0.18	1.96	1.09	0.10	1.89	1.84

		VPP Results					
	TWS (kts)	AWS (kts)	AWA (degrees)	heel (degrees)	BS (kts)	yaw (degrees)	rudder (degrees)
High TWS	16.0	21.45	21.9	19.68	6.04	6.29	0.77
Mid TWS	7.0	10.60	28.37	9.02	4.46	4.25	-1.77
Low TWS	3.5	5.51	31.75	4.48	2.73	3.48	-6.48

	VPP result # of standard deviations from the mean of measured data						
	TWS	AWS	AWA	heel	BS	yaw	rudder
High TWS	0.19	0.95	0.04	-0.53	4.53	0.95	-0.17
Mid TWS	0.53	0.33	0.26	-0.17	-0.25	0.87	-1.27
Low TWS	1.58	1.39	-0.13	-1.03	-1.00	0.46	-2.73

Data from Table 1 was collected from the output from the instrumented J70. The measured data for the Mid TWS is shown in Figure 25 as an example of the recorded measured data used to create Table 1.

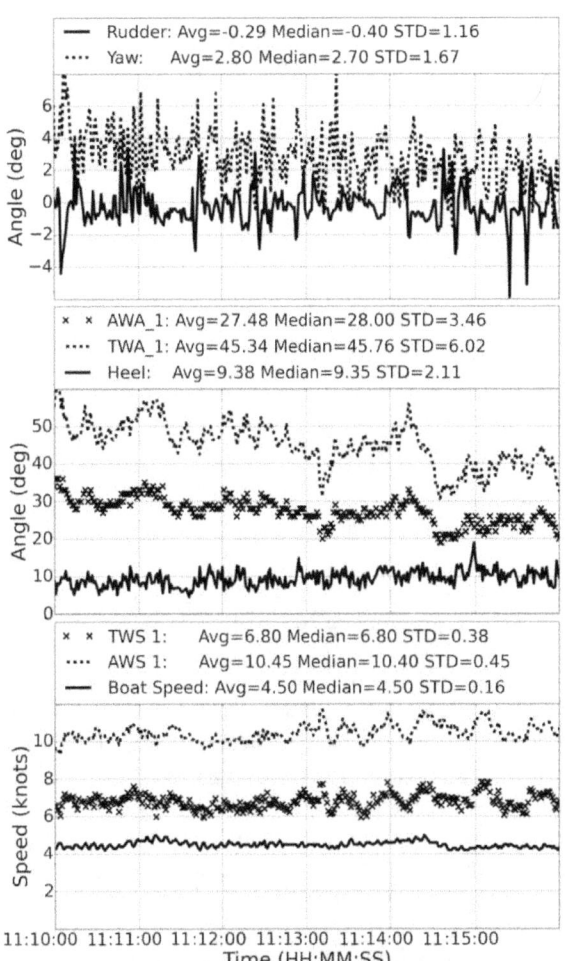

Figure 25 - Measured Data Samples

Boat Speed as a function of TWS: Results & Analysis

The relationship of BS as a function of TWS for the three analyzed data points is shown in Figure 26. For the Low TWS data point, despite having an extra 0.3 kts of TWS, the VPP predicts only 2.7 kts of BS while the measured data was 2.83 kts with a standard deviation of 0.1 kts. The predicted BS is slightly lower than the measured result. For the Mid TWS, the VPP predicted value falls within the measured standard deviation for BS and TWS. The High TWS data point has more than 10% less BS than is predicted by the VPP, which correlates to over 4.5 standard deviations.

30

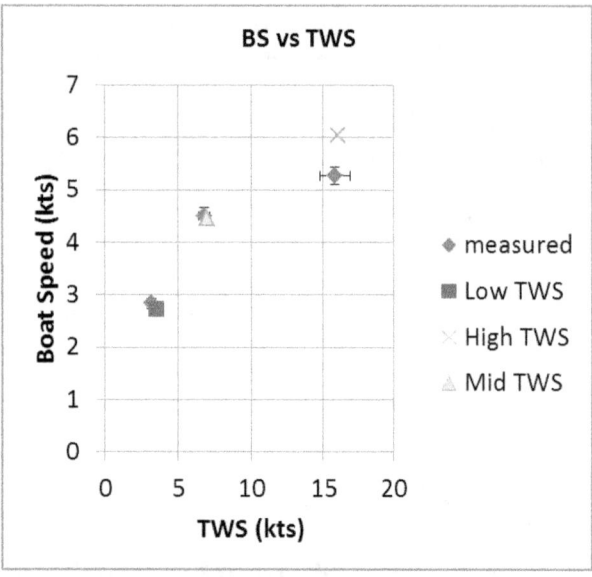

Figure 26 - BS vs TWS

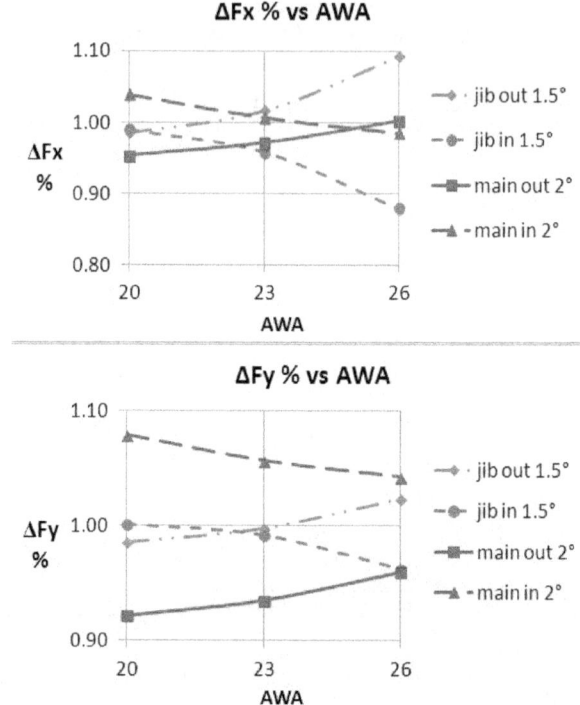

Figure 27 – Effect of potential trim error on sail forces. Changes in Fx and Fy with trim are shown, Mx and Mz changes track closely with changes in Fy

The under-prediction at the Low TWS could be related to current, instrument error, CFD error of either over-prediction of hydrodynamic forces or under-prediction of aerodynamic driving force, or error in the flying shape trim. As will be shown in the discussion of yaw, the VPP results tend to point higher than the actual sailed data to achieve the same TWS and AWA. By pointing higher, the sailboat tends to have force in the sails which reduces heel and BS but increases yaw.

The over-prediction of BS for the High TWS is likely due to the assumption of steady state flow. At 15.8 kts there were waves during the measured data collection, so the flat water assumption for the VOF simulations ceased to be true, which would add extra drag that was not calculated in the RANS VOF solutions. Furthermore, since the dynamic effects affecting righting moment in the VPP were ignored, the boat is simulated as having larger righting moment than is real and performance is therefore increased at high TWS.

One potential source of error is flying shape trim, which could be off by +/- degrees because of slight alignment errors of the camera. This could correlate to a difference of up to 10% of total driving force shown in Figure 27.

Another potential source of error is that the hydrodynamic RANS VOF simulations ignored the effects of pitch, which could alter the resistance of the hull in the VPP.

The analysis of only three VPP driven data points is a limited sample size. To depict that the correct trends are captured by these 3 points, Figure 28 depicts more sailing data points that could be analyzed to fill in additional RANS based VPP data points. The point cloud of sailing points was generated from examining data and selecting cases that have steady data for at least 15 seconds. The VPP results at the Low and Mid TWS fall within the point-cloud, but the High TWS VPP point is still an over-prediction. Seven data points were chosen to illustrate the trends for other variables in more detail, and the respective trendlines are shown.

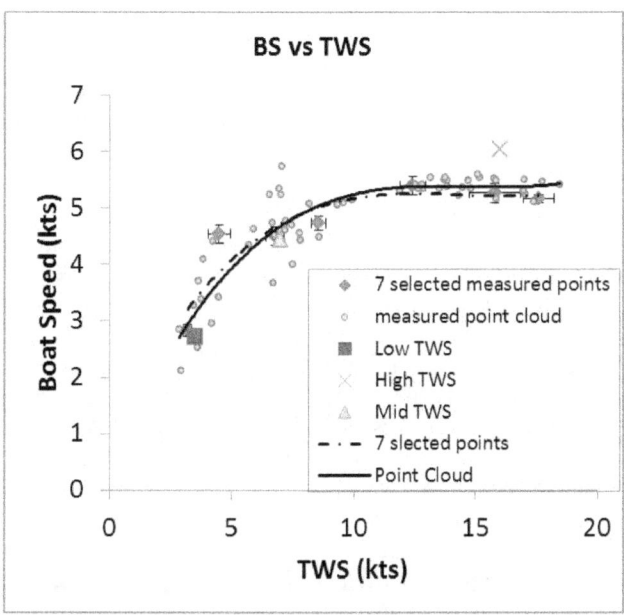

Figure 28 - 7 measured data points of boat speed vs TWS with VPP output

The implication of Figure 28 is that the measured BS plateaus around 10 kts of TWS at 5.3 kts BS, as the boat becomes righting moment challenged and as the unsteady effects of waves become more significant. The implication of the measured data is that the VPP data also likely will plateau, but closer to 6 kts BS due to more ideal conditions. More data points around a TWS of 10 kts should be analyzed to evaluate BS and the effect of waves on BS should be evaluated empirically or with unsteady RANS VOF CFD. Furthermore, the flying shape trim and shape potential error should be reduced to ensure proper sail forces are applied.

Heel, AWA, and Yaw vs TWS

The heel and AWA vs TWS between the VPP prediction and measured data matches within the standard deviation of the measured results as shown in the upper part of Figure 29, except at the lowest speed where heel is predicted to be 1.03 standard deviations below.

Heel is under-predicted from the measured data, which implies that the forces on the sail are lower than the real sailing data. The under-prediction of heel could be caused by the VPP pointing higher, while the measured sailing data may have been collected at broader True Wind Angles as implied by the lower part of Figure 29. Heel could also be under-predicted since the dynamic righting moment effects of the hull and appendages are not included in the VPP.

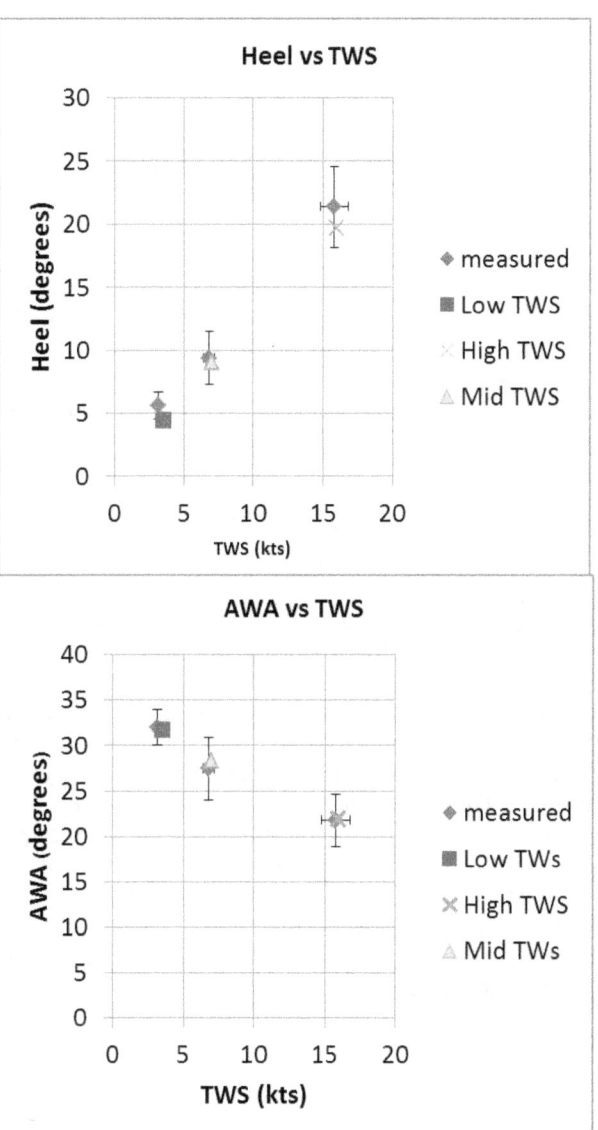

Figure 29 - Heel vs TWS & AWA vs TWS

The yaw vs AWS is over predicted in the VPP compared to the measured data, but does fall in the standard deviation of the measured data. The standard deviation of yaw is large and further instrumentation is necessary to validate the VPP. The yaw as a function of TWS increases nearly four times more than the yaw vs TWS of the mean of the measured data as shown by Figure 30.

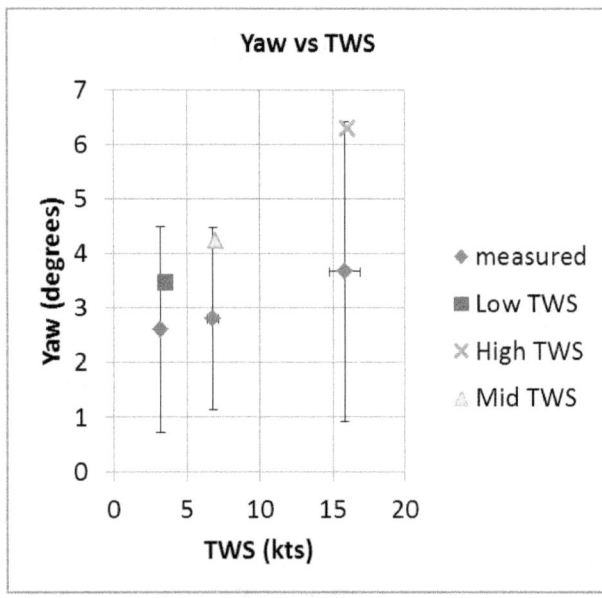

Figure 30 -Yaw vs TWS

If the seven measured data points are added to the analysis, instead of only three used for VPP generation, the spread in measured yaw data becomes wider, as shown by Figure 31. The best fit linear trend-line for yaw vs TWS for the measured data has a steeper slope than the VPP results, however, the VPP data points still have more yaw than the trend-line for the TWS range examined. A quadratic best fit line is also fit to the measured data which implies that the measured data may increase more quadratically than linearly with TWS.

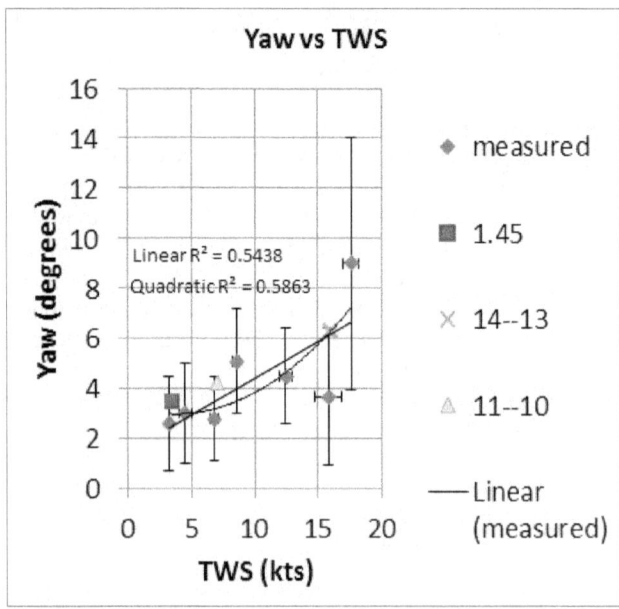

Figure 31 - Yaw vs TWS with more measured data

At the high TWS speed with significant waves, the boat would be sailed to maintain good VMG; therefore the boat

would be sailed at a higher AWA to maintain BS. The hydrodynamic forces assume flat water and to achieve equal AWA the VPP can point the boat higher for a given AWA and TWS, which leads to higher yaw but better VMG. More data points are necessary to further depict the relationships.

Rudder relationships

The rudder values between the measured data and the VPP results have a similar trend but different levels of magnitude. Figure 32 depicts the trends of seven measured data sets with the three VPP data points for rudder as a function of TWS. In general, as TWS is increased the more weather-helm is required (positive rudder is weather helm/negative rudder is lee helm). The measured data-point at 17.6 kts TWS and -5 degrees rudder occurred at a point when the boat was pinching significantly at which point the boat was out of balance and the skipper had to apply lee helm. At other measured data points, the helm is much more neutrally balanced and as TWS increases, the weather-helm increases.

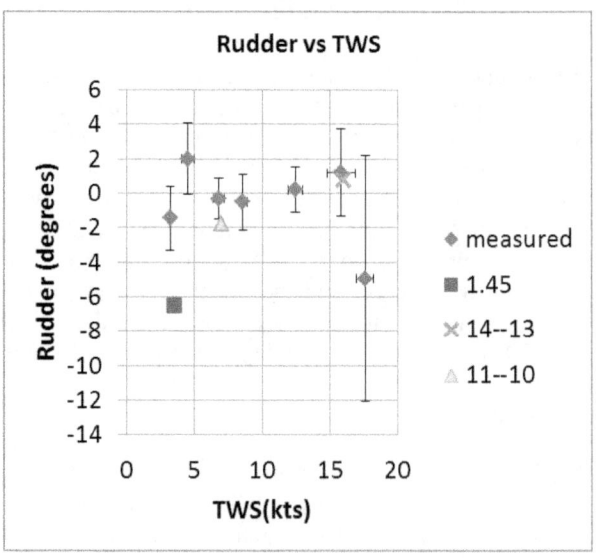

Figure 32 - VPP and 7 measured data point illustrating trend that as TWS increases more weather-helm (positive rudder) is needed

Figure 33 displays the rudder required to balance the yaw moment and side force as a function of TWS for both the VPP output and the corollary measured sailing data. At the Low TWS speed, the VPP predicts that the rudder angle is 2.73 standard deviations below the average sailing data. For the Mid TWS and High TWS, the rudder angle is 1.27 standard deviations and 0.17 standard deviations below the measured data respectively. Further instrumentation and more steady sailing data is required to better validate the rudder angles, since both yaw and rudder have large standard deviations.

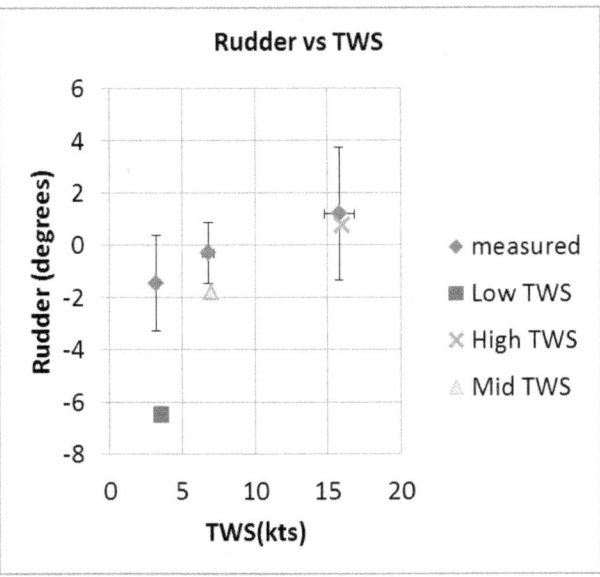

Figure 33 - Measured and VPP results for rudder as a function of TWS

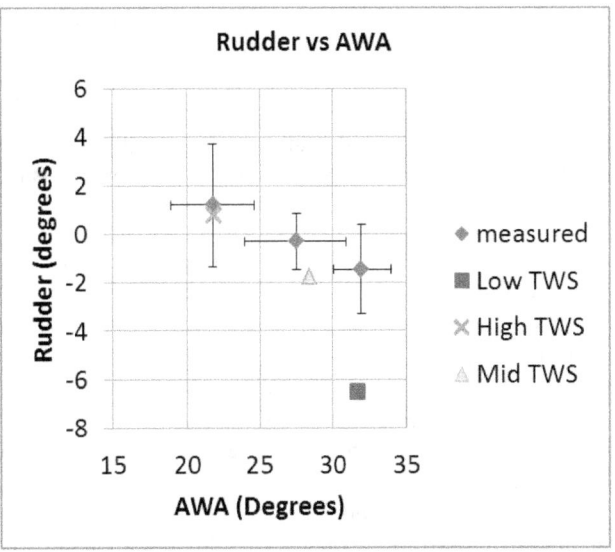

Figure 34 - Rudder vs AWA

The over-prediction of lee-helm for the Low TWS could be caused by the overall relatively low performance of the VPP prediction at this point. The RANS VOF CFD was not analyzed below Fr= 0.2 and the convergence at this point required correction as aforementioned. More RANS VOF points could be analyzed in the range of this BS and at 5 degrees heel to reduce interpolation and directly analyze the results. The VPP over-predicts yaw as previously discussed, and to balance the yaw moment and side-force more lee helm is required in the VPP results. As the true wind speed increases and the helm becomes more neutral, both measured and predicted, the VPP prediction of rudder correlates better with measured data.

Figure 34 depicts the rudder prediction and measured data as a function of AWA. The figure depicts that when sailing upwind as the AWA is increased less weather-helm is required, however, at broader AWA (Low TWS), the boat is out of balance. This further depicts that the very light and unstable conditions for the Low TWS case significantly adversely affected the performance of the boat. The VPP had significant rudder error at the Low TWS speed potentially because the VPP estimates higher pointing, inducing more yaw; whereas the real-world sailor decided to sail broader TWA to maintain BS and pinch less for the same TWS and AWA.

The reason for the varying magnitude of rudder angle could be related to application of rudder forces and moments as a correction to save costly RANS VOF data points, the way the measurements were taken, or a combination of both. During testing, the instrument that was the most problematic was the potentiometer used for the rudder variable. The rudder was centered for over twenty minutes before and after sailing and there was marked float, of up to two degrees, between the initial center reading and the final reading of the day. This implies that the rudder measurements could have up to a magnitude of 2 degrees in error, which could lead the correlations to match better or worse. The error could have occurred from electrical power fluctuations, mechanical oscillation in the bracket or from damage to the sensor. More testing and an improved sensor design are needed to validate the model.

CONCLUSIONS

A J70 has been instrumented, sailed, and sailing performance has been compared to a RANS based VPP. In general good correlation is seen between the RANS based VPP and the real world sailing data. The VPP predicts higher BS at high TWS conditions, which is likely due to the flat water and ideal conditions that the VPP is run under, ignoring the dynamics of real world sailing. In general, yaw is over-predicted. Since TWS and AWA were used as the criterion to match the VPP to measured results, the boat in flat water conditions tends to point higher and yaw more to achieve the same results as the measured data.

The data acquisition system combines a number of standard marine sensors including a sonic anemometer on the bow sprit and masthead, a GPS, a digital compass, an accelerometer and a gyroscope with custom sensors that measure rudder and trim angles as well as a custom sail shape acquisition system. Flying shapes were generated for the RANS VPP by recording images, correcting for lens distortion, and generating STL files for input into the steady state CFD. Forces on the sails, hull and mast are evaluated by using OpenFOAM's steady state solver SimpleFOAM with a wind gradient applied at the inlet. The hydrodynamic forces are evaluated for flat water using OpenFOAM's steady state multiphase solver, LTSInterFOAM. VPP results are generated from the algorithm which builds response surfaces from the CFD results and balances the forces.

With further development the instrumented J70 can be used as a platform for validating a RANS driven VPP or as a design optimization tool. The instrumented platform on the J70 and the corollary RANS based VPP could be implemented on other boats as well. Automation could streamline the optimization process.

Additional instrumentation and testing is required to validate the RANS VPP predictions. Shortcomings of the comparison include limited data, the assumption of steady state in unsteady conditions, and the accuracy of instruments. The dynamic effects on righting moment should be included in the VPP force balance. The rudder sensors robustness and precision needs to be improved for further studies. Furthermore, current and wave effects should be evaluated. More data in more wind conditions should be collected and compared to. Wave effects should be recorded and evaluated either empirically or using a transient moving mesh simulation with waves.

One way to further improve the system is to perform a test with two different instrumented J70s and determine how the VPP would perform in a controlled sailing environment. By using multiple boats to collect data, each would have another boat to compare performance to and would better replicate racing conditions.

REFERENCES

Bohm, C. *A Velocity Prediction Procedure For Sailing Yachts With a Hydrodynamic Model Based on Integrated Fully Coupled RANSE-Free_Surface Simulations*. (Doctoral dissertation) Retrieved from Repository.tudelft.

Doyle, T., Knight, B., "RANS Based VPP Method for Mega-Yachts", Fifth High Performance Yacht Design Symposium. Auckland, New Zealand, 9-11 March, 2015. 267-276. Print.

Kaltschmitt, M., Streicher, W., Wiese, A., "Renewable energy: technology, economics, and environment." Springer, 2007. 55. Print.

Korpus, R. "Performance Prediction without Empiricism: A RANS-Based VPP and Design Optimization Capability", 18th Chesapeake Sailing Yacht Symposium, Annapolis, MD, 2007.

Le Pelley, D. J., and O. Modral. "V-Spars: A Combined Sail and Rig Shape Recognition System Using Imaging Techniques", Proc. Third High Performance Yacht Design Symposium. Auckland, New Zealand, 2-4 December, 2008.

"Lens Distortion." Plug-in Lens distortion. Gimp. Web. 25 Nov. 2015. http://docs.gimp.org/en/plug-in-lens-distortion.html

THE 22nd CHESAPEAKE SAILING YACHT SYMPOSIUM
ANNAPOLIS, MARYLAND, MARCH 2016

Experimental and numerical trimming optimizations for a mainsail in upwind conditions

Matthieu Sacher[1], Naval Academy Research Institute – IRENav, France
Frédéric Hauville, Naval Academy Research Institute – IRENav, France
Régis Duvigneau, INRIA – Sophia Antipolis, France
Olivier Le Maître, LIMSI – CNRS, France
Nicolas Aubin, Naval Academy Research Institute – IRENav, France
Mathieu Durand, K-EPSILON, France

ABSTRACT

This paper investigates the use of meta-models for optimizing sails trimming. A Gaussian process is used to robustly approximate the dependence of the performance with the trimming parameters to be optimized. The Gaussian process construction uses a limited number of performance observations at carefully selected trimming points, potentially enabling the optimization of complex sail systems with multiple trimming parameters. We test the optimization procedure on the (two parameters) trimming of a scaled IMOCA mainsail in upwind conditions. To assess the robustness of the Gaussian process approach, in particular its sensitivity to error and noise in the performance estimation, we contrast the direct optimization of the physical system with the optimization of its numerical model. For the physical system, the optimization procedure was fed with wind tunnel measurements, while the numerical modeling relied on a fully non-linear Fluid-Structure Interaction solver. The results show a correct agreement of the optimized trimming parameters for the physical and numerical models, despite the inherent errors in the numerical model and the measurement uncertainties. In addition, the number of performance estimations was found to be affordable and comparable in the two cases, demonstrating the effectiveness of the approach.

NOTATION

C_X	Drive coefficient	$[-]$
C_Y	Side coefficient	$[-]$
F_X	Drive force	$[\text{N}]$
F_Y	Side force	$[\text{N}]$
ρ	Air density	$[\text{kg/m}^3]$
S	Sail area	$[\text{m}^2]$

[1]matthieu.sacher@ecole-navale.fr

V_∞	Reference velocity	$[\text{m/s}]$
q_∞	Dynamic pressure	$[\text{Pa}]$
L_{sheet}	Main sheet length	$[\text{mm}]$
L_{car}	Main car length	$[\text{mm}]$
T	Sheet tension	$[\text{N}]$
ε	Measurement noise	$[-]$
σ_ϵ	Noise variance level	$[-]$
$\mathcal{P}$	Objective function	$[-]$
$\sigma_\mathcal{P}$	Uncertainty of the objective function	$[-]$
$\mathbf{x}$	Optimization parameters	$[-]$
$\mathbf{x}_*$	Optimal parameters of AEI	$[-]$
$\mathbf{x}_{**}$	Effective optimal parameters	$[-]$
$\mathbf{x}_{\text{opt}}$	Optimal parameters of $\mathcal{P}$	$[-]$
Θ	Vector of hyper-parameters	$[-]$
C_F	Covariance function	$[-]$
$\mathbf{C}$	Covariance matrix	$[-]$
$\mathcal{L}$	Log-marginal likelihood	$[-]$
EI	Expected Improvement	$[-]$
AEI	Augmented Expected Improvement	$[-]$
$\hat{y}$	Prediction mean	$[-]$
$\hat{\sigma}_y^2$	Prediction variance	$[-]$

INTRODUCTION

Researches on sailing yachts have fostered the development of advanced methods dedicated to the prediction and improvement of racing yacht performance. The performance is usually analyzed using so-called Velocity Prediction Programs (VPPs) (Oossanen, 1993), which solve equilibrium equations (balancing hull, appendages and sails loads) to determine several performance indicators, such as Boat Speed (BS) or Velocity Made Good (VMG). The different loads accounted by the VPPs can be based on empirical formulas, experimental data or numerical simulations (Hansen et al., 2003, Korpus, 2007). However, due to the complexity and multi-physic characters of the yachts dynamics, performance studies often consider the hydrodynamic (Huetz and Guillerm, 2014) and aerodynamic (Augier et al., 2012, Menotti et al., 2013, Trimarchi, 2012) aspects separately.

Here, we focus on the aerodynamics, optimizing the performance of a sail system, but the numerical procedure developed below can be used to perform hydrodynamic optimization or even fully coupled yacht performance optimization. Sail systems are subjected to very complex phenomena, such as nonlinear Fluid-Structure Interaction (FSI) effects and instabilities. Moreover, the modeling of real sailing conditions is still an open research problem due to the large uncertainties in wind and sea states. To our knowledge, the optimization of sails has thus been limited so far to idealized situations. For instance, sail shape optimizations (without accounting for the FSI problem) are performed in (Rousselon, 2008), while the trimming of two-dimensional sails is numerically considered in (Chapin et al., 2008). Regarding FSI in three spatial dimensions, the authors in (Ranzenbach et al., 2013) mention an optimization of the trimming of sails, but within an inviscid flow approximation and not much details are provided on the optimization procedure used.

The present work aims at pursuing these efforts toward the development of efficient numerical optimization procedures capable of dealing with complex sail systems, with realistic physical models (*e.g.* nonlinear FSI and turbulent flows) and a large number of optimization variables (*i.e.* trimming parameters). Denoting $\mathbf{x} \in \Omega$ the optimization variables, the optimization problem can be written as

$$\mathbf{x}_{\mathrm{opt}} = \arg\min_{\mathbf{x} \in \Omega} -\mathcal{P}(\mathbf{x}),$$

where $\mathbf{x}_{\mathrm{opt}} \in \Omega$ are the sought optimal parameters and $\mathcal{P} : \Omega \mapsto \mathbb{R}$ is a measure of performance. The main difficulty preventing the straightforward application of standard optimization procedures to sail systems is related to the cost of estimating of the performance $\mathcal{P}$ at tentative values $\mathbf{x}$ of the parameters. Indeed, the estimation of $\mathcal{P}(\mathbf{x})$ involves the resolution of a nonlinear FSI problem, which requires several convergence iterations between the nonlinear elastic and flow solvers. Also, adjoint based techniques are hardly amenable to FSI problems involving coupled nonlinear solvers; this fact precludes the use of efficient gradient-based iterative methods in favor of optimization algorithms such as the simplex based (Nelder and Mead, 1965) or evolutionary (Bäck and Schwefel, 1993, Hansen, 2006) methods. However, depending on the considered problem, these so called gradient-free algorithms are known to require a large number of evaluations of $\mathcal{P}(\mathbf{x})$, making applications to sail systems very costly as a single evaluation may routinely require several hours of CPU on modern parallel computers.

Based on these observations, we advocate the use of meta-model approaches to mitigate the large computational cost of optimizing the trimming parameters of sail systems. Specifically we rely below on Gaussian Process (GP) approximations for the mapping $\mathcal{P} : \Omega \mapsto \mathbb{R}$. This statistical approach uses a coarse set of performance evaluations at some parameters values $\mathbf{x} \in \Omega$ to infer a GP $\mathcal{G}(\mathbf{x}) \approx \mathcal{P}(\mathbf{x})$. One can then apply its favorite optimization procedure to $\mathcal{G}(\mathbf{x})$ to obtain the corresponding approximation of $\mathbf{x}_{\mathrm{opt}}$. The surrogate-based optimization procedure is embedded in an iterative scheme, where new evaluations of the performance at carefully selected new points $\mathbf{x}$ are introduced in order to refine the GP approximation in regions of Ω susceptible to include the optimum. The GP approach is then expected to improve the optimization by a) requiring an overall lower number of performance evaluations, compared to direct gradient-free approaches, and b) enabling the use of efficient global optimization tools. In addition, the GP construction provides a natural way to estimate convergence on the approximation of $\mathcal{P}$ and then to characterize the accuracy on the retrieved optimum.

Another interest in considering an optimization based on GP meta-model is that it naturally accommodates for errors and noise in the performance evaluation. This specificity is exploited in the present work to perform the optimization of an actual physical sail system, consisting in a scaled IMOCA mainsail in upwind conditions. The objective is to find the optimal trimming of the sail, for a performance criterion combining the drive and side aerodynamic force coefficients. Here, the GP-based optimizer used values of $\mathcal{P}(\mathbf{x})$ measured in the wind tunnel of the Yacht Research Unit (Auckland), for the sequence of trimming points requested by the iterative optimization procedure. Because of the imperfections in the experimental apparatus and inherent noise in the measurements, the estimates of $\mathcal{P}(\mathbf{x})$ were subjected to significant errors, that would have compromised the convergence of descent methods (Saul'ev and Samoilova, 1975) without using a GP reconstruction.

In addition to evidence the robustness of GP based optimization, this experiment is used as a reference to assess the relevance of an optimization relying on numerical resolutions of the FSI problem to compute the performance. To this end, the experimental sail system was measured (dimensions, mechanical characteristics of mast and boom, ...), and wind tunnel inflow conditions recorded, to create a numerical model of the experiment. State of the art FSI solvers is then used for the resolution of the resulting numerical model at the sequence of trimming points requested by the GP-based optimizer. The numerical resolution involves a nonlinear structural solver with a mesh deformation utility (K-FSI tools) developed by K-EPSILON, coupled with the finite volume turbulent flow solver FINE^{TM}/Marine from Numeca Software. The Unsteady Reynolds-Average Navier-Stokes Equations (URANS) turbulence model is used in these numerical experiments.

The paper is organized as follows. The first Section briefly reviews the construction of the Gaussian Process approximation and the selection of the new parameters in the iterative optimization procedure. We then detail the experimental set-up and the results of the corresponding optimization procedure in the second Section. The numerical modeling of the experiment and optimization results are reported in third Section. Finally, conclusions of this work and direction for future developments are provided in the fourth Section.

GP MODEL BASED OPTIMIZATION

In this Section we start by briefly summarizing the construction of a Gaussian Process to model a function from noisy measurements. More details on GP models can be found for instance in (Gibbs, 1997, Rasmussen and Williams, 2006). We then describe the GP model optimization procedure (Duvigneau and Chandrashekar, 2012), detailing the selection of successive optimal candidates.

Gaussian Process Model

Consider a dataset $\mathbf{X}_n = (\mathbf{x}_1 \cdots \mathbf{x}_n)^\mathrm{T}$ of n training inputs vectors $\mathbf{x}_i \in \Omega \subset \mathbb{R}^d$. Each element $\mathbf{x}_i \in \mathbf{X}_n$ is associated to an observation (or measurement) $y_i \in \mathbb{R}$ which is assumed to be dependent on a latent function $f(\mathbf{x})$ through

$$y_i = f(\mathbf{x}_i) + \varepsilon_i, \quad i = 1, \ldots, n, \quad (1)$$

where ε_i is a random measurement error (*i.e.* the measurement noise). In this work, the ε_i are assumed *independent* and to follow the same (centered) Gaussian distribution:

$$\varepsilon_i \sim \mathcal{N}\left(0, \sigma_\epsilon{}^2\right), \quad (2)$$

where $\mathcal{N}\left(\mu, \Sigma^2\right)$ denotes the Gaussian distribution with mean μ and variance Σ^2. Thus, $\sigma_\epsilon{}^2$ is referred to as noise variance. The objective is therefore to model the latent function $f(\mathbf{x})$ on the basis on the noisy observations y_i.

The latent function is considered as a realization of a zero-mean multivariate Gaussian process F, with unknown covariance function C_F, that is $F \sim \mathcal{N}(0, C_F)$, with

$$C_F(\mathbf{x}, \mathbf{x}') \doteq \mathbb{E}\left\{F(\mathbf{x}), F(\mathbf{x}')\right\}, \quad (3)$$

where $\mathbb{E}\{\cdot\}$ denotes the expectation operator.

The covariance function of F must be specified. In this work, we consider the Matérn class (Stein, 2012) of stationary covariance functions having for one-dimensional generator

$$M_\nu(r, l) = \frac{2^{1-\nu}}{\Gamma(\nu)} \left(\frac{\sqrt{2\nu}r}{l}\right)^\nu K_\nu\left(\frac{\sqrt{2\nu}r}{l}\right). \quad (4)$$

Here $r = |x - x'|$, ν and l are two positive parameters, and K_ν is the modified Bessel function of the second kind. We shall further restrict ourselves to covariances with $\nu \to \infty$, leading to the squared exponential covariance family with generator

$$M_\infty(r, l) = \exp\left(\frac{-r^2}{2l^2}\right). \quad (5)$$

The multidimensional counterpart is obtained by tensor product of the one-dimensional generator. The final expression of the covariance function for the GP approximation is

$$C_F(\mathbf{x}, \mathbf{x}'; \mathbf{\Theta}) = \theta_1 \prod_{i=1}^{d} \exp\left(\frac{-(x_i - x_i')^2}{2l_i^2}\right) + \theta_2. \quad (6)$$

In the expression (6) of c, $\mathbf{\Theta} = \{\theta_1, \theta_2, l_1, l_2, \ldots, l_d\}$ is a vector of hyper-parameters. The first hyper-parameter, θ_1, scales the distance-dependent correlation, while θ_2 is an offset from zero. The other parameters l_i are the anisotropic correlation lengths associated to the d directions of Ω. From the parametrized covariance function $C_F(\mathbf{x}, \mathbf{x}'; \mathbf{\Theta})$ we derive the covariance matrix $\mathbf{C}(\mathbf{\Theta}) \in \mathbb{R}^{n \times n}$ of the observation points in $\mathbf{X}_n$. The covariance matrix $\mathbf{C}(\mathbf{\Theta})$ has for entries

$$C_{i,j}(\mathbf{\Theta}) \doteq C_F(\mathbf{x}_i, \mathbf{x}_j; \mathbf{\Theta}), \quad 1 \leq i, j, \leq n. \quad (7)$$

Given the n noisy observations y_i, collected into the vector $\mathbf{Y}_n = (y_1 \cdots y_n)^\mathrm{T}$, the predicted observation $y(\mathbf{x})$ at a new point $\mathbf{x} \in \Omega$ is given by the joint Gaussian distribution

$$\begin{pmatrix} \mathbf{Y}_n \\ y(\mathbf{x}) \end{pmatrix} \Bigg| \mathbf{X}_n, \mathbf{\Theta} \sim \mathcal{N}\left(\mathbf{0}, \begin{bmatrix} \mathbf{C} + \sigma_\epsilon{}^2\mathbf{I} & \mathbf{k}(\mathbf{x}) \\ \mathbf{k}^\mathrm{T}(\mathbf{x}) & \kappa(\mathbf{x}) + \sigma_\epsilon{}^2 \end{bmatrix}\right). \quad (8)$$

In (8), the dependence of $\mathbf{C}$ on the hyper-parameters has been removed to simplify the notation, and

$$\kappa(\mathbf{x}) \doteq C_F(\mathbf{x}, \mathbf{x}; \mathbf{\Theta}), \quad (9)$$
$$\mathbf{k} \doteq \left(C_F(\mathbf{x}, \mathbf{x}_1; \mathbf{\Theta}) \cdots C_F(\mathbf{x}, \mathbf{x}_n; \mathbf{\Theta})\right)^\mathrm{T}, \quad (10)$$

while $\mathbf{I}$ is the identity of $\mathbb{R}^n$. Using the conditional rules of a joint Gaussian distribution (Rasmussen and Williams, 2006, Von Mises, 1964), it comes

$$y(\mathbf{x})| \mathbf{Y}_n, \mathbf{X}_n, \mathbf{\Theta}, \sigma_\epsilon{}^2 \sim \mathcal{N}\left(\hat{y}(\mathbf{x}), \hat{\sigma}_y^2(\mathbf{x})\right). \quad (11)$$

The *best* prediction of $y(\mathbf{x})$ is the mean $\hat{y}(\mathbf{x})$ of the distribution; the prediction variance $\hat{\sigma}_y^2(\mathbf{x})$ quantifies the uncertainty in the prediction. The second order properties of the prediction $y(\mathbf{x})$ can be explicitly expressed as

$$\hat{y}(\mathbf{x}) = \mathbf{k}^\mathrm{T}(\mathbf{x}) \left(\mathbf{C}(\mathbf{\Theta}) + \sigma_\epsilon{}^2\mathbf{I}\right)^{-1} \mathbf{Y}_n, \quad (12)$$
$$\hat{\sigma}_y^2(\mathbf{x}) = \kappa(\mathbf{x}) + \sigma_\epsilon{}^2 - \mathbf{k}^\mathrm{T}(\mathbf{x}) \left(\mathbf{C}(\mathbf{\Theta}) + \sigma_\epsilon{}^2\mathbf{I}\right)^{-1} \mathbf{k}(\mathbf{x}). \quad (13)$$

The hyper-parameters $\mathbf{\Theta}$ and noise variance $\sigma_\epsilon{}^2$ are unknown *a priori* and need to be learned from the data. They can be determined by maximizing the log-marginal likelihood (Rasmussen and Williams, 2006) given by

$$\mathcal{L}(\mathbf{\Theta}, \sigma_\epsilon{}^2) = -\frac{n}{2}\log(2\pi) - \frac{1}{2}\log\left|\mathbf{C}(\mathbf{\Theta}) + \sigma_\epsilon{}^2\mathbf{I}\right|$$
$$-\frac{1}{2}\mathbf{Y}_n^\mathrm{T}\left(\mathbf{C}(\mathbf{\Theta}) + \sigma_\epsilon{}^2\mathbf{I}\right)^{-1}\mathbf{Y}_n. \quad (14)$$

The optimal hyper-parameters and noise variance are then found by minimizing $\mathcal{L}$ with respect to its arguments. An evolution strategy algorithm (Hansen, 2006) is used for this purpose.

Once the optimal hyper-parameters are determined, the GP model can be used to predict values at new points using (12) and (13). The most computationally consuming part of the GP construction is the assembly of the (full) matrix $\left(\mathbf{C} + \sigma_\epsilon^2 \mathbf{I}\right)$, and the evaluation of its determinant and inverse required in the definition of the log-marginal likelihood and for new predictions. This can be done efficiently by LU decompositions.

Figure 1 illustrates for a one-dimensional function the effect of the observation noise σ_ϵ on the constructed GP model. The constructions use 6 observations points depicted with circles in the plots and the covariance hyper-parameters are determined by maximizing the log-marginal likelihood. However, in the first case, shown in Figure 1(a) a value $\sigma_\epsilon = 0$ is imposed, while in the second case in Figure 1(b) the noise level is also optimized. In addition to the observations, the two plots report the mean of the GP models with classical $\pm 3\hat{\sigma}$ uncertainty range. It is seen from Figure 1(a) that when the measurements are assumed to be noise-free ($\sigma_\epsilon = 0$), the resulting GP model is interpolating the data, *i.e.* the variance of the prediction is zero at the data points. However, the mean of the GP model exhibits significant oscillations such that over-fitting can be suspected. On the contrary, optimizing the noise level σ_ϵ results in a mean process free of spurious oscillations but that is no more interpolating, as it can be appreciated from Figure 1(b). The averaged distance of the best prediction to the observations is $\sim \sigma_\epsilon$.

Optimization strategy using Gaussian Process models

The GP model (build, previously) can be used to determine the next control parameters $\mathbf{x}_*$ to be included in the data base. Deterministic optimization approaches would classically choose $\mathbf{x}_*$ as the best control parameters, that is the minimizer of the meta-model over Ω. However, GP models are random and not even bounded, so the definition of the best control parameters needs to be clarified in this context. This is classically achieved by introducing an appropriate (deterministic) merit function, and combining the expected prediction $\hat{y}$ and its variance $\hat{\sigma}_y^2$, requiring $\mathbf{x}_*$ to be the maximizer of this merit function. In fact, the merit function should balance a selection of $\mathbf{x}_*$ yielding the minimal expected prediction $\hat{y}$ (optimality) with a selection of $\mathbf{x}_*$ in areas of large variance $\hat{\sigma}_y^2$ to reduce the GP model uncertainty. A complete summary of various merit functions proposed in the literature is provided in (Jones, 2001). In this work, we use the Augmented Expected Improvement (AEI) merit function (Huang et al., 2006) which is an extension of the popular Expected Improvement (EI) (Jones et al., 1998) in the case of noisy estimations. The AEI merit function $AEI(\mathbf{x})$ estimates the expected increase in the performance,

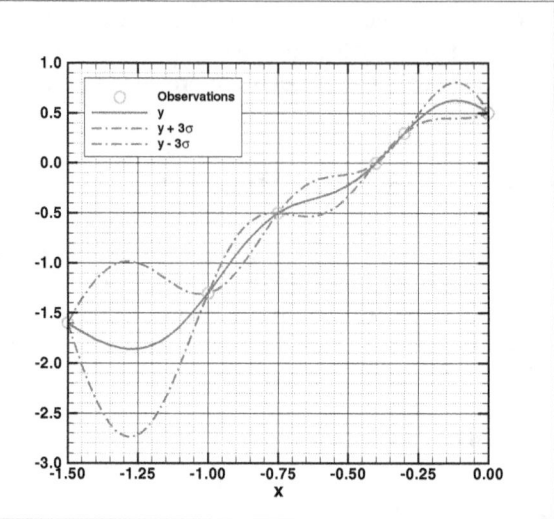

(a) Assuming $\sigma_\epsilon = 0$.

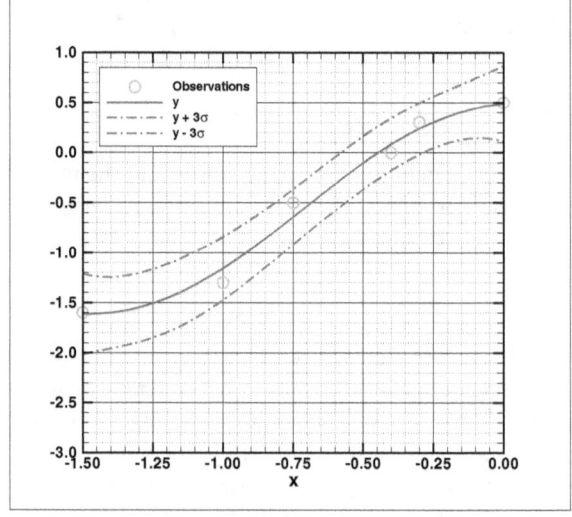

(b) Optimal $\sigma_\epsilon = 0.143$.

Figure 1: Effect of σ_ϵ on the GP model.

taking into account the noise in the observed values and penalizing areas where the variance $\hat{\sigma}_y^2$ is small. It writes as

$$AEI(\mathbf{x}) = EI(\mathbf{x}) \left(1 - \frac{\sigma_\epsilon}{\sqrt{\hat{\sigma}_y^2(\mathbf{x}) + \sigma_\epsilon^2}}\right), \quad (15)$$

with the Expected Improvement defined by

$$EI(\mathbf{x}) = \hat{\sigma}_y^2(\mathbf{x})\left[u(\mathbf{x})\Phi\left(u(\mathbf{x})\right) + \phi\left(u(\mathbf{x})\right)\right],$$
$$u(\mathbf{x}) = \frac{\hat{y}(\mathbf{x}_{**}) - \hat{y}(\mathbf{x})}{\hat{\sigma}_y^2(\mathbf{x})}, \quad (16)$$

where Φ and $\phi = \Phi'$ denote respectively the cumulative distributions (Erf-function) and density of the standard Gaussian distribution, and $\mathbf{x}_{**}$ is the effective best solution:

$$\mathbf{x}_{**} \doteq \arg\min_{\mathbf{x} \in \mathbf{X}_n} \left[\hat{y}(\mathbf{x}) + \hat{\sigma}(\mathbf{x}) \right]. \qquad (17)$$

When the optimum $\mathbf{x}_*$ of the AEI is determined, the corresponding performance $\mathcal{P}(\mathbf{x}_*)$ is evaluated and is included in the database. A new GP model can then be reconstructed using the extended data base, leading to a new maximizer $\mathbf{x}_*$ of the updated AEI, and so on. The iterations carry on until some convergence criterion is satisfied or the resources allocated to the optimization procedure have been exhausted (*e.g.* reaching a prescribed number of performance evaluations). Classical convergence criteria compare the distance between two successive iterates, in terms of optimal parameters $\mathbf{x}_*$ or/and performance prediction $\mathcal{P}(\mathbf{x}_*)$. Overall, we remark that each iteration requires the resolution of two optimization problems (one for the covariance parameters and one for the AEI) and one evaluation of the performance. In practice, all results reported in this work were obtained using the nonlinear non-convex black-box optimization library based on the Covariance Matrix Adaptation Evolution Strategy (Hansen, 2006, Hansen and Ostermeier, 2001).

EXPERIMENTAL OPTIMIZATION

This Section concerns the sail trimming optimization performed in the wind tunnel of the Yacht Research Unit (YRU) (Flay, 1996) at the University of Auckland.

Experimental setup

The sail model is inspired by an IMOCA 60-foot design mainsail at 1:13 scale. It was designed and produced by the sail-makers of INCIDENCES SAILS company. The sail has a surface area of 1 m², for an height of $h = 2$ m, and is supported by a rig consisting of a flexible circular section carbon mast (constant diameter 14 mm), clamped at its base and without spreader, backstay or forestay. The sail and rig are set in the open jet test section of the YRU wind tunnel, see Figure 2(a). The test section is 7.2 wide and 3.5 m high.

Three stepper motors and a control card control remotely the main sheet length (L_{sheet}) and main car position (L_{car}) as shown in Figure 2(b). In the following, L_{sheet} and L_{car} are the only trimming parameters to be optimized. Note that the remote system allows for changing these trimming parameters without switching off the wind tunnel flow and making a new tare of the measurement instruments. A precision of ± 2 mm on the imposed trimming parameters was estimated through repeated measurements.

A six-components force balance, located under the floor of the wind tunnel, was used to measure the aerodynamic forces. The X-direction corresponds to the model longitudinal forward direction (*i.e.* thrust force direction), while the Y-direction is defined as the positive port-side (*i.e.* side force direction) and the Z-direction is the vertical. After careful calibration and testing, the balance precisions in

(a) Mainsail in the YRU wind tunnel.

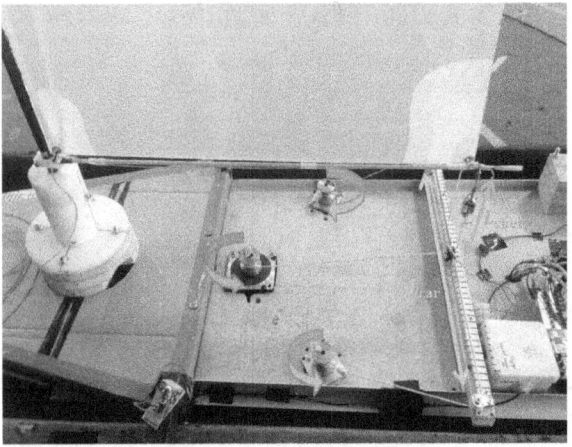

(b) Close view on the trimming system.

Figure 2: Experimental setup.

the X, Y and Z directions were estimated to be ± 0.09 N, ± 0.11 N and ± 0.27 N respectively. An additional load sensor, with 5 daN range, was used to record the load in the sheet with a precision of ± 0.02 N. Flying shapes were also recorded with a V-SPARS acquisition system (Le Pelley and Modral, 2008), tracking the position of five dark red stripes across the sail (see Figure 2(a)).

The wind tunnel inflow velocity was measured and found to have an apparent wind speed (AWS) of 3.5 ± 0.15 m/s for an apparent wind angle (AWA) of 40 ± 2 deg. The corresponding Reynolds number, based on the reference chord length $c = S/h = 0.5$ m, is Re $= 1.2 \times 10^5$. A multi-hole pressure probe (Cobra Probe) was used to measure profiles of the flow velocity at several locations inside the wind tunnel. These measurements were repeated with and without the sail model in the test section to verify its effect on the flow field. Typical profiles are shown in Figure 3; it can be seen that the inflow has no twist.

The optimization problem is finally defined as the maximization with respect to $\mathbf{x} = (L_{\text{sheet}}, L_{\text{car}})$ of the performance $\mathcal{P}(\mathbf{x})$ taken as a composite function of the thrust C_X

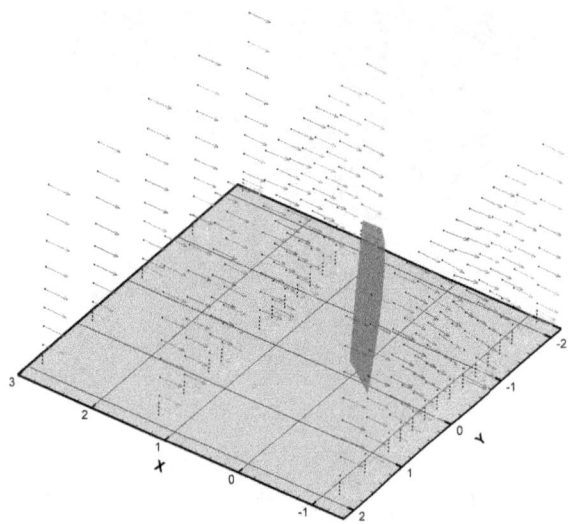

Figure 3: Velocity profiles with sail model.

and side C_Y aerodynamic coefficients:

$$\mathcal{P}(\mathbf{x}) = C_X + 0.1 C_Y. \tag{18}$$

The aerodynamic coefficients are deduced from the aerodynamic forces through normalization by the reference force being $q_\infty S$, where q_∞ is the reference dynamic pressure measured in the wind tunnel (precision ± 1 Pa). In (18), the coefficient 0.1 penalizes the side force to account for the resulting hydrodynamic drag and leeway that would be detrimental to the performance. The optimization of the trimming parameters then follows the procedure illustrated in Figure 4. The primary loop, *"Sampling loop"*, generates an initial Latin Hypercube Sample (LHS) set of 10 trimming parameters (McKay et al., 2000). For each initial sample, the experimental model is remotely set to the corresponding trimming values of L_{sheet} and L_{car}. After the transient flow is over, the aerodynamic loads reported by the balance are averaged over an acquisition time of 30 s to smooth-out the remaining noisy fluctuations in the signals. When the (time-averaged) loads are collected for all the initial samples (blue loop in Figure 4), the GP based optimization is carried out on the initial data set (red block *"Trimming optimization"*). The optimization provides a new trimming point, $\mathbf{x}_*$, maximizing the AEI merit function.

In the *"Optimization loop"*, the experimental apparatus is remotely tuned to the new trimming point $\mathbf{x}_*$. The aerodynamic loads are then averaged over 30 s and the new estimate of $\mathcal{P}(\mathbf{x}_*)$ is included in the database. The Gaussian Process model is updated consequently, generating a new trimming points $\mathbf{x}_*$. These steps (red loop in Figure 4) are repeated until the convergence of the trimming parameters, which is considered achieved when the algorithm proposes two successive trimming points within a distance less than 1 mm (*i.e* the precision on the enforcement of the trimming parameters in the experimental apparatus).

Below, we present the results of the trimming optimiza-

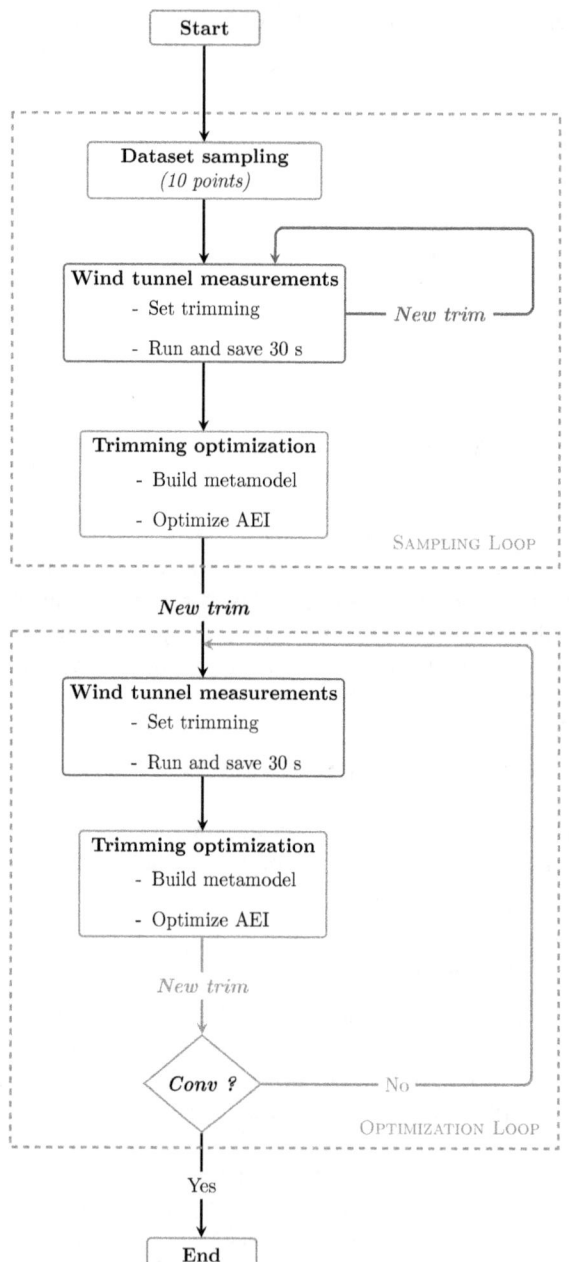

Figure 4: Trimming optimization procedure.

tion, contrasting two cases. In the first case, the measurement noise is set to $\sigma_\epsilon = 0$ (noise-free situation) so the constructed GP models are interpolating the experimental data. In the second case, the actual measurement noise is determined experimentally and it is subsequently used in the GP model construction.

Experimental optimization with $\sigma_\epsilon = 0$

The results of the experimental trimming optimization in the case where σ_ϵ is set to zero is now presented.

The GP model for the experimental $\mathcal{P}(\mathbf{x})$ function, af-

41

ter 36 iterations of the optimization algorithm, is reported in Figure 5. Specifically, Figure 5(a) depicts the color contours of the GP model mean as a function of the trimming parameters L_{car} and L_{sheet}, while Figure 5(b) shows the standard deviation of the GP model. The black dots are the data points where the performance was experimentally estimated. Regarding the location of the observation points, we notice a large dispersion, highlighting the lack of convergence in the successive tentative optimal trimming candidates $\mathbf{x}_*$. In fact, the algorithm has explored the parameter domain Ω without discovering a particular sub-domain of Ω likely to contain the global optimum. This can also be appreciated from the mean field of the GP model, in Figure 5(a), which, although smooth, presents at least 2 local minima. The presence of multiple local minima is in fact spurious and induced by the interpolating nature of the GP model for $\sigma_\epsilon = 0$: the model is fitting the experimental noise. This can also be appreciated from the standard deviation field reported in Figure 5(b), which is zero at the observation points, denoting an inappropriate level of confidence in the GP model approximation of $\mathcal{P}(\mathbf{x})$ at these locations. Further, departing from the observation points, the variance of the GP model prediction quickly increases (observe, in particular, the standard deviation field in the neighborhood of isolated data points) and becomes large. As a result of the over-confidence in the model at measured points and high variance (low confidence) in unexplored areas, the optimization process is led by the AEI merit function to propose new candidates in relatively less populated areas.

Figure 6 depicts the measured values of $\mathcal{P}(\mathbf{x}_*)$ at the successive tentative optima as selected along the iterations of the algorithm (the first 10 iterations correspond to the initial Latin Hypercube Sampling of Ω, and are not actual iterations of the algorithm). The plot shows that the measured performance is not converging and it sustains large fluctuations having the same magnitude as for the initial random sample: the complete absence of an improvement trend in the successive measurement of $\mathcal{P}(\mathbf{x}_*)$ is characteristic of the failure of the present approach. This unsuccessful test highlights the negative effect of not accounting for the noise in the estimates of $\mathcal{P}(\mathbf{x})$: it prevents the GP model to discover trends in the actual performance function from the noisy observations.

Experimental optimization with noise

In a second experiment we set $\sigma_\epsilon^2 = 0.027^2$. This value is not determined as part of the optimization of the log-marginal likelihood in (14). Instead, σ_ϵ is directly estimated from the experimental apparatus, using repeated measurements at the same trimmings. The same 10 previous initial LHS points are used to determined the first GP model, with the prescribed value of σ_ϵ. Then the optimization proceeds and a different sequence of proposed optima is generated as the GP models differ from the previous experiment.

In particular, the optimization now converges in 33 iter-

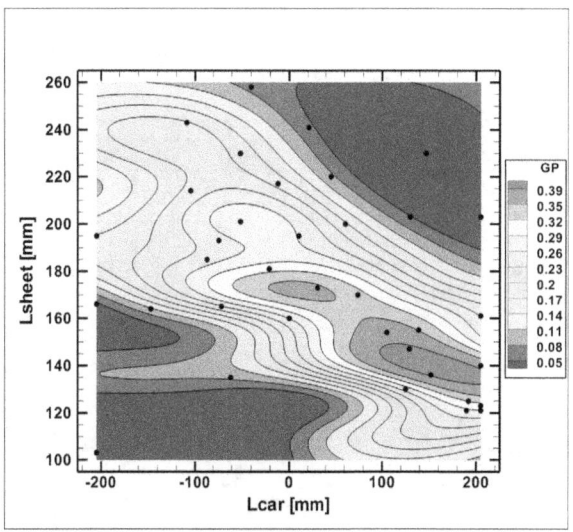

(a) GP model expectation.

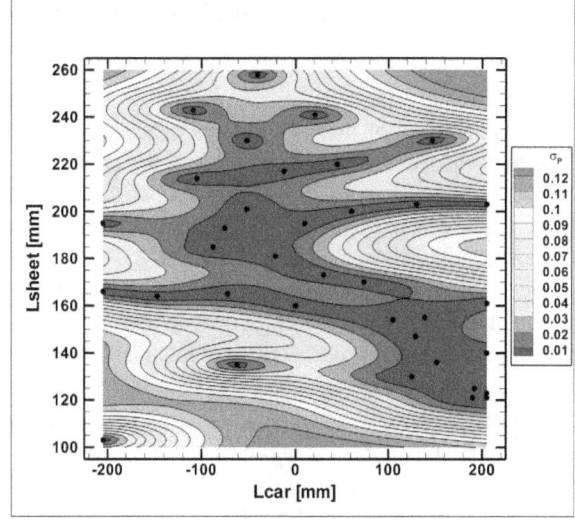

(b) GP model standard deviation.

Figure 5: GP model of the experimental performance $\mathcal{P}(\mathbf{x})$ ($\sigma_\epsilon = 0$).

ations, the last two proposed optima being in sufficiently close distance (less than 1 mm). The convergence of the sequence of proposed optima can be seens in Figure 7, where the mean and standard deviation of the GP model of $\mathcal{P}(\mathbf{x})$ at convergence are plotted. In contrast to the case with $\sigma_\epsilon = 0$, the clustering of the successive proposed optima is clearly visible. Also, the mean of the GP model in Figure 7(a) remains smooth and now exhibits a single well-defined global maximum. The converged optimal trimming is found for $L_{\text{sheet}} = 133$ mm and $L_{\text{car}} = 138$ mm corresponding to a predicted performance $\mathcal{P}(\mathbf{x}_{\text{opt}}) = 0.397$. The standard deviation of the GP model, depicted in Figure 7(b), is seen to be minimal in the neighborhood of the optimum, though assuming values $\gtrsim \sigma_\epsilon$. Other regions of Ω far from the optimum are not explored by the optimization process, although

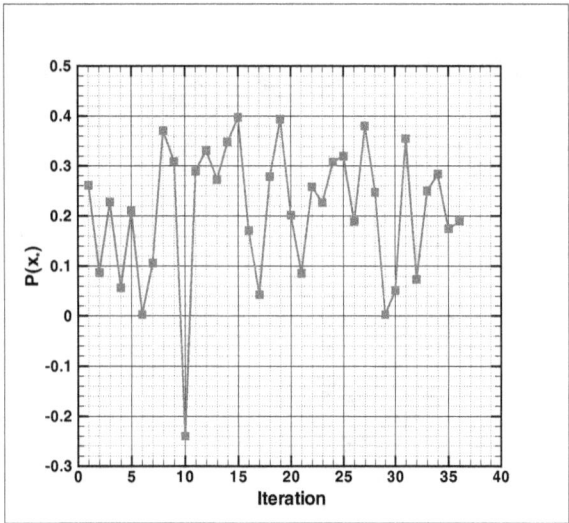

Figure 6: Experimental measurements of the performance for the sequence of proposed optima. Case of $\sigma_\epsilon = 0$.

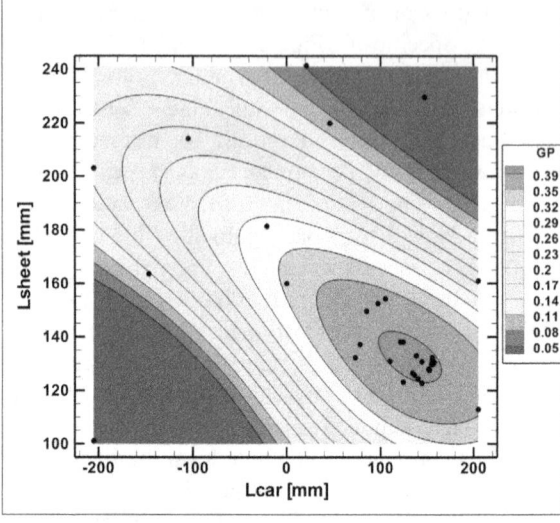

(a) GP model expectation.

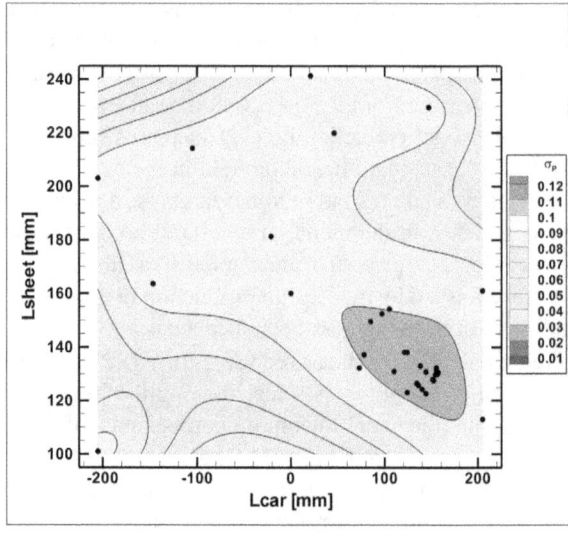

(b) GP model standard deviation.

Figure 7: GP model of the experimental performance $\mathcal{P}(\mathbf{x})$ ($\sigma_\epsilon = 0.027$).

the variance can be large.

The convergence of the optimization procedure can also be appreciated from the plot of Figure 8 which should be contrasted with the results shown Figure 6. It shows a clear improvement of the measured performance (after the first 10 random points). In fact after iteration 25, the remaining fluctuations in the measured performances can be essentially attributed to the measurement noise. These remaining fluctuations, with amplitude $\sim \sigma_\epsilon$, are much less significant compared to the previous case.

In summary, the global optimal trimming parameters are determined despite the noise in the measurements, thanks to the non-interpolating nature of the GP approximation which smooths out the noise. Further, the optimum is found in few iterations only, owing to the AEI merit function which is able to disregard non-interesting areas of Ω, even if they carry large prediction variance.

NUMERICAL OPTIMIZATION

A numerical model of the wind tunnel and sail model has been created to reproduce the previous experimental optimization problem. The objective is to assess the capabilities of the optimization method, when applied to a coupled FSI software, and compare the resulting optimum with the experimental one.

Numerical model

We briefly present the structural and fluid solvers used for the resolution of the FSI problem. Steady solution of the FSI problem are sought by means of a quasi-steady approach.

For the structural model of the sail we rely on the ARA software developed by K-EPSILON. The code ARA considers different structural elements (*e.g.* Timoshenko beams,

cables and Constant Strain Triangles (CST) membrane elements of various types) for the static or dynamic simulation of sail boat rigs in large displacement regime (Augier, 2012). The structural model for the simulations presented hereafter is illustrated in Figure 9; it uses dimensions and mechanical characteristics (for the mast, boom, and sail fabrics) measured on the experimental model.

The solver ARA is coupled to the ISIS-CFD software (from FINE^{TM}/Marine) which solves the Navier-Stokes equations in the flow domain. ISIS-CFD is based on finite volume methods accommodating both structured and unstructured meshes; it also proposes several turbulence models and boundary condition. For the present computations, we consider a parallelepiped computational domain, with spatial extension $7.5h$, $12h$ and $1.8h$ in the X, Y and Z di-

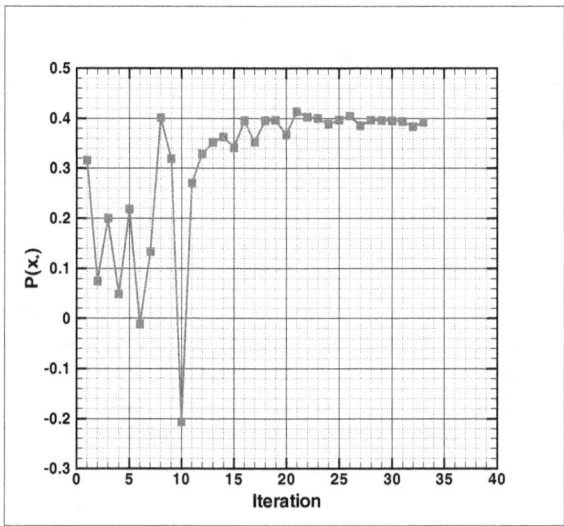

Figure 8: Experimental measurements of the performance for the sequence of proposed optima. Case of $\sigma_\epsilon = 0.027$.

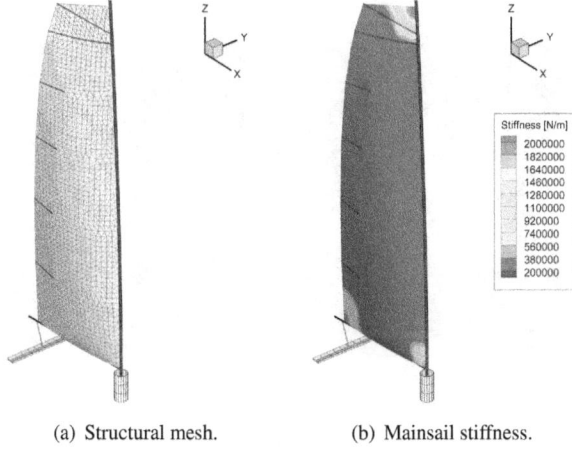

(a) Structural mesh.   (b) Mainsail stiffness.

Figure 9: Numerical model of the sail.

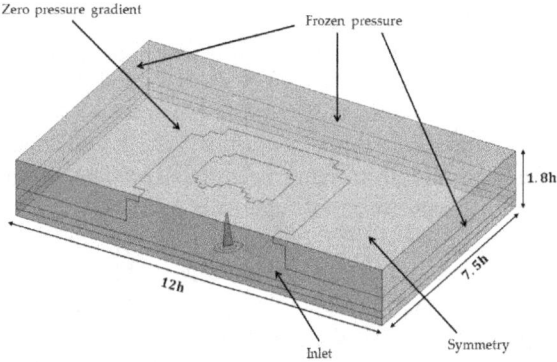

Figure 10: Boundary conditions for the flow solver.

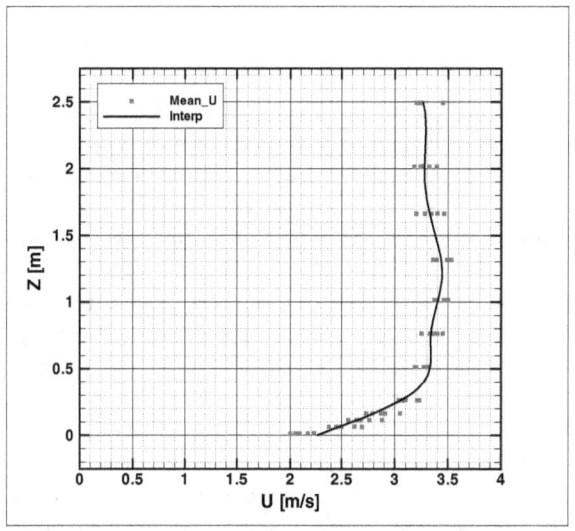

Figure 11: Inflow velocity profile.

rections respectively. These dimensions were selected on the basis of previous numerical experiments (Viola et al., 2013). The boundary conditions, applied on the different faces of the computational domain, are schematically illustrated in Figure 10.

The fluid domain is meshed using HEXPRESS^{TM}, a semi-automated mesh generator. Note that the mast is not meshed in the fluid solver. Regarding the turbulence model, the $SST\ k-\omega$ model (Menter et al., 2003) was selected with wall function boundary conditions (Kalitzin et al., 2005). This choice requires a sufficiently fine mesh over the sail surface and at the bottom of domain (sea level) to correctly capture the vertical profile of the inflow velocity. The later is estimated from the experimental profiles previously reported in Figure 3, and its shown in Figure 11.

For the FSI simulations, the mesh of the fluid domain has to be deformed to adapt to the changes in the bound-

ary geometries. A mesh deformation propagation method, proceeding from the deformable boundaries toward the inside of the fluid domain, was developed at K-EPSILON for this purpose (Durand et al., 2014). The FSI problem is then solved coupling ARA and ISIS-CFD solvers with a quasi-monolithic algorithm (Durand, 2012), which is an implicit coupling procedure adapted to a partitioned solver. Briefly, the resolution of the structural problem is nested inside the iterations on the nonlinear steady flow solution. This approach preserves the convergence and stability properties of the monolithic approach. More details on the solvers and the coupling algorithms can be found in (Durand, 2012, Roux et al., 2002, 2008).

A convergence analysis was conducted in order to select spatial discretization capturing correctly the physics of the FSI problem, while maintaining a reasonable computational cost permitting the optimization of the trimming parameters. In particular, different fluid meshes with up to 4.3 million finite volumes were considered. Eventually, a discretization of the fluid domain with roughly 1.5 million finite volumes and a sail discretization with 2 700 membrane elements was selected for the computations presented below. The computations were carried out on a 64 CPUs cluster; an averaged computational time of 5 h was reported for solving individual FSI problems. From the FSI solution, the associated aerodynamic forces acting on the sail are computed in the same reference frame as in the experimental setup, and the performance in (18) is finally returned to the optimizer. Except for the determination of $\mathcal{P}(\mathbf{x})$, the flowchart of the numerical optimization procedure is identical to that of the experiment in Figure 4. However, for the numerical optimization, σ_ϵ is directly inferred from the data when minimizing (14). In addition, the parameter domain for the numerical case has been increased to encompass higher values of L_{car}.

Numerical optimization results

Figure 12 depicts the mean value of the GP model based on the numerical evaluation of the performance. The GP model is reported at the end of the optimization procedure, which has converged in 34 iterations. The black dots correspond again to the sequence of optimization points $\mathbf{x}_*$ in the data set. We first remark the smoothness of the mean GP model which exhibits a single global optimum, as for the experimental case (considering measurement noise). In fact, the inferred $\sigma_\epsilon = 0.022$ has a value close to the experimental one. Further, in the range $L_{car} \in [-210, 150]$ mm, the terminal GP model is seen to be in good agreement with its experimental counterpart shown in Figure 7(a). However, the valley containing the numerical minimum is larger and flatter, compared to the experimental one, and the numerical optimum appears at a value of a L_{car} value larger than for the experimental case. The variance of the GP model prediction, σ_y^2, exhibits a structure similar in shape and magnitude to the experimental case in Figure 7(b) (not shown).

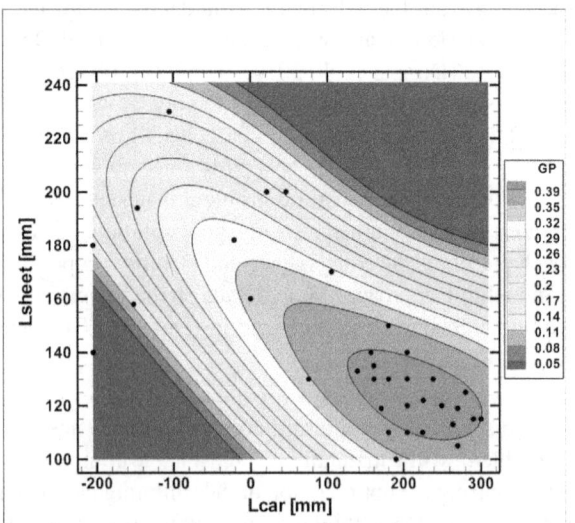

Figure 12: Mean GP model of the numerical performance (inferred $\sigma_\epsilon = 0.022$).

To understand the differences between the numerical and experimental optima, we first compare in Figure 13 and Figure 14 the flyings shapes for two trimming parameters (L_{sheet}, L_{car}) equal to $(160, 0)$ and $(133, 138)$ respectively (lengths in mm). For the case with centered car, $L_{car} = 0$, a small wrinkle is visible in the experimental flying shape (see Figure 13(a)). At the optimal experimental trimming point $(L_{sheet}, L_{car}) = (133, 138)$, shown in Figure 14(a), the wrinkle in the experimental flying shape is even more pronounced. This is in contrast with the corresponding numerical flying shapes, shown in Figures 13(b) and 14(b) respectively, which present no such wrinkle. Modeling errors and experimental uncertainties are deemed responsible for this difference. In particular, the absence of wrinkle in the numerical solution could be mostly due to an incorrect prescription of the tensions in the two full battens of the sail. Another important source of discrepancy between the flying shapes are the boundary conditions of the numerical wind tunnel and effects of confinement, which were shown to have a significant impact on the computed aerodynamic forces (Viola et al., 2013).

(a) Experimental.

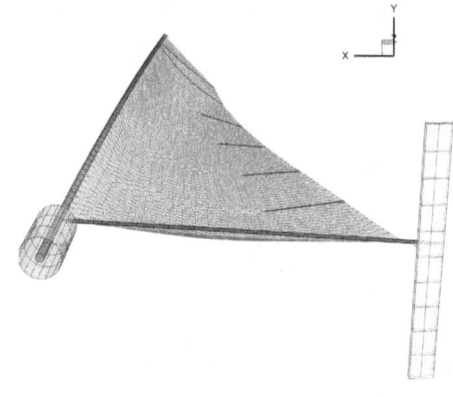

(b) Numerical.

Figure 13: Comparison of experimental and numerical flying shapes for trimming parameters $L_{sheet} = 160$ and $L_{car} = 0$.

To complete the comparison of the experimental and numerical optimizations, we report in Table 1 the computed location of the two optima and the corresponding best predic-

(a) Experimental.

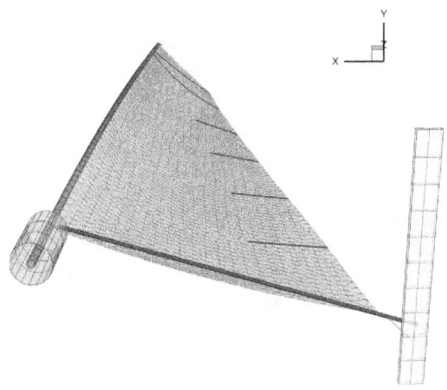

(b) Numerical.

Figure 14: Comparison of experimental and numerical flying shapes for trimming parameters $L_{\text{sheet}} = 133$ and $L_{\text{car}} = 138$.

tion of the performance. We observe that the experimental and numerical optimal trimming are significantly different for the optimal L_{car}. This difference can be explained by the numerical performance function that is particularly flat along the L_{car} direction around the optimal point: variation of L_{car} around the optimum weakly affects the predicted performance. This is consistent with the predicted performances reported in the last column of Table 1, that are in close agreement despite the differing trimming parameters.

	L_{sheet} [mm]	L_{car} [mm]	Pred. $\mathcal{P}(\mathbf{x}_{\text{opt}})$
Exp.	133	138	0.397 ± 0.027
Num.	122	226	0.413 ± 0.024

Table 1: Comparison of the experimental and numerical optima.

A more detailed investigation of the measured and computed fluid forces at the optima, reported in Table 2, reveals that the thrust coefficients are in fact equal up to the third significant digit. On the contrary, the side force coefficients and the sheet tensions exhibit larger discrepancies. The differences in the sheet tensions can be directly attributed to the different optimal L_{car}. The higher disagreement in the

side force coefficients is not surprising. In our experience the side forces are very sensitive to model errors.

	C_X	C_Y	T [N]
Exp.	0.497 ± 0.012	-1.026 ± 0.015	14.9 ± 0.02
Num.	0.497	-0.803	20.1

Table 2: Comparison of aerodynamic coefficients and sheet tension at the experimental and numerical optima.

CONCLUSION AND DISCUSSION

We have proposed to use a Gaussian Process model to enable the optimization of the trimming parameters of a complex nonlinear sail systems. The approach has been first tested on the trimming of an experimental model sail in the Yacht Research Unit wind tunnel. The experiments have validated the approach and have shown its robustness against noisy estimates of the performance. For this two parameters problem, the experimental optimal point was found within few iterations of the algorithm. These tests have validated the proposed optimization method and have demonstrated its robustness against experimental variabilities, as well as the key role of the noise parameter.

A detailed numerical model of the wind tunnel experiment has been established, considering the full Fluid-Structure Interactions problems. The numerical model is based on a turbulent flow solver coupled with a nonlinear elastic solver, using a mesh deformation method. The numerical and experimental optima were found to be consistent, given all the modeling and measurement errors. In particular, it is found that the predicted performances and fluid forces are in much better agreement than the optimal parameters. In fact, it can be reasonably claimed that the differences are consistent with the current predictive capabilities of state of the art FSI solvers, and that the GP approximation does not introduce noticeable errors in the optimization procedure. Specifically, we have shown that, from a limited set of computations, the GP model is able to reconstruct accurately the numerical estimate of the performance. Therefore, for the present tests, only a better numerical modeling of the experimental set-up would help reducing the observed discrepancies. Possible avenues in this direction are: improved battens modeling, better boundary conditions for the flow, accounting for mast/flow interaction, more advanced turbulence model, ... At a more fundamental level, the question of the treatment of experimental uncertainties remains critical. On this aspect, we are considering sensitivity and uncertainty quantification (Le Maître and Knio, 2010) studies to account for the experimental uncertainties. In particular, the results presented in this work point to the need for an appropriate characterization of the optima and the assessment of their robustness to uncertainties and modeling errors. Future works will develop these aspects along with the deployment of the GP-based optimization method to problems involving large numbers of trimming parameters (full yacht rig).

ACKNOWLEDGEMENTS

The authors wish to acknowledge the people of the Yacht Research Unit for their welcoming and help during the experiments in the wind tunnel. Contributions of INCIDENCES SAILS and NUMECA companies are also acknowledged. This work was partially funded by the European Union's Seventh Program for research, technological development and demonstration under grant agreement No PIRSES-GA-2012-318924, and from the Royal Society of New Zealand for the UK-France-NZ collaboration project SAILING FLUIDS (see www.sailingfluids.org). This work was also supported by the "Laboratoire d'Excellence" LabexMER (ANR-10-LABX-19) and co-funded by a grant from the French government under the program "Investissements d'Avenir".

REFERENCES

B. Augier. *Etudes expérimentales de l'interaction fluide-structure sur surface souple: application aux voiles de bateaux*. PhD Thesis, Universite de Bretagne Occidentale, 2012.

B. Augier, P. Bot, F. Hauville, and M. Durand. Experimental validation of unsteady models for fluid structure interaction: Application to yacht sails and rigs. *Journal of Wind Engineering and Industrial Aerodynamics*, 101:53–66, 2012.

T. Bäck and H. P. Schwefel. An overview of evolutionary algorithms for parameter optimization. *Evolutionary computation*, 1(1):1–23, 1993.

V. G. Chapin, R. Neyhousser, G. Dulliand, and P. Chassaing. Design optimization of interacting sails through viscous cfd. In *INNOVSail, Innovation in high performance sailing Yacht*, Lorient, 2008.

M. Durand. *Interaction fluide-structure souple et légère, application aux voiliers*. PhD Thesis, Ecole Centrale de Nantes, 2012.

M. Durand, A. Leroyer, C. Lothodé, F. Hauville, M. Visonneau, R. Floch, and L. Guillaume. Fsi investigation on stability of downwind sails with an automatic dynamic trimming. *Ocean Engineering*, 90:129–139, 2014.

R. Duvigneau and P. Chandrashekar. Kriging-based optimization applied to flow control. *International Journal for Numerical Methods in Fluids*, 69(11):1701–1714, 2012.

R. G. J. Flay. A twisted flow wind tunnel for testing yacht sails. *Journal of Wind Engineering and Industrial Aerodynamics*, 63(13):171 – 182, 1996. ISSN 0167-6105. Special issue on sail aerodynamics.

M. N. Gibbs. *Bayesian Gaussian Processes for Classification and Regression*. PhD thesis, University of Cambridge, 1997.

H. Hansen, P. S. Jackson, and K. Hochkirch. Real-time velocity prediction program for wind tunnel testing of sailing yachts. *Proc. The Modern Yacht, Southampton, UK*, 2003.

N. Hansen. The cma evolution strategy: a comparing review. In *Towards a new evolutionary computation*, pages 75–102. Springer, 2006.

N. Hansen and A. Ostermeier. Completely derandomized self-adaptation in evolution strategies. *Evolutionary Computation*, 9(2):159–195, 2001.

D. Huang, T. T. Allen, W. I. Notz, and N. Zeng. Global optimization of stochastic black-box systems via sequential kriging meta-models. *Journal of global optimization*, 34(3):441–466, 2006.

L. Huetz and P. E. Guillerm. Database building and statistical methods to predict sailing yacht hydrodynamics. *Ocean Engineering*, 90:21–33, 2014.

D. R. Jones. A taxonomy of global optimization methods based on response surfaces. *Journal of global optimization*, 21(4):345–383, 2001.

D. R. Jones, M. Schonlau, and W. J. Welch. Efficient global optimization of expensive black-box functions. *Journal of Global optimization*, 13(4):455–492, 1998.

G. Kalitzin, G. Medic, G. Iaccarino, and P. Durbin. Near-wall behavior of rans turbulence models and implications for wall functions. *Journal of Computational Physics*, 204(1):265–291, 2005.

R. Korpus. Performance prediction without empiricism: A rans-based vpp and design optimization capability. In *The 18th Chesapeake Sailing Yacht Symposium, SNAME*, 2007.

O. P. Le Maître and O. M. Knio. Spectral methods for uncertainty quantification. *Scientific Computation. Springer, New York*, 2010.

D. J. Le Pelley and O. Modral. V-spars: A combined sail and rig shape recognition system using imaging techniques. In *Proc. 3rd High Performance Yacht Design Conference Auckland, New Zealand, Dec*, pages 2–4, 2008.

M. D. McKay, R. J. Beckman, and W. J. Conover. A comparison of three methods for selecting values of input variables in the analysis of output from a computer code. *Technometrics*, 42(1):55–61, 2000.

W. Menotti, M. Durand, D. Gross, Y. Roux, D. Glehen, and L. Dorez. An unsteady fsi investigation into the cause of the dismating of the volvo 70 groupama 4. In *INNOVSail, Innovation in high performance sailing Yacht*, page 197, Lorient, 2013.

F. R. Menter, M. Kuntz, and R. Langtry. Ten years of industrial experience with the sst turbulence model. *Turbulence, heat and mass transfer*, 4:625–632, 2003.

J. Nelder and R. Mead. A simplex method for function minimization. *Computer Journal*, 7(4):208–313, 1965.

P. V. Oossanen. Predicting the speed of sailing yachts. *SNAME*, 101:337–397, June 1993.

R. Ranzenbach, D. Armitage, and A. Carrau. Mainsail planform optimization for irc 52 using fluid structure interaction. In *The 21st Chesapeake Sailing Yacht Symposium, SNAME*, 2013.

C. E. Rasmussen and C. K. I. Williams. *Gaussian processes for machine learning*. MIT Press, 2006.

N. Rousselon. Optimization for sail design. In *modeFRONTIER Conference*, 2008.

Y. Roux, S. Huberson, F. Hauville, J. P. Boin, M. Guilbaud, and B. Malick. Yacht performance prediction : Towards a numerical vpp. In *1st High Performance Yacht Design Conference Auckland, 4-6 December*, Auckland, 2002.

Y. Roux, M. Durand, A. Leroyer, P. Queutey, M. Visonneau, J. Raymond, J. M. Finot, F. Hauville, and A. Purwanto. Strongly coupled vpp and cfd ranse code for sailing yacht performance prediction. In *3rd High Performance Yacht Design Conference Auckland, 2-4 December*, pages 215–225, Auckland, 2008.

V. K. Saul'ev and I. I. Samoilova. Approximation methods for the unconstrained optimization of functions of several variables. *Journal of Soviet Mathematics*, 4(6):681–705, 1975.

M. L. Stein. *Interpolation of spatial data: some theory for kriging*. Springer Science & Business Media, 2012.

D. Trimarchi. *Analysis of downwind sail structures using non-linear shell finite elements*. PhD Thesis, University of Southampton, 2012.

I. M. Viola, P. Bot, and M. Riotte. Upwind sail aerodynamics: A rans numerical investigation validated with wind tunnel pressure measurements. *International Journal of Heat and Fluid Flow*, 39:90–101, 2013.

R. Von Mises. *Mathematical Theory of Probability and Statistics*. Academic press, 1964.

Fully Integrated Fluid-Structural Analysis for the Design and Performance Optimisation of Fibre Reinforced Sails

Sabrina Malpede, SMAR Azure, United Kingdom, sabrina@smar-azure.com
Donald MacVicar, SMAR Azure, United Kingdom, donald@smar-azure.com
Francesco Nasato, SMAR Azure, United Kingdom, francesco@smar-azure.com
Paolo Semeraro, Banks Sails, Italy, paolosemeraro@bankssails.it

ABSTRACT

This paper presents an advanced and accurate integrated system for the design and performance optimisation of fibre reinforced sails, commonly named string sails, developed by SMAR Azure. This integrated design system allows sail designers not only to design sail shapes and the reinforcing fibre paths, but also to validate the performance of the flying sail shape and have accurate production details including the overall sail weight, material used, which means costs, and length of the fibre paths, which means production time.

The SMAR Azure design and analysis method, extensively validated and used to optimise several racing and super-yacht sailing plans, includes a computationally efficient structural analysis method coupled with a modified vortex lattice method, with wake relaxation, to enable a proper aeroelastic simulation of sails in upwind conditions. The structural analysis method takes into account the geometric non-linearity and wrinkling behaviour of membrane structures, such as sails, the fibre layout, the influence of battens, trimming loads and interaction with rigging elements, e.g. luff sag calculation on a headstay, in a timely manner.

Specifically, this paper presents an optimisation of a real fibre reinforced membrane sailplan of an aluminium super yacht, carried out in collaboration with Paolo Semeraro (Banks Sails Europe). The optimisation process of the fibre layouts led to a sensible reduction in maximum stress, strain and displacement compared to the initial designs, keeping the same fibre weight or slightly increasing it. The results have been confirmed in the sailing tests, although no exact measurements have been performed.

NOTATION

AWA	Apparent Wind Angle (°)
AWS	Apparent Wind Speed (kn)
BS	Boat Speed (kn)
CE_z	Height of the Centre of Effort (m)
CST	Constant Strain Triangular
FEM	Finite Element Model
FSI	Fluid Structure Interaction
HM	Heeling Moment (Nm)
VLM	(Modified) Vortex Lattice Method
NR	Newton Raphson
RANS	Reynolds Averaged Navier Stokes
TWA	True Wind Angle (°)
TWS	True Wind Speed (kn)
VPP	Velocity Prediction Program
WTT	Wind Tunnel Test
C_P	Pressure coefficient (-)
δ	Displacement (m)
δ_{max}	Maximum displacement (m)
ε_{max}	Maximum strain (-)
F_E	External loads (N)
F_I	Internal reaction loads (N)
K_E	Elastic stiffness matrix (N.m^{-1})
K_G	Geometric stiffness matrix (N.m^{-1})
p	Static pressure (Pa)
p_∞	Asymptotic static pressure (Pa)
Q_∞	Asymptotic dynamic pressure (Pa)
σ_{max}	Maximum stress (Pa)
V	Velocity (m s^{-1})
V_∞	Asymptotic velocity (m s^{-1})

INTRODUCTION

This paper presents the performance optimisation of a real fibre reinforced membrane sail plan of an aluminium super yacht, achieved through the advanced integrated system developed by SMAR Azure for the simulation of the structural behaviour of fibre-membrane sails. The structural simulation of sails is one of the challenging problems of current marine engineering due to its strong nonlinearities. The integrated system for the design and performance optimisation of fibre-reinforced sails provides an accurate solution method that could be easily used in a timely manner in the everyday work of the sail designers.

The optimisation has been carried out in collaboration with Paolo Semeraro (Banks Sails Europe), who designs and produces the MEMBRANE™ and BFAST™ string sails, the latter with Marco Semeraro. Both BFAST™ and Banks Sails have been using the SMAR Azure technology for almost a decade. Notwithstanding the long experience of Mr. Semeraro in using the technology, given the sailplan size and detailed customer requirements, which included improved durability, strength, reliability and smooth use of in-boom furling, this project was carried out in-cooperation with the SMAR Azure technical team.

Chapter 1 presents the integrated fluid structural analysis method, explaining the main features of the aerodynamic analysis (performed with a modified version of the vortex lattice method), of the structural analysis (a nonlinear finite element method adopting the constant strain triangular membrane elements, CST), and of the aeroelastic analysis, which iterates between the aerodynamic analysis and the structural analysis (fluid structure interaction) to achieve the balance of the internal and external forces.

Chapter 2 presents the fibre layout optimisation of the sailplan of a 43 m, 240 t (tonne) aluminium super yacht, carried out in collaboration with Mr. Semeraro. No aerodynamic optimisation was performed on the sail shapes provided. The analysis and optimisation of the fibre layout of the full mainsail, of the reefed mainsail and of the staysail are presented.

Positioning the battens on a super yacht mainsail is a design challenge, due to the strict requirements of the furling system and the varying thickness of the sail. Chapter 3 presents the algorithm developed for positioning the battens on the mainsail, relevant for the in boom furling system, in order to avoid the battens to overlap each other and the sail to fit inside the boom.

Chapter 4 presents details of the sail production and the results of the sailing test. The sails was made with the MEMBRANE™ and BFAST™ technology, developed by Mr. Semeraro, who supervised the sailing test in a gentle breeze and smooth sea conditions.

1. FLUID-STRUCTURAL ANALYSIS METHOD

Sails are thin flexible structures whose shape depends on wind forces and material properties; therefore the optimisation of the sail shape is an aeroelastic problem, whereby the sailing forces are evaluated on the flying sail shape, in specified sailing and trimming conditions.

The SMAR Azure technology is a fully integrated system for the design and optimisation of sailplans. The system includes a complete toolset for the design of any type of sail, either panelled or fibre-membrane sail, and an integrated static-aeroelastic analysis, composed by the aerodynamic and the structural analysis, repeated iteratively to define the flying sail shape. The analytical methods are summarised in the following sections. Additional information and validation are available in [1], [2] and [3].

1.1 Aerodynamic analysis method

The evaluation of the aerodynamic loads on the sailplan is carried out using a modified version of the Vortex Lattice Method (VLM), which represents the most widely used potential method for solving the inviscid sail aerodynamic problem. In 1968 Milgram started to apply this method to sail aerodynamic analysis [4]. After his basic development, the method has been developed in several ways and is demonstrated to be an effective tool for sail analysis [5], [6]. Low costs in terms of computational and human resources make VLM popular and efficient with respect to RANS techniques and Wind Tunnel Tests (WTT). The implemented Modified Vortex Lattice Method (MVLM) converts the sail surface into a vortex sheet (see Figure 1 andFigure 3), which means that the surface has zero thickness (in this paper the mast aerodynamic influence is ignored) and the jump in velocity across it is equal to the local strength of the vortex sheet. The local vorticity is evaluated by imposing the condition that there is no flow through the surface of the sail and, hence, the velocity field is tangential to the surface. Furthermore, to obtain a unique solution, it is assumed that the flow separates from the sail surface at the trailing edge – the so-called Kutta condition. Essential assumptions are:

- Flow field is irrotational.
- No flow separation on the sail surface.
- Zero pressure gradient across the wake.

The irrotational condition can be considered satisfied because sails can be modelled as thin lifting surfaces. Such sails are characterised by high Reynolds number (between 10^5 and 10^7) and low speed flow (up to 15 m/s in normal sailing conditions). The second assumption is more problematic due to the fact that sails are cambered lifting surfaces working at high angle of attack and a wrong trim can generate even leading edge separation. Thus, the method accuracy decreases for high angles of attack [1]. The third assumption is performed through the wake relaxation process.

The MVLM calculates the jump of the tangential flow velocity over the sail. Thus, the aerodynamic loads are calculated via the Bernoulli principle for the pressure coefficient:

$$C_P = \frac{p - p_\infty}{Q_\infty} = 1 - \left(\frac{V}{V_\infty}\right)^2 \qquad (1)$$

where C_P is the pressure coefficient, p is the static pressure, Q_∞ is the dynamic pressure and V is the velocity. Subscript ∞ denotes asymptotic flow.

The MVLM doesn't account for the suction peak at the leading edge since the sails are normally trimmed such that the stagnation point is situated on the leading edge and therefore no leading edge force is created (the suction peak occurs when the flow is forced to bend sharply around the leading edge, which causes high flow velocities and low pressure) [6].

1.2 Structural analysis method

The structural analysis code developed by SMAR Azure evaluates the deformed shape of sails by considering the local structural properties of the material, the influence of battens and the local reinforcements. It is applied to panelled sails or fibre-membrane sails.

The structural analysis code is a nonlinear finite element method. In particular, a direct stiffness method is used to solve the structural problem [1] [2] [3] [7]. Since a nonlinear problem is solved, Newton-Raphson's method is used to find the deformed equilibrium state of the structure. Geometric and wrinkling nonlinearities are taken into account. The stress-strain relation of the sailcloth material is considered as linear instead: the approximation is valid for most of the sailcloth materials and fibres up to 1% strain, which is the reference strain to determine the cloth or fibre stiffness. Considering one step of the Newton-Raphson's (NR) algorithm, the structural system is linear and its behaviour is described by the equation:

$$(K_E + K_G)\,\delta = \Delta F = F_E - F_I \qquad (2)$$

where K_E and K_G are the elastic and geometric stiffness matrixes, δ is the displacement term, F_E and F_I are the external loads and the internal reaction loads, respectively.

The global stiffness matrices K_E and K_G are assembled adding the contribution of every finite element. Constant strain triangular membrane elements (CST) are used to model the sailcloth. The sailcloth can be modelled as made of an isotropic, orthotropic or anisotropic material. In case the real fibre layout has to be taken into account, a stacking procedure (classical lamination theory) is used to compute the anisotropic stiffness matrix of the laminate. Battens can also be included as beam elements. The wrinkling behaviour of the sailcloth is taken into account by a dedicated model that avoids compression stresses [3].

1.3 Aeroelastic analysis method

The structural analysis is coupled with the modified vortex lattice method in order to obtain a proper fluid-structure interaction (FSI) simulation. The implemented sail static aeroelastic analysis code [1] iteratively performs the actions described below. It:

- generates the vortex lattice on the deformed sail shapes, as computed by the structural analysis;
- performs the aerodynamic analysis in the original sailing and trim conditions to calculate the wind loads;
- performs the structural analysis applying the wind loads on the deformed FEM of the sails to compute the new deformed sail shapes and stress distribution.

The convergence of the iterative process is achieved when the flying sail shape is in static equilibrium with the external loads in the particular sailing and trimming conditions set by the designer. Finally, the aeroelastic analysis provides the complete set of aerodynamic and structural results on the converged flying sail shape.

2. FIBRE LAYOUT OPTIMISATION

This chapter presents the optimisation of a real fibre reinforced membrane sailplan of an aluminium super yacht, carried out in collaboration with Paolo Semeraro (Banks Sails Europe). The optimisation problem is represented by an objective function (the finite element solver), which gives results in terms of stress, strain and displacement, based on the fibre layout and on the sailing conditions. The fibre layout is determined by the shape of each path and by the number of fibres, which translates in the fibre weight. Upon client's requirement, the fibre layout optimisation study aimed to optimise the durability of the sails. In order to ensure durable sails, the optimisation intended to reduce the maximum stress and strain to an acceptable value, keeping the same weight of the sails or increasing it slightly, in a multi-objective optimisation.

Since the durability of the sails is linked to the level of stress and strain, the acceptable values for the maximum stress and strain in the reference conditions have been set to sensibly smaller values than the fibre breaking tenacity and elongation, respectively. Therefore, although the breaking tenacity of the Dyneema® SK90 and Twaron® D2226 is 3800 MPa and about 2900 MPa respectively, the maximum stress in the fibre optimisation has been set to 500 MPa. Similarly, the maximum strain has been set to 1%, although the elongation at break of the Dyneema® SK90 and Twaron® D2226 is 3.5% and 2.8% respectively. The mechanical properties of the fibres, Dyneema® SK90 and Twaron® D2226, are listed in Table 1 andTable 2, while the fibre layout details are reported in Chapter 4. No limitation has been set for the maximum displacement of the sails (the maximum distance between the relative points of the undeformed design shape and the final flying shape).

Property	Unit	Typical value
Yarn count	dtex	1040
	denier	936
Tenacity	cN/dtex	39.5
	g/den	44.7
	GPa	3.8
Modulus	cN/dtex	1435
	g/den	1625
	GPa	140
Elongation	%	3.5

Table 1 - Dyneema® SK90 mechanical properties

Property	Unit	Twaron D2226 1210 dtex	Twaron D2226 1610 dtex
Effective linear density	dtex	1300	1710
Breaking strength	N	265	340
Breaking tenacity	MPa	2956	2883
Elongation at break	%	2.80	2.80
Modulus	GPa	100	100

Table 2 - Twaron® D2226 mechanical properties

The structural analysis of the original fibre layout provides the input for the optimisation, which is performed by:

- aligning the fibres along the principal stress directions, editing the shape of the fibre groups;
- spreading the fibres toward the most stressed areas;
- varying the number of the fibres in the fibre groups.

The next sections provide a detailed description of the optimisation process and results for the mainsail in the full and reefed configurations and for the staysail. The sailing conditions for the aerodynamic analysis and fibre layout optimisation have been determined in agreement with Mr. Semeraro, based on the estimation of the superyacht upwind performance (no VPP data available).

2.1 Full mainsail

The optimisation of the fibre layout of the full mainsail has been carried out for the loads evaluated on the sail set composed by the full mainsail and yankee (forward headsail, see Figure 1), in the following sailing conditions:

- TWS = 10 kn
- TWA = 60°
- BS = 8 kn
- Heeling angle = 5°
- Yankee sheeting angle = 14.5°
- Full mainsail sheeting angle = 8°

The sailing conditions correspond to an apparent wind of 16 kn (AWS) at 34° (AWA). The twist of the yankee has been increased from 24° maximum, as in the original design, to 30° maximum in order to trim the sail properly, accordingly to the specific sailing conditions. The aerodynamic analysis results of the sailplan in the specified sailing conditions are:

- Forward driving force = 11630 N
- Lateral force = 29240 N
- Height of the Centre of Effort (CE$_z$) = 19.6 m
- Heeling moment (HM) = 573 kNm

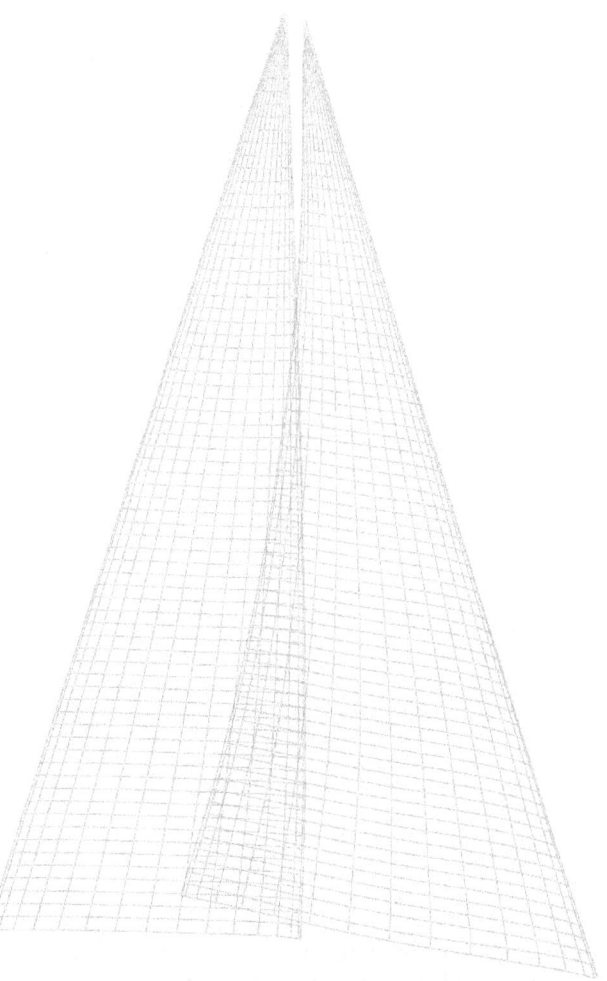

Figure 1 - Vortex lattice of the full mainsail and yankee sailplan.

The structural analysis, performed applying the aerodynamic loads on the original full mainsail layout, reveals the loads at the corners of the sail in the specific sailing conditions:

- Head load = 26000 N
- Clew load = 28000 N
- Tack load = 4200 N

The maximum stress, strain and displacement on the original design are reported in Table 3. The maximum stress and strain are at an acceptable level, therefore the optimisation focused on the further reduction of those parameters without increasing the fibre weight of the sail (35.7 kg, considering only the fibres of the sail).

Fibre layout	σ_{max} [MPa]	ε_{max} [%]	δ_{max} [m]	Weight [kg]
Original	176	0.19	0.274	35.7
Optimised	137	0.20	0.269	35.7

Table 3 - Maximum stress, strain, displacement and weight of the full mainsail fibre layout.

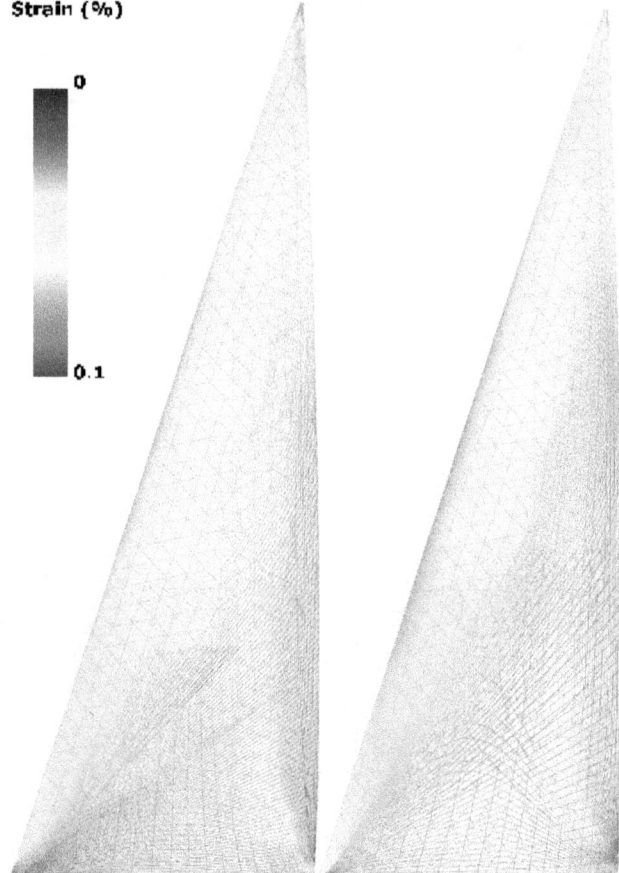

Strain (%)

0

0.1

Figure 2 - Full mainsail fibre layout and strain map (0.1% threshold): original (left) and optimised (right). Reef fibre groups are hidden. Notice the reduced maximum strain on the head of the optimised fibre layout.

The original and optimised fibre layouts and the strain of the fibres are displayed in Figure 2. The resulting parameters on the optimised fibre layout are displayed in Table 3, compared to the original parameters: the maximum stress and displacement are reduced by 22% and 2% respectively. The maximum strain is slightly increased by approximately 5% (still 20% of the maximum acceptable strain).

2.2 Reefed mainsail

The mainsail fibre layout was also verified for the reefed condition. The reefed mainsail fibre layout optimisation has been carried out according to the wind loads on the reefed mainsail and staysail sailplan (see Figure 3), in the following sailing conditions:

- TWS = 25 kn
- TWA = 45°
- BS = 12 kn
- Heeling angle = 15°
- Staysail sheeting angle = 13°
- Reefed mainsail sheeting angle = 0°

The sailing conditions correspond to an apparent wind of 35 kn (AWS) at 31° (AWA). The twist of the staysail has been increased from 14° maximum, as in the original

design, to 30° maximum in order to trim the sail properly, accordingly to the specific sailing conditions. The aerodynamic analysis results of the sailplan in the specified sailing conditions are:

- Forward driving force = 32630 N
- Lateral force = 108000 N
- Height of the Centre of Effort (CE_z) = 16.9 m
- Heeling moment (HM) = 1950 kNm

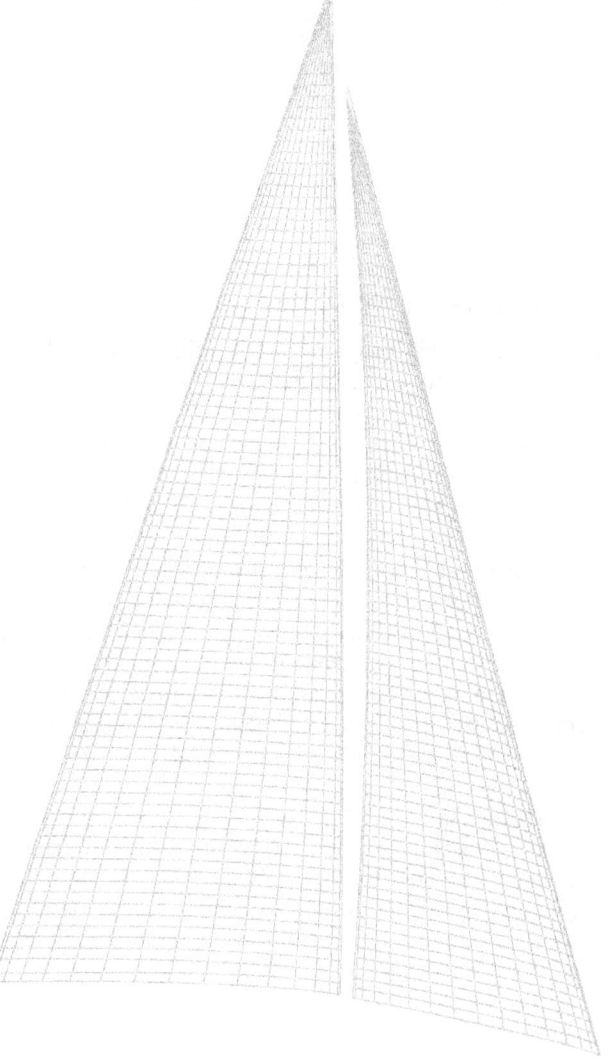

Figure 3 - Vortex lattice of the reefed mainsail and staysail sailplan

The structural analysis, performed applying the aerodynamic loads on the original reefed mainsail layout, reveals the loads at the corners of the reefed mainsail in the specific sailing conditions:

- Head load = 95000 N
- Clew load = 11000 N
- Tack load = 3900 N

The maximum stress, strain and displacement on the original design are reported in Table 4. The maximum stress and strain, specifically, are not acceptable, therefore the optimisation focused on the reduction of those

angle from the proposed values is not known. Further validation and development to the batten positioning algorithm might be necessary.

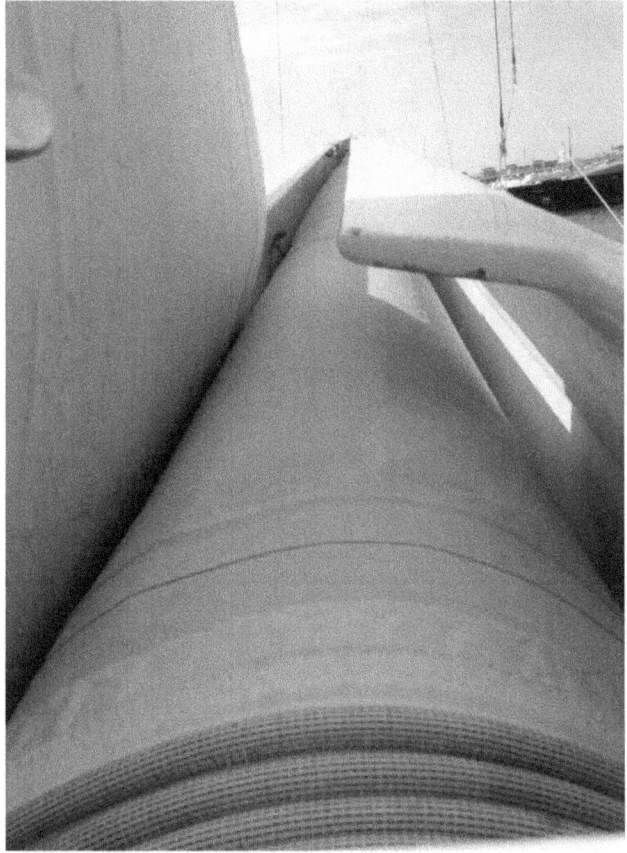

Figure 8 - The mainsail furled smoothly in the boom.

4.2 Sailing test

Sailing tests have been supervised by Mr. Semeraro in a gentle breeze and smooth sea conditions. Those conditions allowed for a proper test of the full mainsail and yankee sailplan (Figure 9). Due to the light wind conditions the reefed mainsail has not been tested and the staysail hoisted with the full mainsail (Figure 10).

The optimised sailplan, as anticipated by the analysis results, holds the desired shape (reduced maximum displacements), as assessed by eye during the sail trials, although no exact measurements have been performed. The test for the durability in terms of reduced maximum stress and strain has not been accomplished; nonetheless further data might be available through the lifespan of the sailplan, assessing if the maximum stress and strain parameters in the fibre layout optimisation process had been properly set.

Figure 9 - Full mainsail and yankee sailing test in gentle breeze and smooth sea conditions.

Figure 10 - Full mainsail and staysail sailing test in gentle breeze and smooth sea conditions.

CONCLUSIONS

About 450 m² of upwind sailing area was analysed and structurally optimised. The analysis of the final fibre layouts reveals a sensible reduction in maximum stress, strain and displacement compared to the initial designs, keeping the same fibre weight or slightly increasing it. A combination of Dyneema® SK90 and Twaron® D2226 fibres were laid in a double lamination process, in order to increase the adhesive bonding between the epoxy matrix and the fibres. A long term vacuumed post-curing period sealed the production phases. A model for the batten positioning to avoid the overlap on the furled mainsail has been developed and tested. The final sail plan, as anticipated by the analysis results, holds the desired shape.

REFERENCES

[1] Malpede, S., Baraldi, A., 'A Fully Integrated Method for Optimising Fiber-Membrane', Proceedings of the 3rd High Performance Yacht Design Conference, Auckland, 2-4 December 2008.

[2] Malpede, S., Nasato, F., 'A Fully Integrated Sail-Rig Analysis Method', Proceeding of the 2nd International Conference on Innovation in High Performance Sailing Yachts, Lorient, 30 June - 1 July 2010, 203-210.

[3] Malpede, S., D'angeli, F., Bouzaid, R., 'Advanced Structural Analysis Method for Aeroelastic Simulations of Sails', Proceedings of the 3rd International Conference on Innovation for High Performance Sailing Yachts INNOV'SAIL, Lorient, France, 26-28 June 2013.

[4] Milgram, J.H., 'The Analytical Design of Yacht Sails', Transactions of the Society of Naval Architects and Marine Engineers, Vol. 76, 1968, 118-160.

[5] Fiddes, S.P., Gaydon, J.H., 'A New Vortex Lattice Method For Calculating The Flow Past Yacht', Journal of Wind Engineering and Industrial Aerodynamics, Vol. 63, Issues 1-3, October 1996, 35-59.

[6] Werner, S., 'Application of the Vortex Lattice Method to Yacht Sails', July 2001.

[7] Levy, R., Spillers, W.R., 'Analysis of Geometrically Nonlinear Structures 2nd Edition', ISBN 0-4020-1654-9, Kluwer Academic Publishers, Dordrecht, Netherlands, 2003

Unsteady Sail Dynamics Due To Bodyweight Motions

Riley R. Schutt, Cornell University, Ithaca, NY, USA
C.H.K. Williamson, Cornell University, Ithaca, NY, USA

Figure 1: View from the top of the mast as 2012 Olympian Sarah Lihan sails upwind in a Standard Laser outfitted with a camera array, inertial measurement unit, GPS, and anemometer.

ABSTRACT

In small sailboats, the bodyweight of the sailor is proportionately large enough to induce significant unsteady dynamics of the boat and sail. Sailors use a variety of techniques to create sail dynamics which can provide an increment in driving force, increasing the boatspeed. In this study, we experimentally investigate the unsteady aerodynamics associated with one such technique, called "sail flicking". We employ a two-part approach.

First, on-the-water experiments are carried out using a Laser class sailboat sailed by Olympic and world championship level sailors. Data collected from an on-board GPS, IMU, anemometer, and camera array are used to measure

characteristic motions of the boat and sail relative to the apparent wind.

Second, laboratory experiments using the characteristic motion of the sail are run in a computer-controlled XY towing tank. We use water as the working fluid. Rather than directly experiment with three-dimensional sail shapes, we represent the primary effects of the sail dynamics using rapidly prototyped two-dimensional flexible sail geometries. Shapes are based on draft stripe shapes from the upper half of the sail. The laboratory experiments approximately match the key non-dimensional parameters of the on-the-water sailing conditions, including the reduced frequency and heave-to-chord ratio. Vorticity measurements and force measurements are used to analyze the driving force and corresponding vorticity flow field generated by the model sail during the dynamic motions.

On-the-water testing shows that the characteristic motion of a sail section during sail flicking is a combination of translation caused by the actions of the sailor and a passive twisting of the sail due to rig flexibility. The translation of a sail section is due to rotation (roll) of the rig around the longitudinal axis of the hull. This is significant because the "heaving" motion representing the sail's translation is, in general, at an angle that is not normal to the apparent wind. There are components of the "heave" motion which are both perpendicular and parallel to the oncoming wind flow. This is distinct from classical aerodynamic studies with heaving motions purely perpendicular to the incoming flow.

In our laboratory experiments, the characteristic flicking motion is applied to a 2D sail angled at 10° to the flow. This angle is maintained throughout the laboratory studies. The sail flicking leads to an increase in lift, decrease in drag, and overall increase in resultant driving force of the boat. The beneficial effect of this dynamic motion becomes greater as the apparent wind angle increases. Each flick of the sail leads to the formation of a strong vortex pair whose impulse is associated with the augmented driving force on the sail.

1   NOTATION

c	Chord (m)
h	2D peak-to-peak heave (m)
h/c	Heave to chord ratio
f	Frequency (Hz)
α	Angle of attack, sheeting angle to the wind (°)
u	Velocity (m/s)
x	X-position (m)
y	Y-position (m)
ρ	Fluid density (kg/m^3)
μ	Fluid dynamic viscosity (kg/(ms))
U_∞	Freestream velocity (m/s)
k	Reduced frequency $\pi f c / U_\infty$
Re	Reynolds number $\rho U_\infty c / \mu$
AWA	Apparent wind angle (°)
TWA	True wind angle (°)
AWS	Apparent wind speed (knots)
TWS	True wind speed (knots)
BS	Boat speed (knots)
C_D	Drag coefficient
C_L	Lift Coefficient
C_{Drive}	Driving force coefficient
C_{Side}	Side force coefficient
C_T	Thrust coefficient $-C_D u / \frac{1}{2}\rho c U_\infty^3$
C_P	Power coefficient $C_L \dot{y} / \frac{1}{2}\rho c U_\infty^3$
η	Efficiency C_T / C_P

2   INTRODUCTION

The performance of racing sailboats is affected by the unsteady aerodynamics of the sails. Gerhardt et al. (2011) show how the propulsive forces of a main and jib pair can be affected by the unsteady motion through waves. Aside from this study, there are surprisingly few previous studies in the open literature which look into the effect of unsteady sail dynamics on sail driving forces. In the present study, we look at boats with a single sailor and a single mainsail. In these boats, unsteady aerodynamics can be caused by boat kinetics due to deliberate bodyweight motions of the sailor (Bethwaite, 1996). Here we investigate the kinetic technique sail flicking.

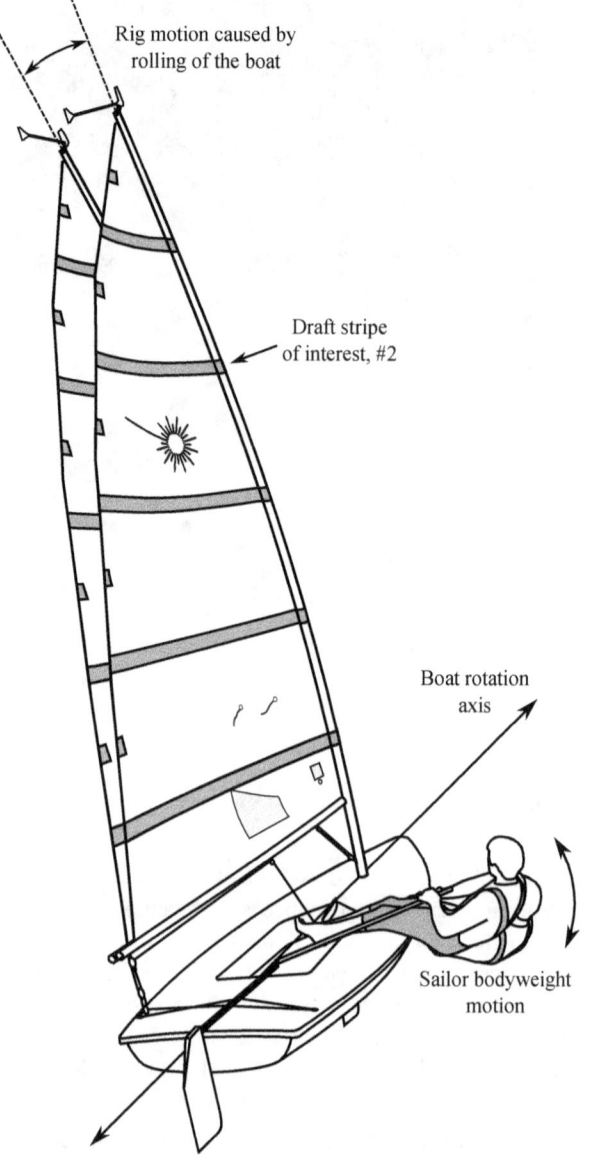

Figure 2: Movement of the sailors bodyweight causes a flicking motion in the rig and sail.

Figure 2 shows that as the sailor moves his upper body up and down, the boat rolls around an axis approximately parallel to the centerline of the hull. This rolling motion of the hull causes the rig to rotate, with individual points on the sail moving in a shallow arc. For a roughly upright rig, the motion of a sail section is approximated as back and forth translation in a horizontal plane (parallel to the water surface). Sail flicking gets its name from the observed motion of the sail, especially the leech. The sail leech "flicks" because of both the motion of the rig, and the dynamic twisting of the sail due to leech tension and sail flexibility under the varying aerodynamic load.

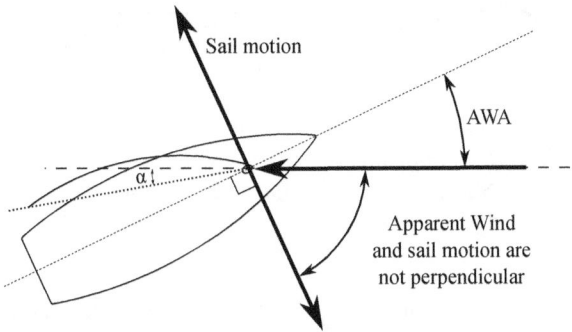

Figure 3: Roll motion around the centerline of the boat results in a 2D heave of the sail section that is not perpendicular to the incoming apparent flow.

In the present study, one of the goals is to understand the vortex dynamics influencing the pressure distribution. This in turn yields the forces on the sail. Rather than modeling a specific rig of a particular sailboat, we initially propose to study the forces and vortex dynamics of a 2D section of the sail (the second draft stripe down from the head of the sail, as experimentally measured on the water). We expect that the vortex dynamics will be qualitatively similar to those around the 3D sail shape at this sail section, which is something we intend to show in further work. This approach to qualitatively understanding the sail flow as a set of 2D flowfields, has been used in the sail aerodynamic research of Gentry (1971) and Gerhardt et al. (2011), and in general unsteady aerodynamics (Heathcote and Gursul, 2007, Izraelevitz and Triantafyllou, 2014). There is no question that the actual flow of a specific sail is more complex, both in the case of a model 3D sail, as well as in the full-scale case. Our qualitative approach does not take account of the full-scale boundary layer effects at much higher Reynolds numbers, the 3D nature of the flow around a triangular sail, or the rolling up of streamwise vorticity near the sail head into a "tip" vortex further downstream.

Aerodynamic literature has a long history exploring the effects of unsteady motion on two-dimensional airfoils. Classically, Katzmayr (1922) showed that an airfoil can produce a net thrust in the direction opposing the free stream when subjected to variations in angle of attack. Garrick (1936) analytically showed thrust can be produced when

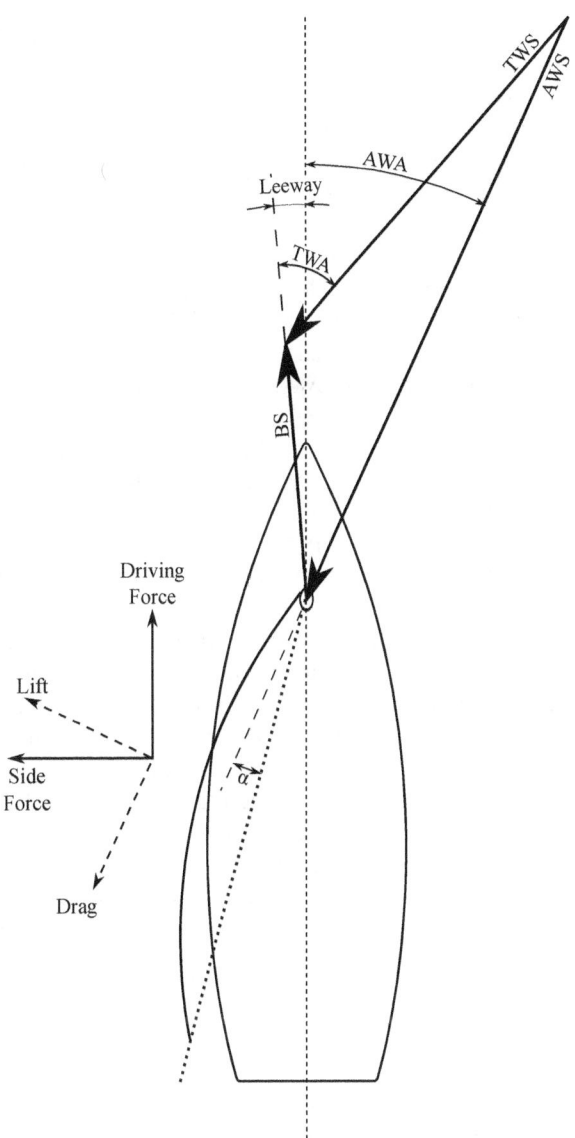

Figure 4: The boat travels with boatspeed, BS, at a leeway angle relative to the boat centerline. The true wind angle, TWA, is the angle between the boat velocity vector and the true wind vector. A vector addition of the true wind and boat velocity yield an apparent wind at angle AWA relative to the centerline of the boat. This apparent wind is the wind felt by the boat and represents the aerodynamic free stream.

oscillating a symmetric thin airfoil perpendicular to the freestream in a simple "heave" motion, which causes a variation in instantaneous angle of attack. In a sailing context, a benefit can be gained by a number of factors, including decreasing sail drag, producing thrust (negative drag), or increasing the lift generated by a sail. Heathcote and Gursul (2007) show that a NACA 0012 airfoil at a positive angle of attack receives augmented lift when oscillated perpendicular to the flow, in a classic "heaving" motion.

A sail experiencing unsteady motion departs from classic unsteady aerodynamic studies in two distinct ways. First, the direction of the sail section motion is not perpendicular to the free stream direction. This is true for boats traveling through waves (Gerhardt et al., 2011), and for the presented flicking case, where motion is necessarily perpendicular to the boat centerline. Figure 4 shows that a vector addition of the boat speed, BS, and true wind speed, TWS, results in an apparent wind vector with speed AWS. In sail aerodynamics, the AWS, which represents the freestream velocity, is at an apparent wind angle relative to the centerline of the boat. Figure 3 shows that, during sail flicking, the motion of the sail section forms an angle with the freestream that is complementary to the AWA. There are components of the sail motion both perpendicular and parallel to the apparent wind.

Figure 6 shows typical boatspeeds and sailing angles for a Laser (Binns et al., 2002). This, combined with our own on-the-water data, is used to determine the appropriate AWA for our laboratory experiments representing different sailing conditions.

The second departure of the current study from classic aerodynamics is that the primary quantity of interest, when considering the two dimensional sail section, is driving force, rather than pure lift or drag. As shown in Figure 4, the driving force is defined parallel to the centerline of the boat, while the lift and drag are defined relative to the freestream. The quantities, driving force coefficient, C_{Drive}, and side force coefficient, C_{Side}, are related to the lift and drag coefficients, C_L and C_D, by the apparent wind angle, and can be calculated with Equations 1 and 2, as follows:

$$C_{Drive} = C_L \sin(\text{AWA}) - C_D \cos(\text{AWA}) \qquad (1)$$

$$C_{Side} = C_L \cos(\text{AWA}) + C_D \sin(\text{AWA}) \qquad (2)$$

Figure 5 shows the relationship between C_{Drive} and AWA for a given C_L and C_D. C_L and C_D remain constant with a change in AWA because the sheeting angle of the sail relative to the apparent wind, α as shown in Figure 3, does not change as the AWA increases. C_{Drive} increases with AWA because the component of the large lift vector, generated by the sail, becomes increasingly aligned with the boat centerline and contributes in greater proportion to the driving force.

2.1 Legality of Forced Unsteady Motions While Racing, RRS Rule 42

Within the Racing Rules of Sailing (RRS), it is rule 42 which governs the permitted propulsion of the racing sailboat (ISAF, 2012). This generally makes actions like sail flicking illegal during a race. The relevant part of rule 42 is given in the quotation below:

> *Crew may adjust the trim of sails and hull, and perform other acts of seamanship, but shall not otherwise move their bodies to propel the boat.*

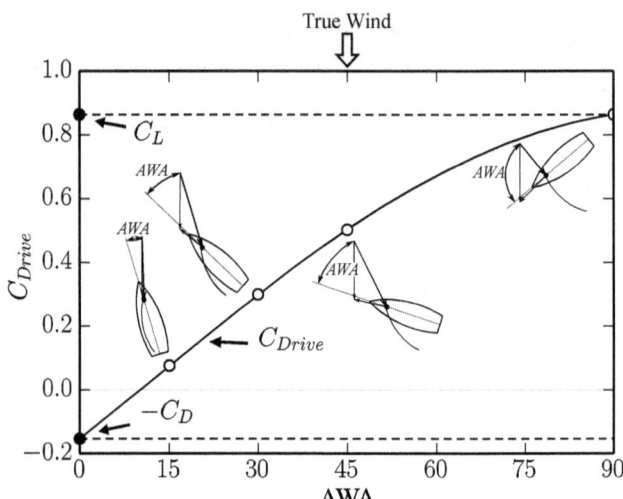

Figure 5: With constant sheeting angle to the apparent wind, the driving force increases with AWA. Lift and drag remain constant. Sketches of the boat at selected AWAs give plausible sailing angles with a true wind coming down the page.

While this limits the application of kinetic techniques, there are some provisions in the RRS which allow them. Appendix P5(a), provides for individual class rules to relax rule 42 above certain wind speeds. In appendix B4(42), rule 42 is modified to allow most kinetic techniques in windsurfing. Further, sailors employ variations of these techniques even when not strictly legal. The necessity of rule 42 is evidence that some kinetic techniques increase the performance of the boat. The aim of the presented work is to explore the underlying unsteady aerodynamic principles involved. The legal or illegal application of these are left to the sailor. World Sailing (ISAF, 2013) has issued an interpretation of rule 42, which classifies the sail flicking discussed in this study most closely as "PUMP 6", described as "Repeated flicks of a sail due to body [motion]".

3 EXPERIMENTAL DETAILS

We use a two-part approach to investigate the unsteady aerodynamics involved in sail flicking. First, on-the-water experiments to collect performance and sail motion data are done using a modified Laser sailboat with a Standard rig, from here referred to simply as a Laser. Second, the characteristic motion extracted from the on-the-water tests is used in the laboratory environment to enable us to study the principle sail-flow interactions that are present during sail flicking.

3.1 On-The-Water Experiments

On-the-water experiments are conducted using a Laser. Non-class legal instrumentation has been added to facilitate the collection of data. Figure 7 shows the location of instrumentation. A similar setup was reported by this research

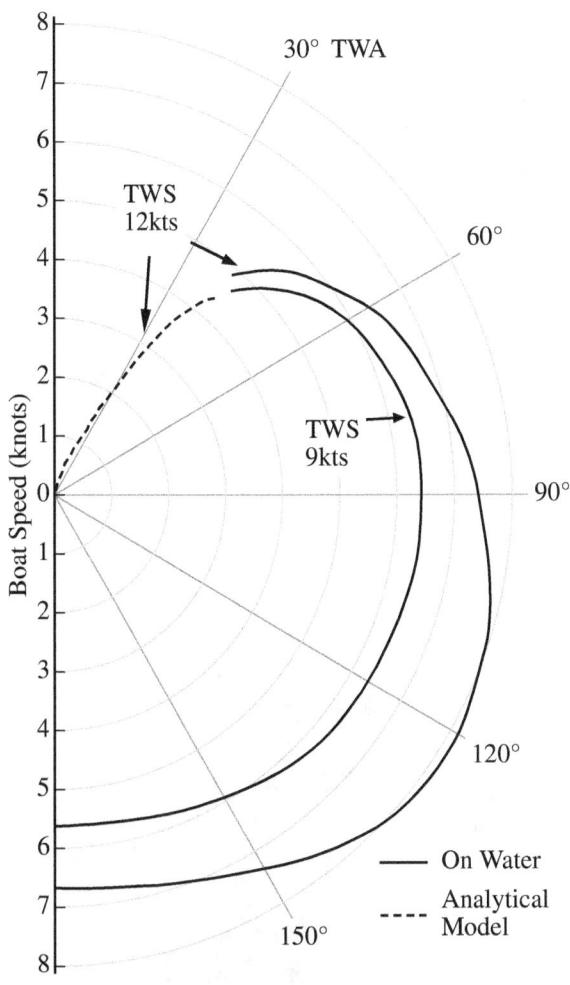

Figure 6: Laser performance polar (Binns et al., 2002).

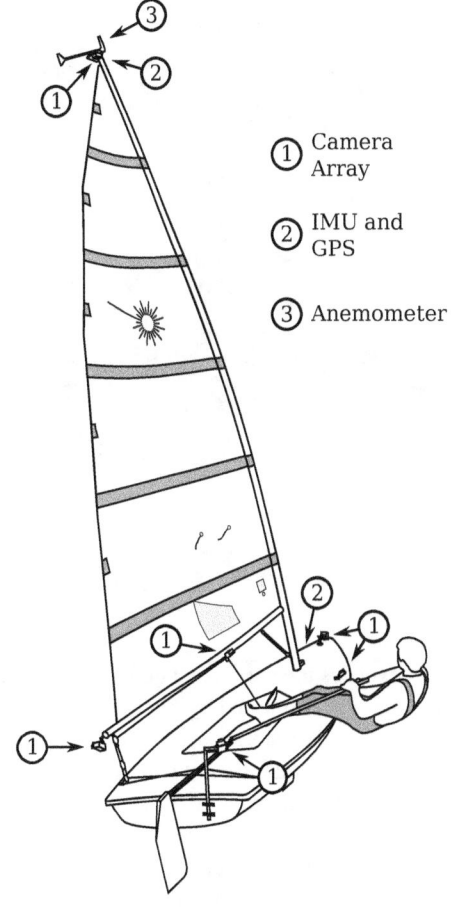

Figure 7: The experimental boat was a Laser Standard outfitted with a camera array, anemometer, Inertial Measurement Units, and GPS.

group previously (Keil et al., 2015, Schutt and Williamson, 2015, Scuttlebutt, 2014).

Two in-house Inertial Measurement Units, IMUs, have been built and are used to measure the rigid body motion of the boat and rig. Figure 8 shows the IMU which is secured to the deck just in front of the mast step. Developing the IMU in-house from proven technologies allows for a cost effective solution while maintaining a sufficient confidence in the accuracy of results. The IMU at it's core uses the ArduPilot Mega 2.5 hardware developed by 3DRobotics Inc.

This board is Arduino compatible and based on Amtel's ATMEGA2560 micro controller. It features the Invensense MPU-6000 (6-DoF accelerometer and gyroscope), Honeywell HMC5883L-TR (3-Axis digital compass), Measurement Specialties MS5611-01BA03 (altimeter/barometer), and uBlox LEA 6 (GPS). Additional hardware has been added to extend the data logging capabilities, provide on board power and charging, and encase the sensors in a waterproof housing. The software is a modified version of the Ardupilot Plane 2.65, with custom changes to logging func-

tionality and time resolution. The modified IMU is capable of recording heading, pitch, and heel at 50Hz and boat speed and bearing at 10Hz for over 7 hours. Our approach, through testing of known motions, shows good representation of rigid body movement. A second IMU uses the same hardware and software and is housed in the mast-top instrumentation assembly shown in Figure 9.

The mast-top instrumentation assembly in Figure 9 is designed to fit over the top of the Laser mast and sail. The assembly secures to the masthead strap in the sail and has been designed to be impact and water resistant, while keeping weight to a minimum. A number of capsizes during on-the-water testing have shown it to be robust and to sufficiently protect the enclosed sensors. The assembly is a combination of rapidly prototyped housings and fittings, plastic, and carbon fiber for impact resistance.

A Kestrel 4500 logging anemometer from Nielsen-Kellerman is used to measure the apparent wind velocity and direction. These variables, and many other atmospheric conditions, are logged at 0.5Hz. Data can be logged continuously at 0.5Hz for 1 hour 46 minutes and 40 seconds.

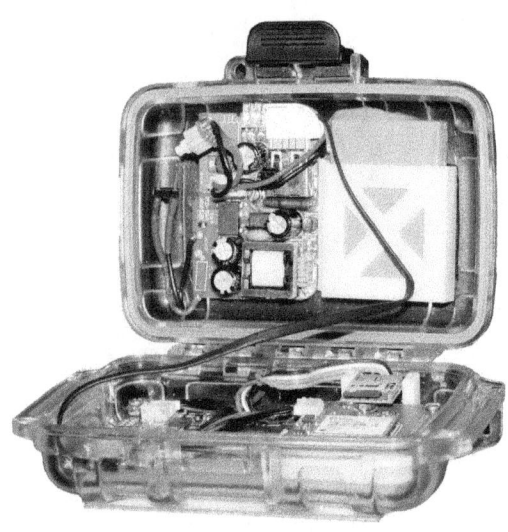

Figure 8: In-house instrumentation containing an Inertial Measurement Unit and GPS in a waterproof housing.

Note that this exceeds the duration stated in the user manual, the correct value was confirmed by Nielsen-Kellerman. The logging frequency and data accuracy of this device when placed at the top of the mast are not sufficient to record transient apparent wind behavior during unsteady sailing maneuvers, or while employing kinetic techniques. However, the log data is more than sufficient to determine the wind conditions experienced by the boat on a minute-to-minute time scale. The anemometer is mounted to a freely rotating wind vane, which maintains an orientation into the apparent flow. This wind vane also serves as a visual aid in seeing the shifts in incoming flow direction at the top of the mast and developing a conceptual understanding of the changes in instantaneous angle of attack experienced by the upper sail section.

An array of 6 GoPro and GoPro2 cameras is used to visually track the motion of the boat and sail. The position of the 6 cameras is shown in Figure 7 and described in Table 1. The field of view of each camera is checked using the GoPro LCD BacPac™. Each camera can record for approximately 2 hours, limited by battery life and memory card capacity. To ameliorate picture quality issues caused by water drops and fogging on the lens, Rain-X is applied to the outside of the waterproof housing, Rain-X Anti-Fog is applied to the inside of the housing, and GoPro Anti-Fog Inserts are placed in each housing. This is a generally successful strategy, with the exception that the curved lens of the down-facing camera at the masthead occasionally collects a water drop in the center of the outer lens after getting wet.

An official Laser Standard (pre-2016 version) sail in racing condition is used for all testing. Draft stripes, as illustrated in Figure 7, have been added to the sail to facilitate the observation of sail section shapes. Stripes are cut from black adhesive-backed Dacron sail material and care is given to

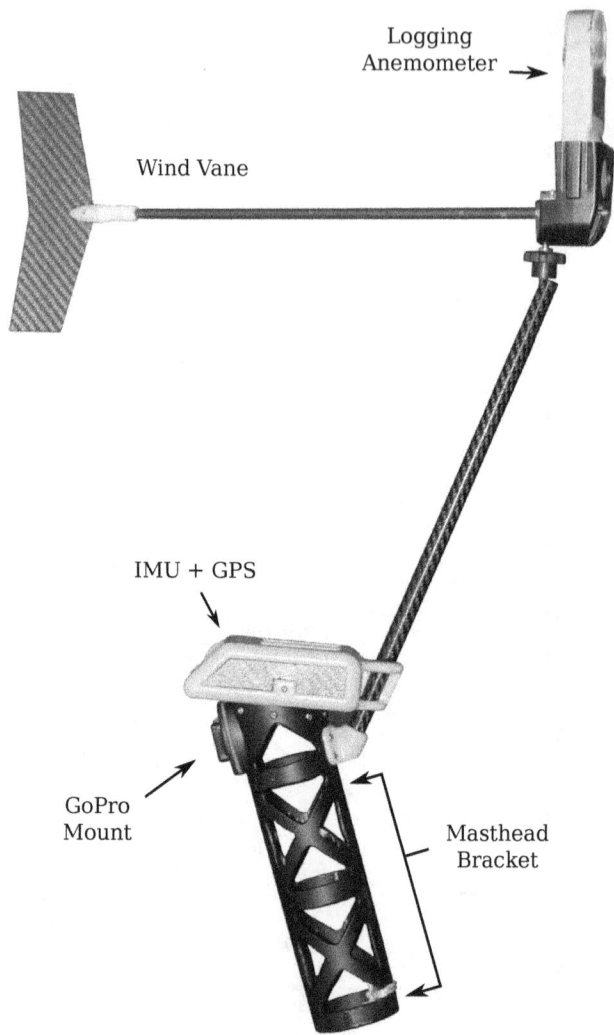

Figure 9: Masthead instrumentation assembly containing an Inertial Measurement Unit, GPS, anemometer, and mount for a GoPro camera.

match tension in the sail cloth during installation. This results in very minimal distortion of the flying shape of the sail due to the stripes. We focus on the second draft stripe from the top, #2 see Figures 2, which is 3.2m above the sail foot and has a 1.35m chord.

On-the-water tests are conducted during 2 hour sailing sessions. This duration is limited by the memory capacity of the Kestrel 4500 and the recording duration of the camera array. The GPS track of typical part of one session is shown in Figure 10. Sessions begin by powering on all equipment and giving initial instructions to the sailor. A chase boat maintains close distance to the test boat, providing testing instruction and recording external video. Each sailing session has multiple sailors that rotate into the test boat. Each sailor is instructed to sail at multiple points of sail. Boat kinetics during tests vary between legal motions, motions that violate rule 42 but might be used in a race, no motion, and

Camera	Location	Orientation/View
1	Masthead, offset to starboard.	Vertically down showing the starboard side of the sail, draft stripes, hull, and sailor while on starboard tack, as seen in Figure 1.
2	Bow, centered.	Portrait orientation looking backwards, showing the entire mast from step to head
3	Starboard rail, forward of sailor.	Looking up and backward at the sail, showing the majority of the sail while close-hauled.
4	Mid-boom, offset to starboard	Vertically up, showing the majority of the sail and draft stripes, with the bottom section of the sail out of the field of view.
5	Sting off boom-end, centered	Looking up along the leech of the sail, showing the entire leech of the sail.
6	Sting off starboard quarter.	Portrait orientation looking forward at the sail, showing the whole sail while close-hauled.

Table 1: Location and orientation of GoPro cameras used in an array to visually track the rig and sail motion and deformation.

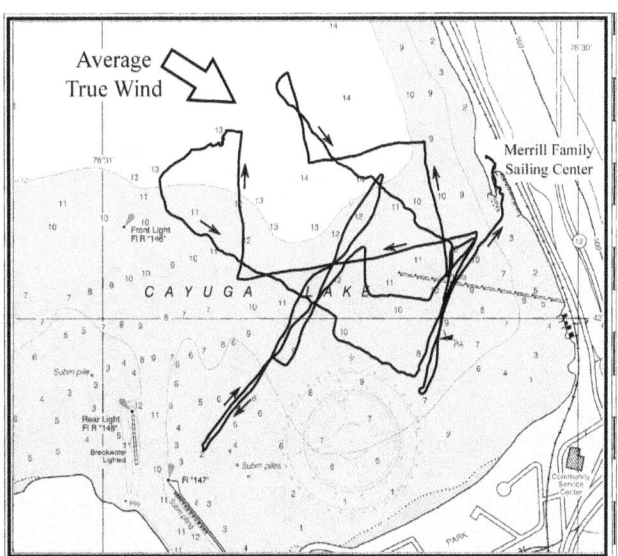

Figure 10: GPS tracks from an on-the-water testing session based out of Cornell University's Merrill Family Sailing Center in Ithaca, NY.

Figure 11: Sail draft stripe shapes are extruded to create geometries which are 3D printed. Variations in thickness between the rigid (left) and flexible (right) sections control chordwise stiffness.

motions that blatantly violate rule 42, but which, in the best judgment of the sailors, will result in the best boat performance.

During on-the-water tests, the exact techniques used, and the proficiency in which they are executed, depend on the sailor helming the test boat. As such, an effort is made to have sailors with a proven track record in high level regattas sail the boat. We are very fortunate to have Sarah Lihan, member of the 2012 US Olympic Team and Phillip Alley, top ten ICSA college singlehanded sailor, and Robbie Gilmore, Irish youth national champion, among others, sail for us.

Tests are conducted at the US Sailing Center in Miami, FL and Cornell University's Merrill Family Sailing Center in Ithaca, NY.

3.2 Laboratory Setup

With the goal of understanding the flow physics around the sail during sail flicking, a series of experiments are performed using 2D sail section shapes in a ~6m computer-controlled XY towing tank in the Fluid Dynamics Research

Laboratories at Cornell University.

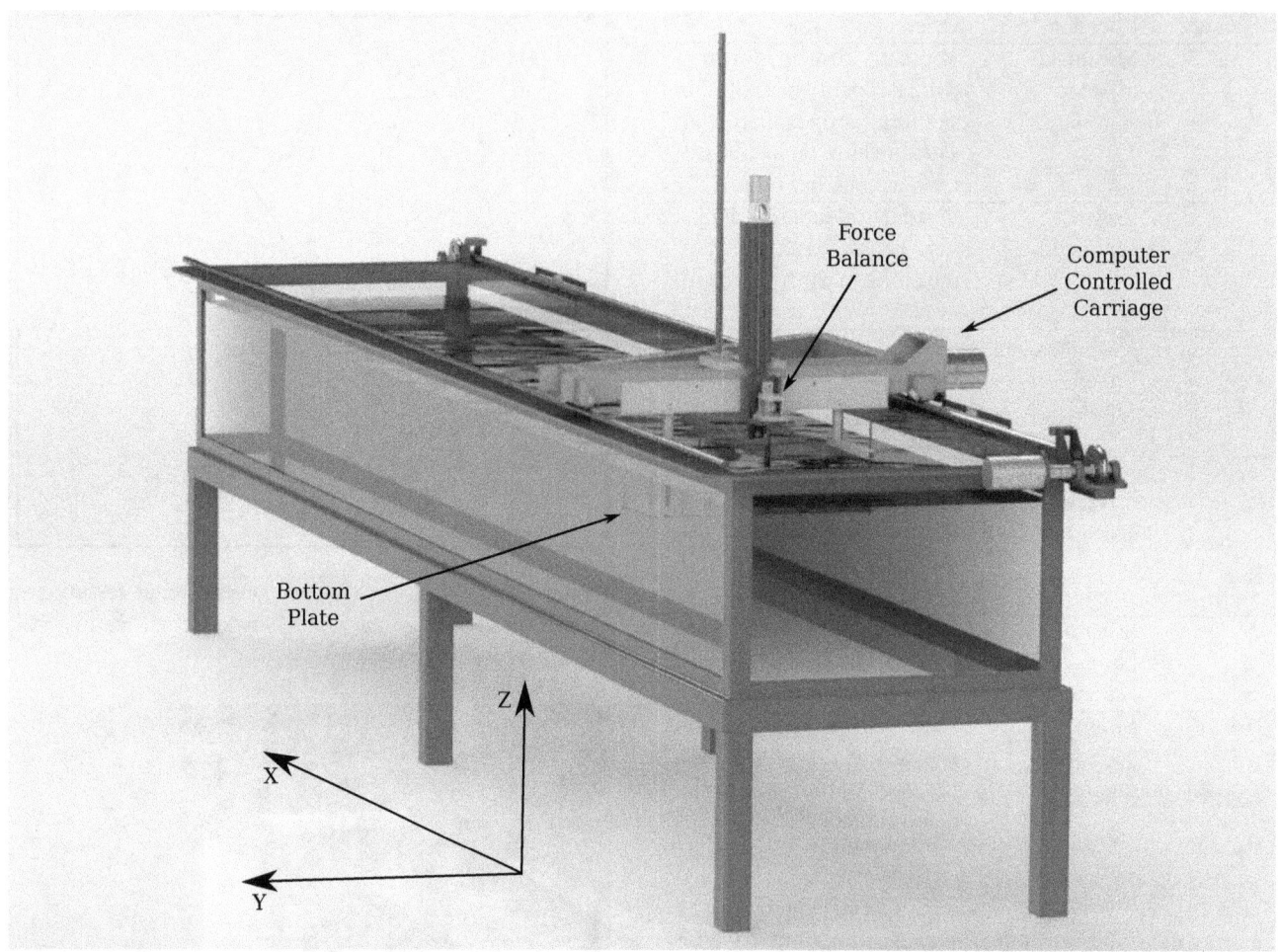

Figure 12: Laboratory experiments are conducted in a ~6m towing tank with computer-controlled X and Y degrees of freedom. The upper boundary condition is the free surface, the lower boundary is an end-plate which moves with the carriage.

All experiments are run with Reynolds Number 10,000 (validations with a NACA 0012 were additionally run at Re = 20,000). Cleaver et al. (2009) show that the dependence on Re is low for a symmetric airfoil oscillating at positive angles of attack. The dependence on non-dimensional reduced frequency, k, and heave to chord ratio, h/c, is significant. Experiments are run for a range of each parameter k = [0.0, 0.8, 1.6, 2.4] and h/c = [0.0, 0.1, 0.2, 0.3]. Experiments cover a range of apparent wind angles, AWA = [0°, 15°, 30°, 45°], which affects the direction of motion of the sail section relative to the freestream. Angle of attack, α, is held constant relative to the freestream at 10°.

Two dimensional section shapes have been extracted from mid-boom, on-the-water video footage of the Laser sail draft stripe #2. The section shape is given thickness and extruded to create a 3D solid model. All section shapes have a 0.1m chord and 0.2m span (0.17m is submerged in experiments). The physical 2D sections are printed using an Objet30 3D printer and FullCure 705 resin. Chordwise flexibility of the section is altered by varying section thickness and leaving strategic voids inside the geometry to allow bending. With water as the working fluid, the desired flexing during ex-

periments is achieved with reasonable printed thicknesses. Two section flexibilities are tested, a "flexible" section and a stiffer "rigid" section.

When printed at thicknesses which are thin enough to allow flexibility, FullCure 705 deforms in warm and humid environments. To reduce this, each foil is sealed with spray epoxy, painted, and then sanded smooth. The final experimental geometries are shown in Figure 11. A female mold for each section shape is printed and all 2D sections are stored in these molds between batches of experiments to ensure that the sections remain in the design shape.

Our XY towing tank setup was previously used to measure forces on a transient cylinder by Stallard et al. (2009). The setup is depicted in Figure 12. X-direction motion of the carriage is actuated by an electric motor driving a cable and pulley system. Y-direction motion is driven by a servo motor and lead screw. A stepper motor driven lead screw was added to the carriage in the vertical Z-direction and is used to raise the 2D sections out of the water between individual experiments. This reduces the amount of submerged time for the 2D sections, and prevents water from infiltrating into the epoxy coated FullCure 705 material. All degrees of freedom are computer controlled by sending the actuators a voltage signal generated by LabView and Python code. Force measurement experiments are run in batch mode, with the computer capable of running a large unattended series of experiments.

To encourage the two-dimensionality of the flow, a different approach is used at each end of the vertical span. As previously described by Khalak and Williamson (1996), in the case of a vertical cylinder, the upper boundary condition is provided by the free surface. This gives a nearly-slip condition and allows vortex lines to pass through the boundary. The deformation of the free surface during experiments is small, resulting in minimal out-of-plane flow. The lower boundary condition is provided by a Lucite bottom plate attached to the X-stage of the towing tank carriage. It travels in the X-direction along with the 2D section during an experiment. The gap between the 2D section and the bottom plate is maintained at less than 0.002m but there is no contact.

Forces in X and Y are measured on the 2D section with a force balance utilizing linear variable differential transducers. Voltages from the transducers are sent to LabView and recorded at 200Hz. Electronic noise and internal vibrations of the force balance are filtered using a low-pass filter. For each experiment with the 2D section submerged in water, a second experiment is performed with the Z-stage elevated and the 2D section in air. This air-run measures the inertial forces generated by the motion of the 2D section. Fluid dynamic forces from the water are found by subtracting away the inertial forces. Each force experiment is repeated 5 times, waiting 20 minutes between experiments. Force data is phase averaged for all periods within a single experiment, and then averaged across the 5 repeated experiments.

Force measurements have been validated against experimental results (Heathcote and Gursul, 2007) and computational viscous Navier-Stokes results (Young and Lai, 2004) of a NACA 0012 oscillating perpendicular to the flow with $h/c = 0.35$ and $\alpha = 0$. Thrust, power coefficient, and efficiency are all shown to agree well in Figure 13. At Re = 10,000 our efficiency is lower than the Re = 20,000 and 30,000 values at $k < 1.5$; however, there is no Re = 10,000 data to compare to at these low reduced frequencies.

The vorticity field in the flow around the 2D sections is measured using particle image velocimetry, or PIV. Water in the towing tank is seeded with 10 μm hollow glass spheres (Potters Industries SPHERICEL 110P8), which are illuminated by a sheet of laser light from a Coherent Innova 90 laser. Pairs of particle images are acquired at 15Hz using a stationary Jai CV-M2CL CCD camera (1600 x 1200 pixels). Particle displacement between images within a pair is analyzed using the PivLab software developed by Thielicke and Stamhuis (2014). Further details on our PIV technique are given in Govardhan and Williamson (2000). Each PIV experiment is repeated four times, twice with the laser sheet entering the tank from the left, and twice from the right.

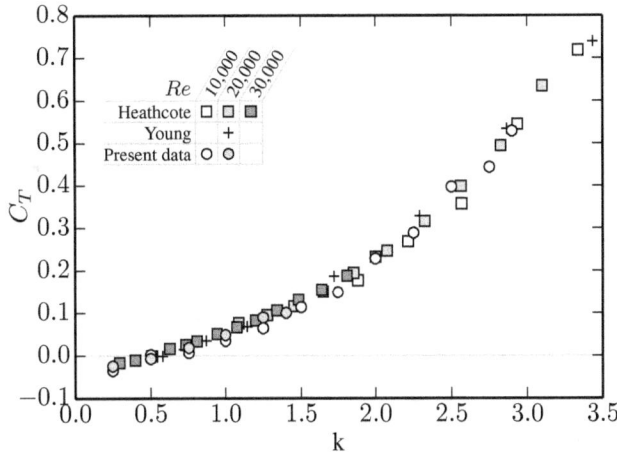

(a)

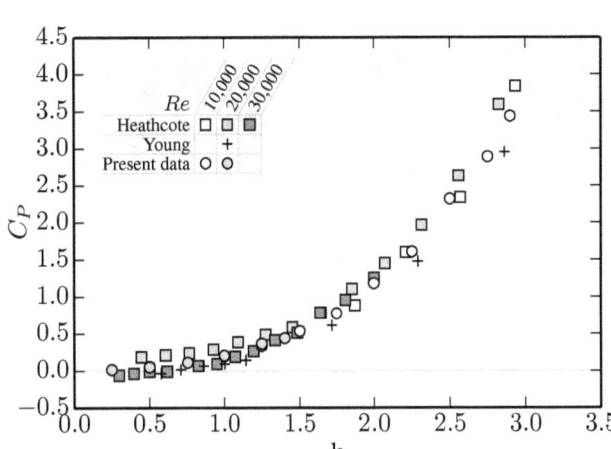

(b)

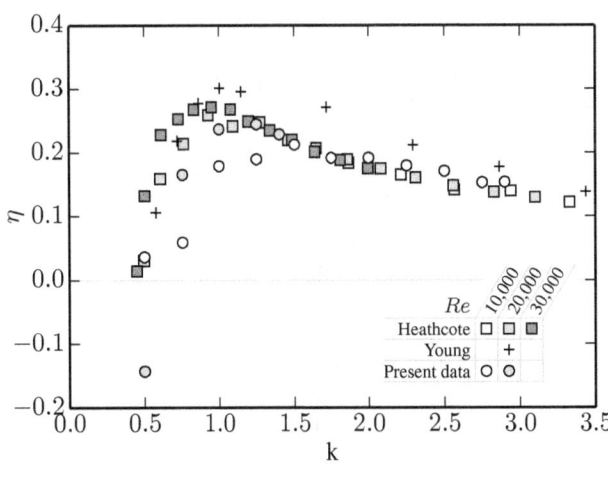

(c)

Figure 13: A heaving NACA 0012 airfoil is compared to Navies Stokes simulations Young and Lai (2004) and previous experimental work (Heathcote and Gursul, 2007). $h/c = 0.35$, $\alpha = 0$.

Flow fields are phase averaged within each single experiment and between all 4 repetitions. This eliminates shadows cast by the 2D section and the supports for the bottom plate, allowing a recreation of the whole vorticity field. The stationary camera captures the evolution of the wake as the 2D section moves down the tank and out of the field of view.

4 SAIL FLICKING

4.1 On-the-Water Experiments in a Laser Standard

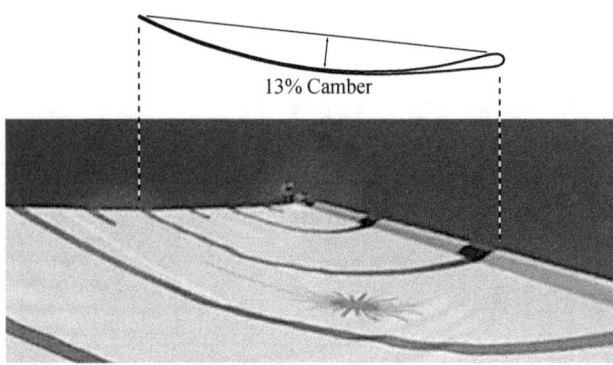

Figure 14: The 2D section shape of draft stripe #2 is extracted from footage of the mid-boom camera during upwind sailing.

Analysis of video footage and IMU data of repeated sail flicking shows typical body pumping and flicking frequencies on the order of 1-1.5Hz and a rolling motion of the boat with peak to peak amplitude between 2° and 5°.

Footage from the mid-boom, camera 4 in Table 1, is used to determine the 2D section shape of draft stripe #2, as shown in Figure 14. For a 2D section at a given height up the mast, the rolling motion becomes an approximately horizontal 2D translation, or "2D heave", transverse to the centerline of the boat. This is coupled with a passive pitching of the 2D section due to the twisting of the sail under aerodynamic load. The magnitude of this pitching varies with leech and vang tension in the rig, but is observed on the order of 5°.

The amplitude of the 2D heaving motion varies along the mast, with the masthead having a large motion compared to an almost zero translation at the mast step. The 2D chord, c, of the sail sections also varies along the mast, ranging from 2.74m at the foot of the sail to 0.10m at the head. Figure 15 shows how the variation of the 2D heave and chord along the mast affects the key non-dimensional parameters, heave-to-chord, h/c, ratio and reduced frequency, k, for a typical light wind sailing condition. Higher heave-to-chord ratios and reduced frequencies in this parameter range are generally associated with larger increases in lift, as discussed in Section 4.2. Consequently, this means that if either h/c or

k are too low, there will not be a significant increase in lift coefficient at that sail section. Figure 15 shows that the draft stripe of interest, draft stripe #2, is located in a region of the sail where h/c and k are both large enough to be effective. At the draft stripe for the displayed wind condition, $k = 1.6$, and $h/c = 0.18$. Parameters in the range $0.0 < k < 2.4$ and $0.0 < h/c < 0.3$ cover sailing conditions from steady sailing without flicking, to vigorous sail flicking in light wind.

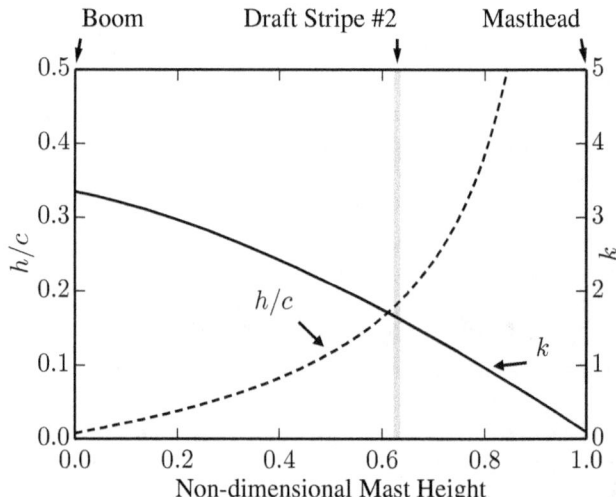

Figure 15: The non-dimensional heave-to-chord ratio, h/c, and reduced frequency, k, vary along the mast due to changes in sail chord length and motion amplitude. Values are shown for a typical light wind sailing condition at AWS=5kts while sailor's bodyweight causes 4° rolling of the boat with a frequency of 1hz.

Visual tracking of the sail luff is used to extract the motion of the 2D sail section. Averaging the motion of the sail section from multiple individual "flicks", each normalized in period and amplitude and filtered, yields the characteristic motion shown in Figure 16 and mathematically described as a Fourier series with coefficients given in Table 2. This is the motion transverse to the centerline of the hull. A visual comparison shows that the characteristic motion is similar to a sine curve, but with a more abrupt return as the sail is forced back to windward. The similarity to a sine curve is also shown by the Fourier coefficients, where the first term, a_1, is an order of magnitude greater than all subsequent terms.

4.2 Laboratory Experiments with a Representative 2D Sail Section

Experiments with 2D sail sections in the laboratory setting yield results that confirm conceptual ideas about effective techniques while sailing. Results provide additional physical insight into the flow dynamics around a sail and show fundamental trends for the unsteady aerodynamics of 2D, airfoil-like, bodies.

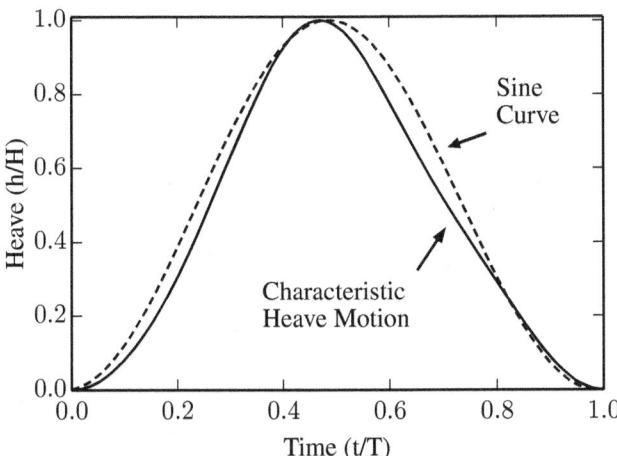

Figure 16: Characteristic upwind flicking motion derived from on-the-water experimental data. Heave is non-dimensionalized dividing by peak to trough distance H, and time is non-dimensionalized dividing by the duration of an individual flick, T.

$$f(x) = \frac{1}{2}a_0 + \sum_{n=1}^{\infty} a_n \cos(nx) + \sum_{n=1}^{\infty} b_n \sin(nx) \quad (3)$$

a_0	a_1	a_2	a_3
0.920	-0.478	0.030	-0.014

b_1	b_2	b_3
0.041	-0.040	0.012

Table 2: The first seven coefficients of the Fourier series of the characteristic flicking motion.

All laboratory experiments were conducted using 2D sail sections with a geometric section shape matching Figure 14 and using the characteristic flicking motion illustrated in Figure 16 and Table 2.

Rigid and Flexible 2D sail sections were used to simulate the effect of different leech tension on sail deformation in the Laser rig. The flexible 2D section is characterized by the ranges of the camber and geometric angle of attack, α_{geo}, of the deformed shapes during experiments, as presented in Table 3. Note that the rigid sail section has some small flexibility, generally causing angle of attack variations of ±0.5°.

4.2.1 Force Measurement

Figure 17 shows the time averaged lift coefficient for the rigid 2D sail section with a flicking motion at an angle to the flow corresponding to AWA = 30° for a range of heave to chord ratios and reduced frequencies. The lift coefficient generated by the 2D sail section increases with both frequency, k, and amplitude, h/c. This is similar to the effect

Section	k	h/c	Camber (%)	α_{geo} (°)
Rigid	0.0	0.0	13.2	10.1
Flexible	0.0	0.0	13.3	9.1
Rigid	2.4	0.1	13.1	9.8
Flexible	2.4	0.1	13.0 to 13.6	6.1 to 9.9
Rigid	2.4	0.2	13.0	9.7
Flexible	2.4	0.2	10.7 to 13.6	3.0 to 10.3
Rigid	2.4	0.3	12.9	9.4
Flexible	2.4	0.3	9.7 to 14.3	0.7 to 11.8

Table 3: Deformations of rigid and flexible 2D sail sections. The minimum to maximum range of camber and α_{geo} due to deflection of the geometry under load during each motion period are given for the flexible section for a selection of experimental parameters.

observed for a symmetric airfoil oscillating perpendicular to the incident flow with $\alpha = 15°$ (Cleaver et al., 2009). When either reduced frequency or heave to chord ratio is too low, the increase in C_L is minimal. In the context of Figure 15 showing the variation of these non-dimensional parameters along the mast, this trend supports the idea of an "active", or most effective, area of the sail during flicking motions.

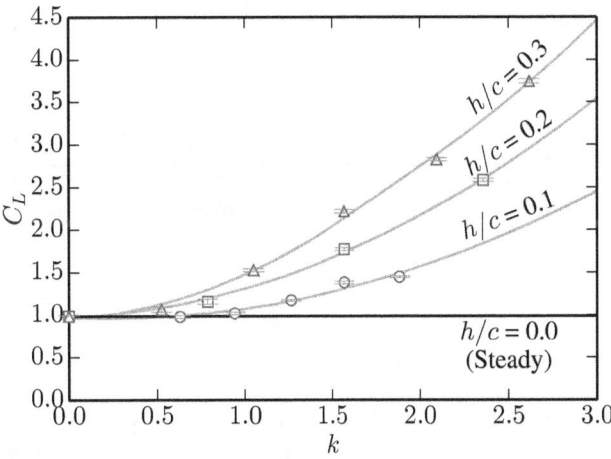

Figure 17: Lift Coefficient, C_L, for the Rigid 2D sail with motion corresponding to AWA=30°. Error bars indicate the 95% confidence interval.

In these experiments, apparent wind angle is complementary to the angle of flicking motion relative to the incident flow, as shown in Figure 3. In the present work, we vary this AWA while keeping angle of attack constant at 10°. Over the tested apparent wind angle range in Figure 18 (0° to 45°), the C_L dependence on AWA is low, with larger effects being attributed to k, h/c, or section flexibility. The rigid section has a higher steady state lift and importantly sees larger increases in lift for equivalent flicking motions.

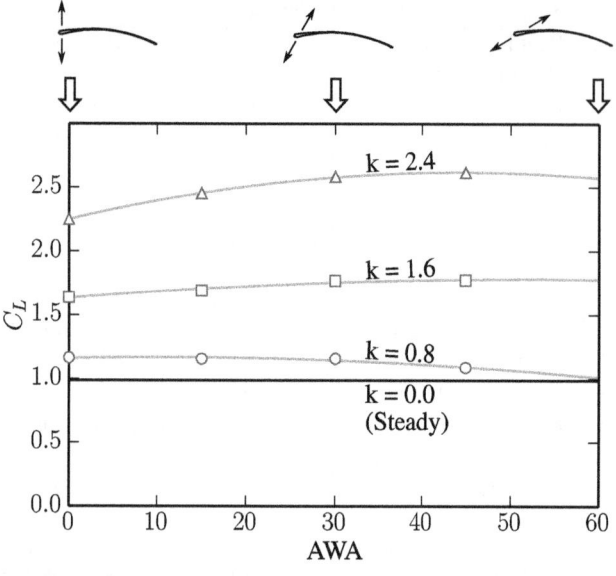

(a) **Rigid 2D sail**

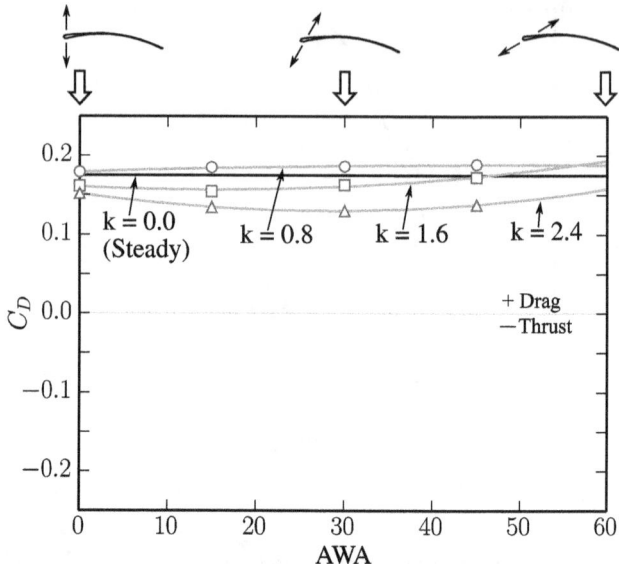

(a) **Rigid 2D sail**

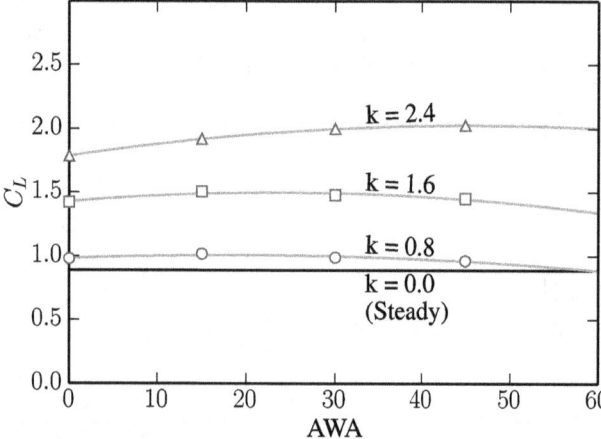

(b) **Flexible 2D sail**

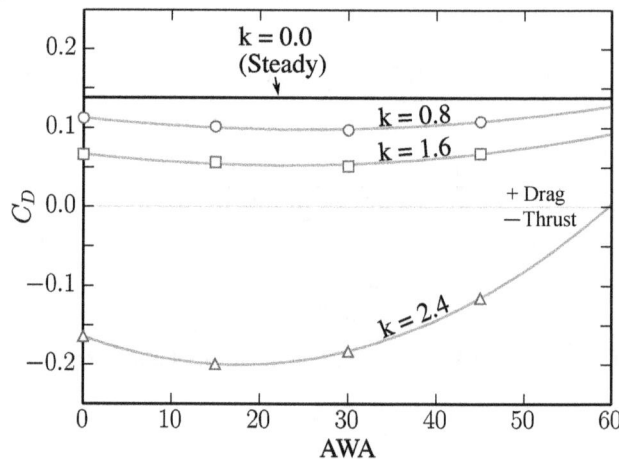

(b) **Flexible 2D sail**

Figure 18: Lift Coefficient, C_L, of the 2D sail geometries at $h/c = 0.2$. A visual representation of the flicking motion angle is shown above.

Figure 19: Drag Coefficient, C_D, of the 2D sail geometries at $h/c = 0.2$.

Figure 19 shows C_D over the same parameter space. A similar low dependence on AWA is observed. Apart from the rigid foil at the lowest reduced frequency, $k = 0.8$, the flicking motion reduces the drag on the section. The flexible foil more effectively reduces drag, and at the most aggressive flicking, $k = 2.4$, the drag becomes negative, representing a thrust into the oncoming flow.

Table 3 shows that at AWA = 0° the flexible section has variation in α_{geo} of 11.1° due to deformation under load. This chordwise flexibility (leading to changes in angle of attack, and thrust generation) is similar to the affect on thrust production described by Heathcote et al. (2007) for a flexible 2D geometry oscillating perpendicular to the flow.

In a sailing context, driving force produced by the 2D sail section is of primary interest. Figure 20 shows that driving force in all cases increases significantly with an increase in apparent wind angle. Equation 1 and Figure 5 show that this is largely due to a greater component of C_L contributing to C_{Drive} at higher AWA. At $k = 1.6$ and 2.4, the augmentation of driving force from flicking, also increases at higher AWA. For $k = 0.8$ the increase in driving force is minimal; because k is normalized using the freestream velocity, U_∞ (here equal to AWS), we expect flicking to be most effective at low wind speeds. This agrees with sailing experience.

All flicking experiments reduce the crossover AWA where C_{Drive} becomes positive. This means a boat can point higher into the wind while maintaining forward progress if the sail is flicked. The flexible sail geometry at $k = 2.4$ has positive driving force at AWA = 0°. This is also when C_D is

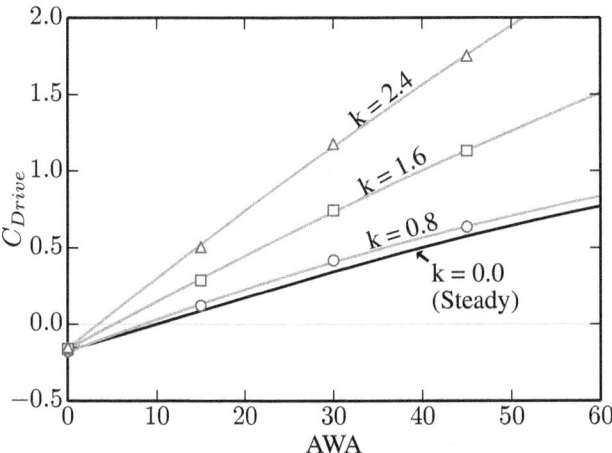

(a) **Rigid 2D sail**

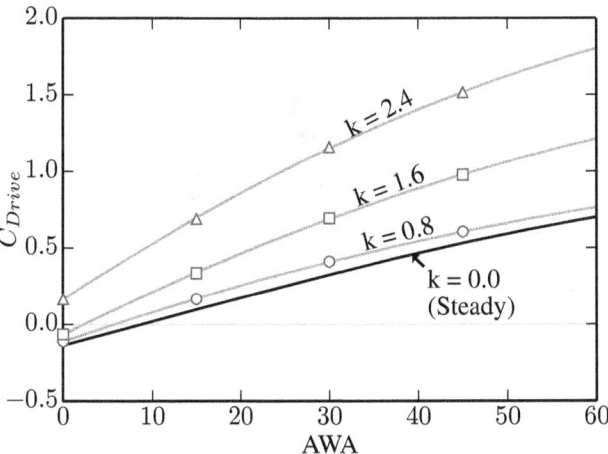

(b) **Flexible 2D sail**

Figure 20: Driving Force Coefficient, C_{Drive}, of the 2D sail geometries at $h/c = 0.2$.

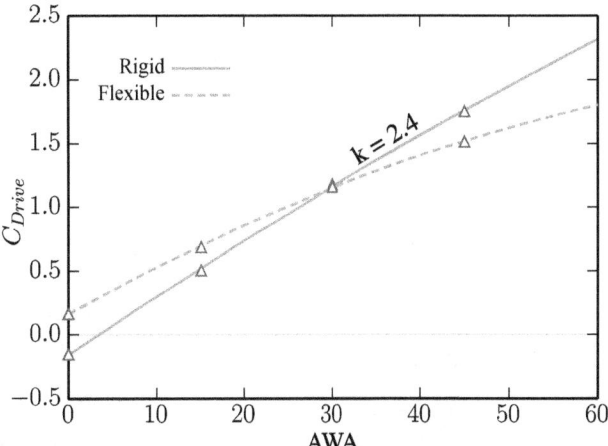

Figure 21: Driving Force Coefficient, C_{Drive}, compared for the rigid and flexible 2D sail geometry at $h/c = 0.2$ and $k = 2.4$.

tion. Figure 22 shows the vortex pattern that develops in the steady flow case when there is no flicking motion. The flow is unstable and the vorticity in the wake rolls up into a "Karman vortex street" characterized by anti-symmetric alternating-sign vortices.

The mean wake of the steady section is deflected slightly to windward relative to the apparent wind vector shown in the wind triangle in Figure 22. This is indicative of lift being generated by the sail section, as the flow passes the sail section it is pushed to the side by the angle and camber of the sail, generating lift toward the opposite side.

Figure 23 shows that much stronger vortices develop in the wake of the rigid sail, flicking with $k = 2.4$ and $h/c = 0.2$. Rather than forming naturally, as in the steady case, the vortex formation is forced by the motion of the sail. A vortex pair is shed from the sail during each "flick" and the frequency of vortex shedding is locked to the frequency of the sail motion.

In the wake, the clockwise vortices (shown in red) are shed as the sail moves to windward. These appear more strongly concentrated than the counter-clockwise (blue) vortices. This is likely a consequence of the many non-symmetric parameters in the experiment: a cambered section, a positive angle of attack, and a characteristic motion that has a more abrupt motion to windward. The stronger concentration of the clockwise vortex causes the vortex pair to rotate clockwise around each other, as observed by noting the relative positions of each pair further downstream in the wake.

The incident turbulence, colloquially "dirty" or "bad" air, impinging on a boat located in the wake of the flicking sail will be much more severe than in the wake of the steady sail.

negative. This corresponds to a "head-to-wind" sailing condition, and suggests that forward progress could be made with sufficiently vigorous sail flicking and rig flexibility.

Figure 21 overlays the $k = 2.4$ results for the rigid and flexible 2D sections. It illustrates that below AWA = 30°, the flexible sail provides more driving force, but that this trend reverses for higher AWA. This is a direct result of the superior drag reduction of the flexible section at low AWA. At higher AWA, this is overcome by the increased lift production of the rigid section. The crossover point occurs at lower AWA for lower reduced frequencies. This suggests that, for optimal increases in C_{Drive} during flicking, sail flexibility (likely controlled by leech tension), should be trimmed to both point of sail and flicking frequency.

4.2.2 Vortex Dynamics Corresponding to Sail Flicking

Utilizing particle image velocimetry, we measure the velocity and vorticity fields in the wake of the rigid 2D sail sec-

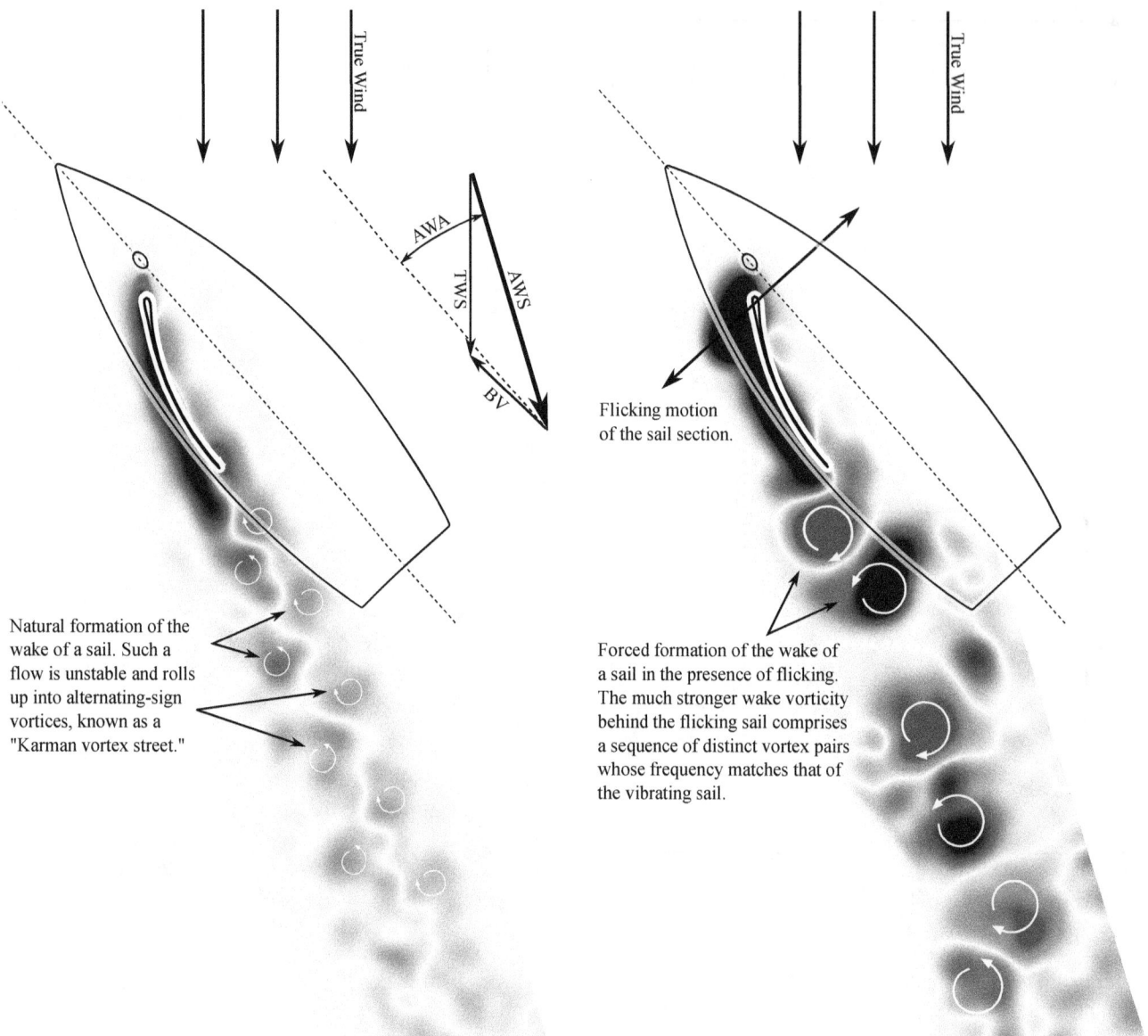

Figure 22: Natural wake vortex formation shown in preliminary analysis of the vortex dynamics in the wake of the rigid sail at steady state, h/c=0.0, k=0.0. Angles are based on a representative upwind VMG sailing condition with AWA=25° and TWA=47°.

Figure 23: Forced wake vortex formation shown in preliminary analysis of the vortex dynamics in the wake of the rigid sail section flicking at h/c=0.2, k=2.4.

5 CONCLUSION

The aerodynamics of sail flicking are investigated with a two part approach. On-the-water tests are conducted to find typical non-dimensional flicking parameters, frequency (k) and amplitude (h/c), and the characteristic motion of the sail during each "flick". Representative laboratory experiments using a 2D sail section show that driving force produced by the sail is augmented, especially at higher reduced frequencies and heave to chord ratios. This suggests that sail flicking is particularly effective in low apparent wind speeds. While flicking, sail sections continue to produce

positive driving force at lower apparent wind angles, even when "head to wind" given the right flicking parameters. Analysis of the vortex dynamics associated with sail flicking show strong vortex pairs shed from the sail during each flick. The presence of these pairs is connected to increases in time-averaged lift on the sail section, and to more severe "dirty air" in the downwind wake of the sail.

6 ACKNOWLEDGEMENTS

We would like to thank members of the Cornell sailing team, Colin Keil, Jennifer Borshoff, Mike Rivlin, Sam Webster, and Philip Alley who contributed to this work both technically and as sailors. Max de Zegher and Theodore Meynard helped in the laboratory. We would also like to thank Sarah Lihan, member of the 2012 US Olympic Team, and Marissa Lihan, Niklas Anderson, and Robbie Gilmore, Irish youth national champion, for sailing our test boat.

Special thanks to Philippe Williamson and the Alley family for use of their personal boats and gorgeous accommodations during our testing. And thanks to the US Sailing Center in Miami, FL and Cornell University's Merrill Family Sailing center in Ithaca, NY, particularly Pat Crowley and Cornell Coach Brian Clancy, for use of their facilities.

REFERENCES

BETHWAITE, F.: *Kinetics*, High Performance Sailing, , Adlard Coles Nautical, London-U.K., 2nd edition, (1996) 320-332.

BINNS, JONATHAN R. AND BETHWAITE, FRANK W. AND SAUNDERS, NORMAN R.: *Development Of A More Realistic Sailing Simulator*, High Performance Yacht Design Conference, 2002.

CLEAVER, D. J., WANG, Z., GURSUL, I.: *Lift Enhancement On Oscillating Airfoils*, 39th AIAA Fluid Dynamics Conference, **4028**, (2009), 1-16.

GARRICK, I. E.: *Propulsion Of A Flapping And Oscillating Airfoil*, NACA Technical Report, **567**, (1936), 419-427.

GENTRY, A.: *The Aerodynamics of Sail Interaction*, Proceedings of the 3rd AIAA Symposium on the Aero/Hydronautics of Sailing, 20th November, Redondo Beach, CA, (1971).

GERHARDT, F. C., FLAY, R. G. J., RICHARDS, P.: *Unsteady aerodynamics of two interacting yacht sails in two-dimensional potential flow*, Journal of Fluid Mechanics, **668**, (2011), 551-581.

GOVARDHAN, R., WILLIAMSON, C.H.K.: *Modes of vortex formation and frequency response of a freely vibrating cylinder*, Journal of Fluid Mechanics, **420**, (2000), 85130.

HEATHCOTE, S., GURSUL, I.: *Flexible Flapping Airfoil Propulsion at Low Reynolds Numbers*, AIAA Journal, **45**(5), (2007), 1066-1079.

ISAF: *Interpretations Of Rule 42, Propulsion*, International Sailing Federation, Southampton-U.K., May, (2013).

ISAF: *The Racing Rules Of Sailing For 2013-2016*, International Sailing Federation, Southampton-U.K., June, (2012).

IZRAELEVITZ, J. S., TRIANTAFYLLOU, M. S.: *Adding in-line motion and model-based optimization offers exceptional force control authority in flapping foils*, Journal of Fluid Mechanics, **742**, (2014), 5-34.

KHALAK, A., WILLIAMSON, C.H.K.: *Dynamics of a hydroelastic cylinder with very low mass and damping*, Journal of Fluids and Structures, **10**, (1996), 455-472.

KATZMAYR, R.: *Effect of Periodic Changes of Angle of Attack on Behavior of Airfoils*, NACA, **147**, (1922), 1-18.

KEIL, C., SCHUTT, R.R., BORSHOFF, J., ALLEY, P., DE ZEGHER, M., WILLIAMSON, C.H.K.: *Aerodynamics of Unsteady Sailing Kinetics*, 68th Annual Meeting of the APS Division of Fluid Dynamics, **60**(21), (2015).

SCHUTT, R.R., WILLIAMSON, C.H.K.: *Unsteady Aerodynamics of "Roll-Tacking" in Olympic Class Sailboats*, 68th Annual Meeting of the APS Division of Fluid Dynamics, **60**(21), (2015).

SCUTTLEBUTT: *Full speed ahead: The physical art of sailing*, Scuttlebutt Sailing News, http://www.sailingscuttlebutt.com/2014/11/24/full-speed-ahead-physical-art-sailing/ , November (2014).

STALLARD, T., TAYLOR, P.H., WILLIAMSON, C.H.K., BORTHWICK, A.G.L: *Cylinder loading in transient motion representing flow under a wave group*, Proc. R. Soc. A, **465**(2105), (2009), 1467-1488.

THIELICKE, W., STAMHUIS, E.J.: *PIVlab Towards User-friendly, Affordable and Accurate Digital Particle Image Velocimetry in MATLAB*, Journal of Open Research Software, **2**(1):e30, (2014).

YOUNG, J., AND LAI, J.C.S.: *Oscillation Frequency and Amplitude Effects on the Wake of a Plunging Airfoil*, AIAA Journal, **42**(10), (2004), 20422052.

THE 22nd CHESAPEAKE SAILING YACHT SYMPOSIUM
ANNAPOLIS, MARYLAND, MARCH 2016

A Comparison of RANS and LES for Upwind Sailing Aerodynamics

Stefano Nava, University of Auckland, Auckland, New Zealand
John Cater, University of Auckland, Auckland, New Zealand
Stuart Norris, University of Auckland, Auckland, New Zealand

INTRODUCTION

Computational methods currently form a significant part of a sailing yacht design project. The fluid-dynamics that characterises a sailing yacht is extremely complex, due to the fact that the yacht is partly immersed in water and partly in air with major three-dimensional and turbulent phenomena. A large number of investigations, using both experimental and numerical methods, have created an extensive knowledge of the physics of a sailing yacht, but at the same time highlighted the extent and complexity of this research field. A better understanding of the physics involved in both the aerodynamics and fluid dynamics of the yacht, together with the need to develop, improve and validate existent numerical models are primary reasons to further extend the work done on this subject.

There have been a number of studies of the aerodynamics of upwind sails. Viola (Viola et al., 2013) modelled the air flow field around a hypothetical AC33 class yacht design, using a steady RANS solver, and highlighted the main flow features and structures involved in upwind sailing aerodynamics. Even though the flow is mainly attached to the sails, numerical results showed areas of leading edge flow separation, especially at the top of the mainsail, where the influence of the downwash generated by the headsail is less predominant. Querard & Wilson (Querard & Wilson, 2007) and Masuyama (Masuyama et al., 2007) showed the need for a high quality, fine mesh in order to capture these flow features and correctly reproduce the pressure distribution on the sail surface. Queutey (Queutey et al., 2015) addressed some misbehaviour of numerical models to geometric incongruities with the experimental benchmark model.

The present work further investigates upwind sailing aerodynamics, analysing the effects of geometric and mesh modifications, in combination with the use of Large Eddy Simulation. LES has proven its ability to model

highly turbulent and separated flow (Sagaut & Mary, 2001) (Sampaio et al., 2014), albeit at the expense of high computational costs. To date there are no published results of LES in sailing aerodynamics, and so the application of the methodology to the modelling of a well documented experiment (Fluck, 2010) allows the testing of LES's capabilities and applicability to this field of research.

A high resolution grid was used so as to capture the finest flow structures trying to correctly reproduce the pressure distribution across the sails, especially in regions of flow separation. Particular effort was made to replicate the experimental test conditions and to evaluate the influence that different set ups have on the computational results. Simulations were performed using RANS and LES on the same mesh, allowing a direct comparison between the methods.

EXPERIMENTAL TEST CASE

The RANS and LES methods were validated against the experiments of Fluck of an idealised upwind sail plan, which were carried out at the Yacht Research Unit Twisted Flow wind tunnel (Fluck, 2010). The pressure distributions on the sails were summarised in (Fluck et al., 2010) and (Viola et al., 2011). Two thick fiberglass sails were used to model a 1/15th scale upwind sail plan of a hypothetical AC33 class yacht design. The distance between the top of the mainsail and the foot of the headsail was 2.25 m with the maximum chord-length of the head and mainsails being 0.7 m and 0.65 m respectively. The sails were approximately 4 mm thick. The mast and hull were not included in the model, which was lifted 600 mm above the floor to avoid the wind tunnel floor boundary layer, and mounted on a square baseplate. Moreover no twist vanes were used to ensure the sails were tested in a straight uniform onset flow (Figure 1).

The head and mainsails were both equipped with pressure taps connected to a pressure transducer box which recorded

Figure 1: Wind tunnel experiment (Fluck, 2010).

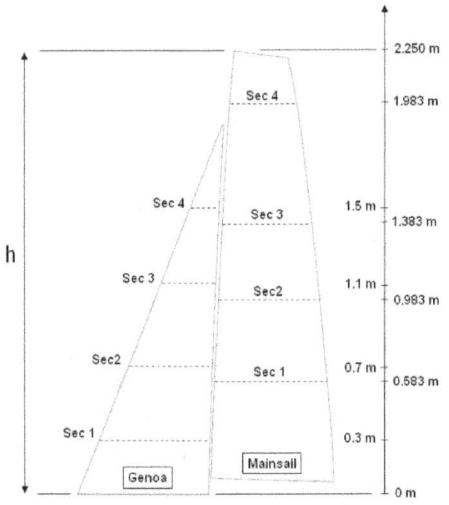

Figure 2: Pressure tap locations (Fluck, 2010).

the static pressure on the sails and the dynamic pressure upstream. The pressure taps were distributed in eight rows, four on the headsail and four on the mainsail, as shown in Figure 2. They are referred to by the sail and the height above datum in mm; for example, the topmost section of the mainsail is MAINSAIL 1983. The sails and base were oriented to give an apparent wind angle of 18.7°, and the two sails were each trimmed in four different positions, giving 16 different sailing configurations. The maximum driving force was generated by the sheeting position termed G3M2 and this has been chosen as the reference case for this investigation.

NUMERICAL METHOD

The computational modelling was performed using the CFD software FLUENT. This software can numerically solve the

Reynolds Averaged Navier Stokes equations used in RANS models (steady RANS were used throughout this study), as well as the spatially filtered Navier Stokes equations used in Large Eddy Simulation. A block structured hexahedral mesh was used, which was generated with ICEM-CFD. This type of mesh was preferred over other types used since it allowed total control over cell dimensions, enabling the specification of different refinement levels in different axes, and it generated a high quality computational grid. The cells dimension in the streamwise direction varied from a maximum of 0.01 m to a minimum of 0.0001 m in some simulations; in the spanwise direction cells were 0.005 m wide near to the sail edges growing up to 0.02 m in the middle part of the sails. Finally the first cell thickness in the normal direction was selected as 0.0001 m in order to ensure a maximum y^+ value of 1 on the sail surface. The RANS calculations were performed using the SST turbulence model using a SIMPLE solver scheme, with second order upwind differencing being used for the momentum equations while the turbulence scalars were discretized using first order upwinding.

The LES calculations used the PISO solver with three iterations per timestep, with the momentum equations being discretised with second order central differences, whilst time stepping used a second order implicit formulation. The timestep size was based on the smallest cell size in order to achieve a Courant number of approximately 1, with the highest values reaching 5 in practice. Finally, the subgrid scale turbulence was modelled using the dynamic Smagorinsky-Lilly model.

The inlet flow turbulence intensity was set to 3% and this resulted in turbulence intensities at the location of the sails of 1% and 0.6% for the RANS and LES models respectively. These values reflect the turbulent intensity of the tunnel which has been measured as approximately 1.5%. The turbulence length scale was set to 0.4 m which is three times the mesh size at the inlet section, allowing enough grid nodes for the LES model to resolve the large scale turbulent structures. All the simulations were run until the equation residual values reached a steady behaviour, and for the LES calculations variables were averaged over the following 4 seconds. The LES simulations were initialised using results from an equivalent RANS calculation. The calculations were performed on an Intel based high performance computational cluster, and the wallclock times for RANS calculations varied between 1 to 6 hours using 64 cores while LES calculations took 30 to 600 hours using 128 cores.

COMPUTATIONAL DOMAIN SET UP

The initial investigation was focused on identifying an optimal computational domain geometry and set up, in order to minimise errors in the numerical simulations. The model centreline was rotated by an angle of 18.7°

to match the nominal experiment value. The effects on the flow field generated by the square baseplate were reproduced by including a square no-slip surface underneath the sails in the computational model. Instead of lifting the baseplate and the sails 0.6 m above the floor as in the experiments, the computational domain was cut at the height of the baseplate with a free-slip boundary condition being applied to the lower surface. This choice has minor influence in the wake development behind the model but reduced the required computational effort. The sails were modelled as thick surfaces in order to reproduce the thickness in the experiment. Although the inclusion of thickness resulted in a small increase in the number of cells in the mesh, it improved the prediction of the location of the stagnation point at the leading edge of the sails.

Different leading edge shapes were tested, including squared, chamfered and rounded geometries. Since no significant differences were observed in the resulting pressure distributions on the sails, a squared geometry was chosen, which simplified the meshing process.

As was found in previous studies (Viola et al., 2013) (Queutey et al., 2015), the choice of domain size and geometry requires careful consideration. The outer box used was 13 m wide and 15 m long with the sails placed in the centre. The height of 3.1 m reproduced the distance from the base to the wind tunnel roof. Since the experiment was carried out in a open jet wind tunnel and the test wind velocity can be achieved with two different wind tunnel configurations (full width and half width), the author investigated the influence of the onset flow type on the flow field around the model. The three different internal configurations shown in Figure 4 were tested. These represented a simple rectangular box with uniform flow and the two possible configurations of the true wind tunnel, with the inlet section at its maximum width of 7 m or at half width.

For the box case, the inlet spanned the entire width of the computational domain, the sides were modelled as free-slip walls and an outlet was placed downstream at the opposite end to the inlet. For the wind tunnel models, the inlet only covered a portion of the front surface, while all the other bounding surfaces were set with an outlet flow boundary condition. The wind tunnel walls were modelled as free-slip walls as well as the roof in each of the three configurations, in order to minimise the computational effort in calculating the flow behaviour close to these surfaces. Finally the sails were given a no-slip wall boundary condition.

This preliminary test was carried out on a 1.7 million cell mesh using the RANS solver. Observation of the streamlines flowing from the inlet section indicated that in the rectangular box case the free stream flow was constrained to follow a straight path by the lateral walls, while, in the wind tunnel models it was free to fan out as soon as it exited the

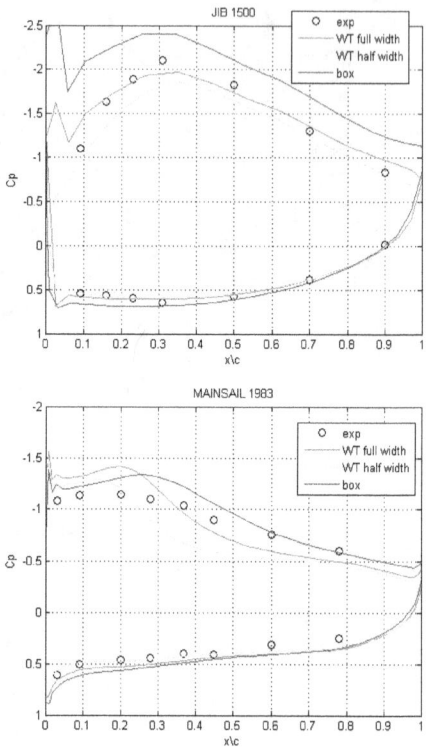

Figure 3: Pressure coefficient distribution along the top sections of the two sails for the three different domain geometries.

inlet section, and it deflected under the influence of the sails flow field. The contours of flow direction in Figure 4 reveal that the deflection of the jet was maximised for the narrower configuration of the wind tunnel, with the downwash at the leading edge of the headsail varying by 5°. As it might be expected, these differences in the inlet flow were reflected in the pressure distribution on the sails. As Figure 3 shows, the box case, which had the lowest downwash and highest apparent angle of attack, had the highest suction on the headsail with values around 15% higher than experiment.

For the full and half width wind tunnel models the suction values decrease being nearly 11% lower than experiments for the half width case. Another clear difference between models is the length of the separation bubble occurring at the top of the mainsail, with differences of approximately 15% of the chord.

It is clear that the modelled results are affected by the sails operating in a jet rather than a uniform onset flow. This is presumably true for the experimental data. The simulation results discussed in the next paragraphs were carried out using the full width wind tunnel configuration.

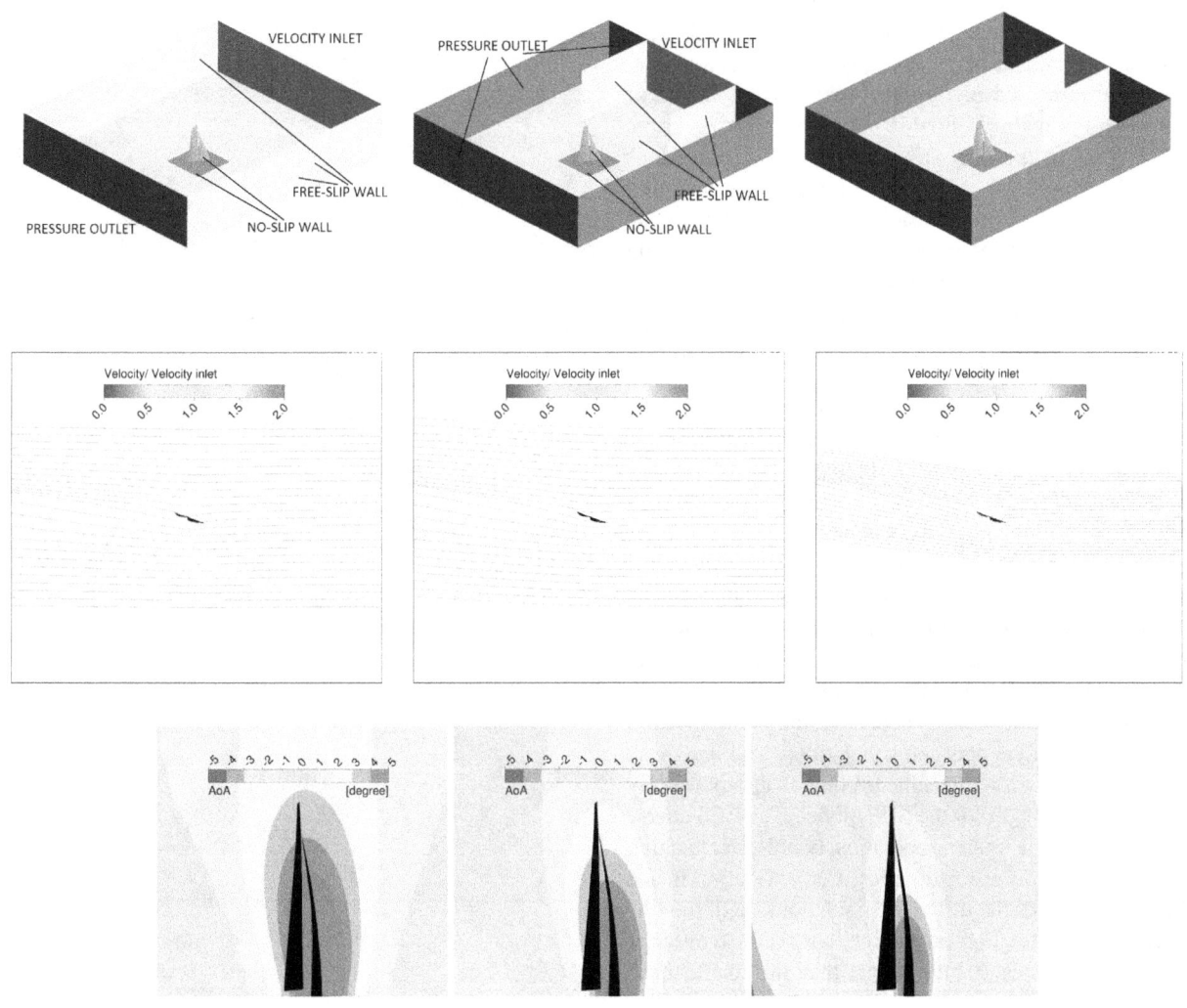

Figure 4: Wind Tunnel configurations (top), flow streamlines released from a horizontal rake 1 m above datum at the inlet (middle), and contours of deviation of flow from the nominal angle of attack of 18.7° on a vertical cross section at the headsail leading edge (bottom). Rectangular box (left), full width wind tunnel (centre) and half width wind tunnel (right).

PRESSURE DISTRIBUTION AND FLOW FIELD

The results shown in this section were calculated on two different meshes composed of 1.7 and 4.7 million cells using RANS and LES. The wall shear streamlines shown in Figure 5 for the leeward surfaces of the sails indicate that the flow was mostly attached with a small region of trailing edge separation. The contours of pressure coefficient show the maximum suction occurs on the headsail, typically occurring at the position of maximum camber. The pressure distribution on the mainsail is more uniform than the jib, partly due to the flatter design of the sail and partly because this sail is immersed in a zone of high pressure that develops on the windward face of the headsail. The influence of the latter on the mainsail flow is noticeable looking at the drastic change in the flow structure between the sections above and below the hounds at the head of the headsail. Indeed the head of the mainsail is not immersed in the above mentioned higher pressure zone and the incoming flow is not affected by the downwash generated by the jib, resulting in a high apparent angle of attack and flow separation.

The predicted amount of separation differs between the calculations. As shown in the pressure coefficient distribution labelled MAINSAIL 1983 in Figure 6, the LES results are generally characterised by a smaller, more defined separation bubble starting at the leading edge.

In the RANS cases the section shown in Figure 6 in the MAINSAIL 1983 pressure plot is fully stalled, with the air backflowing from the trailing edge to the leading edge. In contrast, the two LES simulations predict the reattachment of the leading edge separation bubble although the location of reattachment differs between simulations. The fine mesh calculation predicts a reattachment point at approximately 40% of the chordlength for the LES model. The LES coarse mesh case shows a longer separation bubble, reattaching at around 0.55c. Moreover the pressure distribution inside the bubble varies for the different simulations, with the fine mesh LES having the highest suction, almost 30% over the experimental results and 20% greater than the coarse mesh simulation.

Significant differences between the RANS and LES solutions can also be observed in the mainsail trailing edge flow. The RANS simulations predict the flow to be fully attached, while the LES shows a a region of separated flow, extending for nearly a quarter of the length of the chord. The differences between the solutions can be observed in the pressure coefficient distribution, with the two LES simulations characterised by a systematic offset compared with the RANS solutions. The latter are in better agreement with the experimental data, suggesting that the flow is actually attached in rear part of the mainsail.

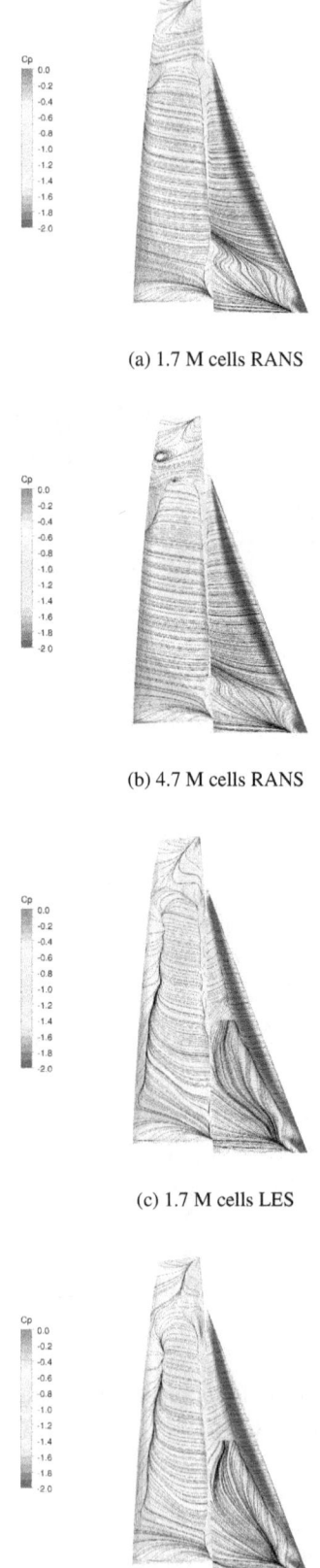

(a) 1.7 M cells RANS

(b) 4.7 M cells RANS

(c) 1.7 M cells LES

(d) 4.7 M cells LES

Figure 5: Wall shear streamlines and contours of pressure coefficient on the sails leeward surface.

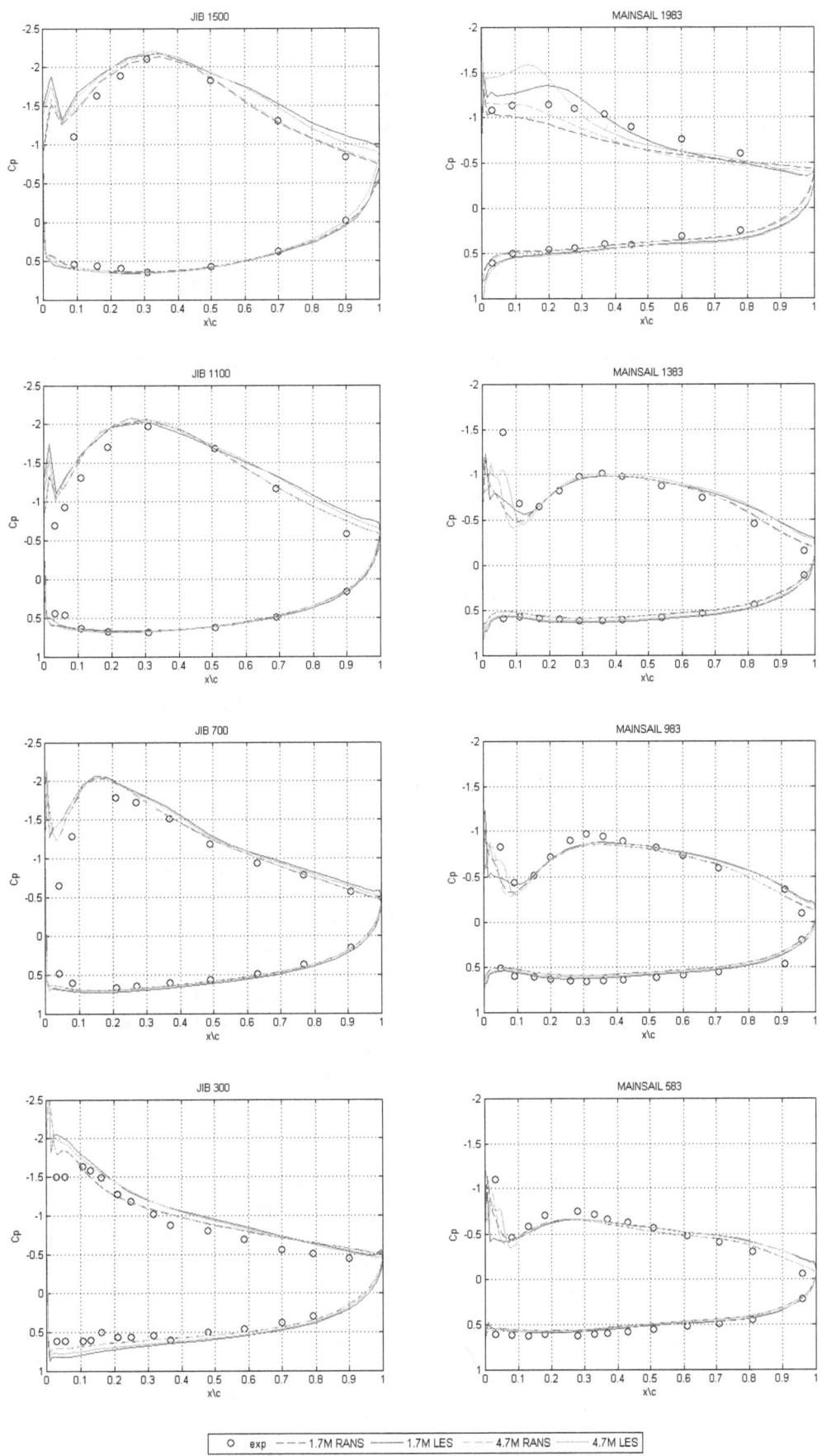

Figure 6: Distribution of pressure coefficient on the four headsail sections (left) and on the four mainsail sections (right).

Looking at the headsail flow, the results are in good agreement between RANS and LES solutions, for both the coarse and fine meshes, at the lower sections of the sail. Indeed the wall shear stress streamlines indicate that the flow near to the surface is close to separation and it moves upward due to the suction at the upper sections which are immersed in lower pressure. Generally, the pressure distributions are very similar for the two lower sections and they follow the experimental trend. However, the predicted flow over the head of the headsail differs for the two models. The RANS simulations show the flow as attached, with the C_p matching experiment. In contrast, the LES simulations show separated flow moving forward from the trailing edge to a point near mid-chord. This is reflected in the pressure distribution on the two upper sections which overpredict the suction approximately 10% compared to the experimental results.

A substantial improvement in the fine mesh LES results at the top section of the mainsail is achieved when decreasing the time step size, in order to ensure a maximum Courant number of 1 everywhere in the computational domain. Indeed, the pressure distribution plot shows (Figure 7) a longer recirculation bubble, which reattaches to the sail surface at approximately 65% of the chord length. The suction values inside the separated flow region are also closer to the experimental data, with the computational results matching the experiments in the first section of the bubble.

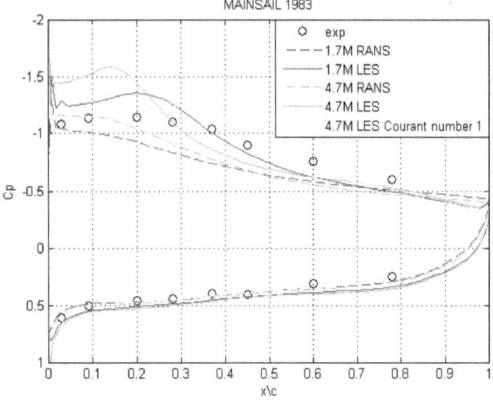

Figure 7: Distribution of pressure coefficient at the top section of the mainsail with different Courant numbers.

LEADING EDGE SEPARATION ON THE MAINSAIL

The results presented in the previous section show that the top of the mainsail is the most difficult region to model. The reason for this is the development of a leading edge separation bubble due to the sharp edge at the luff of the sail. The flow in this region is mostly planar, with a small component of vertical velocity, hence the following discussion is based

on the analysis of a horizontal section at the height of 1.983 m. The flow is visualised using surface streamlines, which are constrained to lie on a surface, neglecting the velocity component normal to the surface plane. The surface streamlines shown in Figure 8 present data for a RANS and a LES computation of the same case (4.7 million cell mesh and maximum Courant number of 1) and show that the flow detaches at the edge of the leeward side of the sail, increases speed along the length of the recirculation bubble and reattaches at around half of the chordlength. Here the surface streamlines divide, with part of the fluid flowing to the trailing edge and part of the fluid going towards the leading edge, feeding the recirculation bubble. The contours in Figure 8 of the surface streamwise velocity indicate that, depending on the structure of the separation bubble, the velocity recovery downstream changes radically.

In the RANS solution the centre of the bubble is placed at 20% of the chordlength while in the LES results it is placed at around 30% of the chordlength. For the RANS model, even though the velocity surface streamlines suggest a reattachment of the flow downstream, careful examination shows that the flow is still stalled, with low velocity values that extend past the trailing edge. In contrast, the LES velocity contours show a rapid velocity recovery with values 3 to 5 times higher than the RANS solution. An explanation for why the recirculation bubble is resolved differently by the two models can be found when looking more in detail at the leading edge of the sail (Figure 8c and 8d).

The LES velocity surface streamlines and the streamwise velocity contours predict the development of a secondary separation bubble not present in the RANS solution. Indeed, the backflowing section of the leading edge separation bubble is characterised by a high flow speed, approximately one quarter of the freestream velocity. This flow decelerates near the leading edge and it separates due to the high positive pressure gradient forming the secondary separation bubble. This phenomena has been studied in detail for a flat plate in the experiments of Crompton & Barret (Crompton & Barret, 2000), which show that the secondary bubble rotates in the opposite direction compared to the primary bubble. A preliminary CFD test carried out on the flat plate experiment highlighted the potential of LES for capturing this particular phenomena and its similarity to the aerodynamics of upwind sails. 3D simulations on a mesh characterised by the same spatial discretisation in terms of y^+, Δx^+ and Δz^+ were carried out, adopting the same time discretisation and model settings. The simulations were carried out with the timestep size resulting in a maximum Courant number of 4. As is shown in Figure 9, the secondary recirculation bubble is captured only by the LES model and the presence of this structure shifts the primary recirculation bubble further downstream, as also predicted in the sails case. Despite the LES model correctly predicting the flow structure, ex-

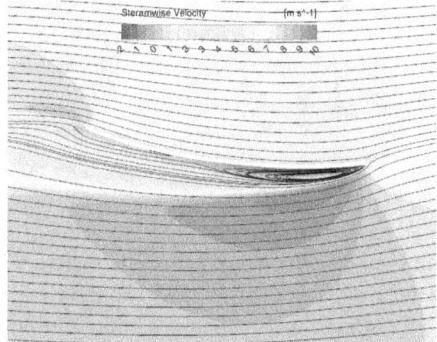

(a) RANS velocity surface streamlines

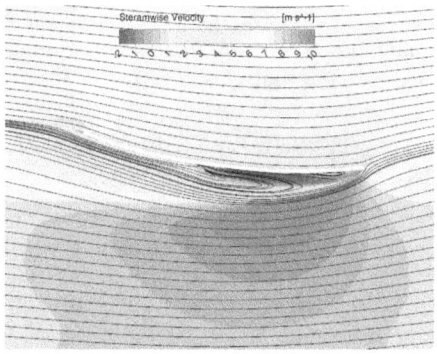

(b) LES velocity surface streamlines

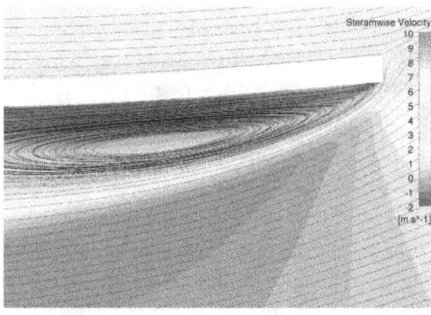

(c) RANS velocity surface streamlines-leading edge detail

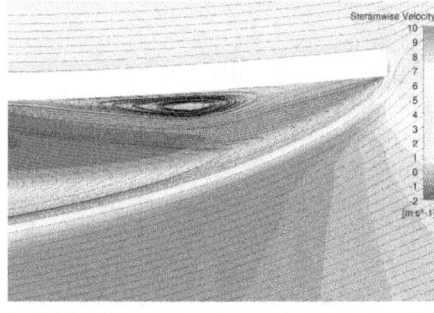

(d) LES velocity surface streamlines-leading edge detail

Figure 8: Surface streamlines at the top section of the mainsail.

amination of the predicted pressure coefficients for both the flat plate and the sail model (Figure 10) show the difficulty of capturing the correct length of the leading edge separation bubble if not adopting a sufficient fine time discretisation. Consequently the pressure distribution in this area is wrongly predicted, with LES always overestimating the suction pressures.

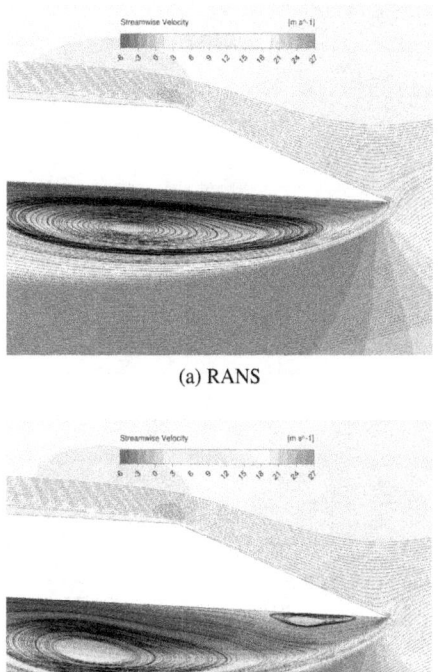

(a) RANS

(b) LES

Figure 9: Velocity streamlines at the leading edge of the flatplate.

EFFECTS OF LEADING EDGE MESH REFINEMENT

The high sensitivity of the leading edge recirculation bubble to the simulation set up has been further investigated with the comparison of the results of LES simulations carried out on four similar meshes and the same Courant number values, with the maximum being around 4. The computational grids differ in the level of refinement in the streamwise direction at the leading edge. Starting from a coarse value of first cell size of 0.01 m (coarse) the mesh has been refined (intermediate) at the leading edge decreasing the value to 0.004 m and then further refined initially to 0.001 m (fine) and subsequently to 0.0001 m (ultrafine).

The plot of the pressure coefficients reveals a clear trend,

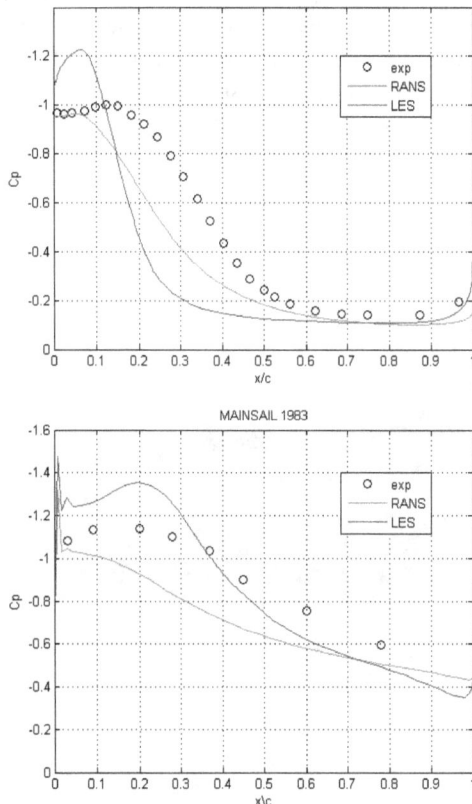

Figure 10: Pressure coefficient distribution at the leading edge of the flat plate (top) and the mainsail (bottom).

with the separation bubble becoming shorter and the pressure suction increasing with refinement at the leading edge. The solution of the coarse mesh simulation seems to provide the best agreement with the experimental pressure distribution, but the detail of the leading edge flow shown in Figure 11a indicates that the mesh is too coarse to correctly capture the flow phenomena structure. The good results achieved in terms of pressure distribution can be explained by the excessive numerical diffusion which has the effect of stretching the recirculation bubble to the correct length. Refining the mesh in that region leads the solver to predict the formation of a secondary recirculation bubble, the centre of which is located at approximately the same position for the intermediate and fine meshes while it moves towards the leading edge when switching to ultrafine mesh. Moreover, the thickness of the bubble appears to be similar for the intermediate and fine meshes, but decrease with the further refinement of the mesh, with the backflow in the secondary recirculation bubble, separating again a few cells from the leading edge, creating another recirculating structure.

The movement of the secondary recirculation bubble towards the leading edge and its shrinkage appear to be the leading factor that determines the inability of the LES model

to correctly predict the primary leading edge separation bubble. The pressure distribution is affected by the incorrect prediction of the leading edge separation bubble (Figure 12), with the suction pressure overestimated by nearly 20% for the original case, 40% for the fine case and 60% for the ultrafine case, in comparison with the experimental data.

CONCLUSIONS

A set of upwind sails have been modelled using RANS and LES and the results compared with existing experimental data. The computational results are shown to be sensitive to the boundaries of the flow domain and the onset flow conditions. The numerical simulations show a large variation in the pressure distribution depending on the choice of the computational domain, and it is presumed that the experimental data are similarly sensitive.

Differences are seen between the results of the RANS and LES models, with the latter predicting flow separation on part of the mainsail trailing edge and at the top region of the headsail.

A substantial difference arises in the structure of the leading edge separation bubble, with only the LES model showing the formation of a secondary recirculation bubble, a structure that has been observed experimentally on flat plates. In both cases the LES model has been shown to be able to predict the correct flow structures while the RANS simulations shows less accurate results.

The structure of the leading edge recirculating region and the pressure distribution associated with it is highly influenced by the mesh level of refinement in the streamwise direction and generally the LES model overestimates the suction in this region. Decreasing the time step size in order to achieve a maximum Courant number of 1, radically improve the results with the model correctly capturing the length of the separation bubble and the associated values of pressure.

Although there is a substantial difference between the flow feature predicted by the solvers, the differences in the pressure distribution is minimum, with both the computational models in good agreement with the experimental data in the attached zone of the flow, while large differences between RANS and LES appear in the areas of flow separation.

Finally, the computational effort required by the LES simulation is much larger than the RANS growing by a factor of 60 for the coarser meshes and by a factor of 200 for the finer meshes. In conclusion no significant differences can be noticed between RANS and LES when simulating attached flow. However LES used with a fine time and spatial discretisation provides an improvement in the accuracy of predicted flow structure and pressure distribution in regions of

(a) Coarse: first cell size 0.01 m

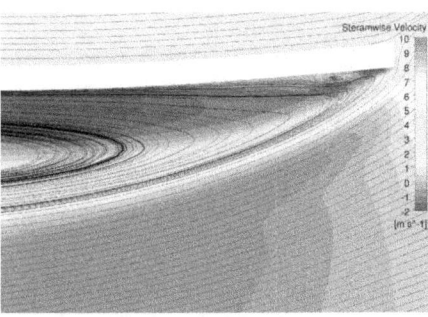

(b) Original: first cell size 0.004 m

(c) Fine: first cell size 0.001 m

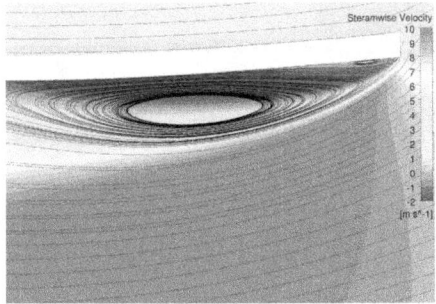

(d) Ultrafine: first cell size 0.0001 m

Figure 11: Velocity surface streamlines at the mainsail top section (MAINSAIL 1983) leading edge.

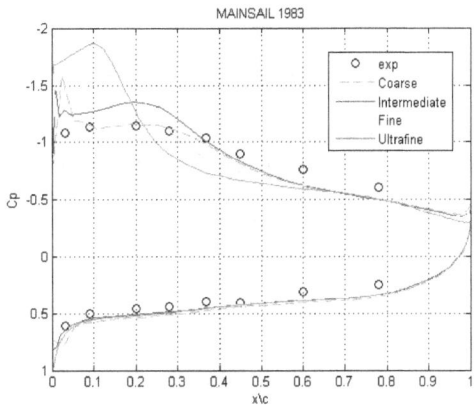

Figure 12: Distribution of pressure coefficient at the top section of the mainsail using different mesh refinement.

flow separation.

REFERENCES

Crompton M., Barrett R.: *Investigation of the separation bubble formed behind the sharp leading edge of a flat plate at incidence*, in the proceedings of the Institution of Mechanical Engineers, Part G, v. 214, pp. 157-176. (2000).

Fluck M.: *Extended Lifting Line Theory applied to two interacting Yacht Sails*, Master Thesis, Institute of Aerodynamics, Technological University of Munich, Germany. (2010).

Fluck M., Gerhardt F.C., Pilate J.P., Flay R.G.J.: *Comparison of potential-flow based and measured pressure distributions over upwind sails*, Journal of Aircraft, Vol.47, pp. 2174-2177. (2010).

Masuyama Y., Tahara Y., Fukasawa T., Maeda N.: *Database of Sail Shapes vs. Sail Performance and Validation of Numerical Calculation for Upwind Condition*, in the proceedings of the 18th Chesapeake Sailing Yacht Symposium (CSYS18), Annapolis, Maryland, USA. (2007).

Querard A.B.G., Wilson P.A.: *Aerodynamic of Modern Square Head Sails: a Comparative Study Between Wind-Tunnel Experiments and RANS Simulations*, in the proceedings of the Modern Yacht, October 11-12, Southampton, UK. (2007).

Queutey P., Guilmineau E., Wackers J., Visonneau M.: *RANS and DES CFD Simulation of Rigid Multiple Sails*, in the proceeding of The 5th High Performance Yacht Design Conference (HPYD5), March 10-12, Auckland, New Zealand. (2015).

Sagaut P., Mary I.: *Large Eddy Simulation of Flow Around a High Lift Airfoil*, Direct and Large-Eddy Simulation-IV, pp. 157-164, Springer Netherlands. (2001).

Sampaio L.E.B., Rezende A.L.T., Nieckele A. O.: *The challenging case of the turbulent flow around a thin plate wind deflector, and its numerical prediction by LES and RANS models*, Journal of Wind Engineering and Industrial Aerodynamics, v. 133, pp. 52-64. (2014).

Viola I.M., Pilate J.P., Flay R.G.J.: *Upwind sail aerodynamics: A pressure distribution database for the validation of numerical codes*, International Journal of Small Craft Technology, Vol.153, pp. 47-58. (2011).

Viola I.M., Bot P., Riotte M.: *Upwind sail aerodynamics: A RANS numerical investigation validated with wind tunnel pressure measurements*, International Journal of Heat and Fluid Flow, February, Vol.39, pp. 90-101. (2013).

Pressure Measurements on Yacht Sails: Development of a New System for Wind Tunnel and Full Scale Testing

Fabio Fossati, Ilmas Bayati, Sara Muggiasca, Ambra Vandone, Department of Mechanical Engineering, Politecnico di Milano, Milano, Italy.
Gabriele Campanardi, Department of Aerospace Science and Technology, Politecnico di Milano, Milano, Italy
Thomas Burch, CSEM, Centre Suisse d'Electronique et de Microtechnique SA, Alpnach Dorf, Switzerland
Michele Malandra, North Sails Group, Carasco, Italy

ABSTRACT

The paper presents an overview of a joint project developed among Politecnico di Milano, CSEM and North Sails, aiming at developing a new sail pressure measurement system based on MEMS sensors (an excellent compromise between size, performance, costs and operational conditions) and pressure strips and pads technology. These devices were designed and produced to give differential measurement between the leeward and windward side of the sails. The project has been developed within the Lecco Innovation Hub Sailing Yacht Lab, a 10 m length sailing dynamometer which intend to be the reference contemporary full scale measurement device in the sailing yacht engineering research field, to enhance the insight of sail steady and unsteady aerodynamics [1].

The pressure system is described in details as well as the data acquisition process and system metrological validation is provided; furthermore, some results obtained during a wind tunnel campaign carried out at Politecnico di Milano Wind Tunnel, as a benchmark of the whole measuring system for future full scale application, are reported and discussed in details.

Moreover, the system configuration for full scale testing, which is still under development, is also described.

NOTATION

ρ	Density of air (kg/m^3)
Cp	Pressure coefficient (-)
p	Actual Measured Pressure (Pa)
p_0	Reference static pressure (Pa)
V_∞	Incoming/apparent wind speed (m/s)
LWL	length water line (m)

INTRODUCTION

The possibility of knowing the effective pressure distribution over the sail plan is of great interest for the aerodynamic and structural design of sails and for the selection and the optimal use of materials an production techniques. Integral measurements alone may not be sufficient in understanding how a sail plan can be optimized on specific purposes, if any information about the complex local fluid-structure interaction are provided.

In the last few years there has been a revival of pressure measurements on yacht sails and recently several contributions can be found in literature aiming to assess sail pressure distribution detection ([2-9]).

The present paper presents an overview of an ongoing joint project developed among Politecnico di Milano, CSEM and North Sails aiming at assessing a new sail pressure measurement system based on MEMS pressure transducers, connected to strips and pads. These devices were designed and produced with the scope to provide differential measurements between the leeward and windward side of the sails.

The project has been developed within the Lecco Innovation Hub Sailing Yacht Laboratory project, a 10 m length sailing dynamometer which longs for being the modern reference full scale measurement setup in the sailing yacht engineering research field, in order to enhance the insight of sail steady and unsteady aerodynamics.

An overview of the Sailing Yacht Lab project is provided in [1]: a brief summary of the origin and building steps of the vessel's design are given, along with a description of principal design and performance criteria. Also the project management and commissioning are described, as well as the measurement capabilities and data acquisition procedure.

Furthermore, an important feature of this project is the availability of measurement systems for pressure distribution acting on the sails at full scale.

In the following, the pressure measurement system is described in details, as well as the data acquisition process and system metrological validation is provided.

The pressure measurement system has also been tested in the wind tunnel using a scale model of sailing yacht and compared with a different pressure measurement system already available at Politecnico di Milano Wind Tunnel. For wind tunnel tests were realized strips and pads adequate for the model sails.

Some results obtained during a wind tunnel campaign carried out during an Offshore Racing Congress project aimed at revising sails aerodynamic coefficients and ORC VPP aerodynamic model are reported and discussed in details.

In conclusion the pressure measurement system designed for full scale testing is described.

PRESSURE MEASUREMENT SYSTEM

The pressure distribution on the sails is carried out by means of MEMS sensors (an excellent compromise between size, performance, costs and operational conditions) and dedicated pressure strips and pads which have been designed and produced aiming to provide the differential measurement between the sail leeward and windward side.

The pressure sensors are designed and built to provide the differential measurement between the measurement point and a reference pressure value which can be supplied by the user.

In the following a detailed description of the pressure scanners will be provided, as well as of the other main components of the system.

Pressure strips scanner description

The scanner CSEM C16 is a miniaturized electronic pressure scanner in a slim, lightweight and waterproof package (Figure 1). It provides 16 differential pressure sensors and a CAN bus interface for the communication. High attention was given to dimensions and shape of the scanner box. The scanner height, of only 6 mm, has minimal impact on the airflow, which makes it possible to place the scanner directly in a custom built sleeve close to the actual measurement section on the sails. Each of the 16 sensors has its own reference input which makes the scanner especially suited for measuring the pressure difference between leeward and windward side on dedicated spots on the sails.

The commercial MEMS pressure dies, integrated in the scanner, are a new generation of piezo-resistive differential low-pressure dies to reach very low full scale ranges below 1000 Pa. Despite the die size of only 2 x 2 x 0.5 mm, that is much smaller than traditional low-pressure dies, it provides improved zero-stability, reduced g-sensitivity and reduced sensitivity to humidity. This added stability permits use with added amplification to achieve accurate performance in ranges much lower than its nominal 1000 Pa rating. The key

specification of the scanner is given in Table 1.

The MEMS sensors are cost efficiently bonded to a FR4 substrate using innovative die bonding techniques based on elastic adhesives (Figure 5). The sensors are packaged in a sensor array with minimal air cavity to ensure optimal performance in combination with the micro-channels of the pressure strips.

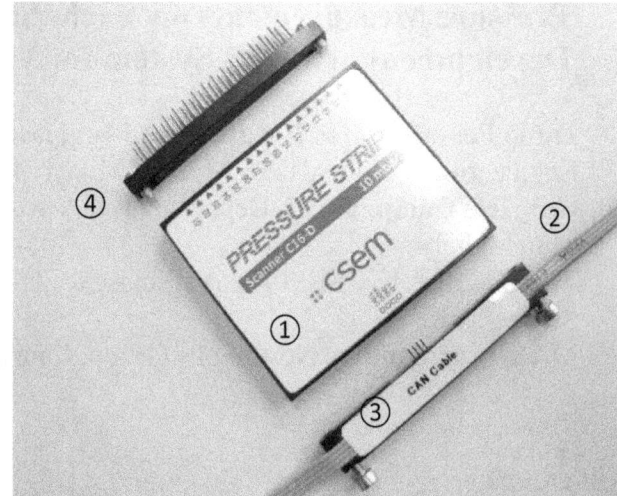

Figure 1 - Pressure Scanner C16 with auxiliary parts. 1) Scanner, 2) CAN Cable, 3) CAN connector, 4) Tube Adapter

Parameter	C16	Unit
FS pressure range	±1000	Pa
Number of pressure inputs	16	
Number of reference inputs	16	
Measurement resolution	0.01	% FS
Static accuracy after zeroing	0.25	% FS
Total thermal error	0.01	%FS/°C
Sample rate	1 - 100	Hz
Input voltage	12	V
Operation current	60	mA
Communication CAN Interface	1	Mbit / s
Maximal CAN cable length	40	m
Internal flash data memory size	8	Mbit
Operating temperature range	-10 to 70	°C
Size	65x55x6	mm
Weight	50	gram

Table 1 - Pressure Scanner Specification

A dedicated pressure flange system makes the scanner compatible with either the pressure strips or with standard tubing. Three different pressure adapters have been developed, which can be screwed to the scanner. The first adapter provides 32 tubes (2 per pressure sensor, one facing to the front side and one to the reference side of the sensor). A second adapter combines all reference inputs to a single tube, in order to connect all MEMS sensors to the same reference (Figure 2 and Figure 3). Finally, the third adapter provides direct access to the pressure strips without the need of any tubes.

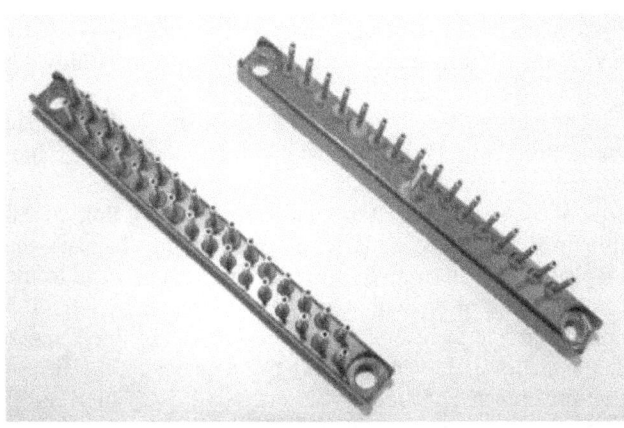

Figure 2 - Pressure tube adapter 2 x 16 tubes (left) and 1 + 16 tubes (right)

Figure 3 - Pressure flange with gasket and threads to connect and seal pressure tube- or strip-adapters

The scanner C16 supports a standard CAN interface (CAN 2.0A) with a proprietary CAN protocol, allowing remote access to the essential commands required when integrating the unit into an instrumentation system. The serial CAN bus topology (Figure 4) allows for up to 128 scanners in a single network. In practice, the number of scanners per network should be below 16 (i.e. 256 pressure sensors) in order to reduce the data traffic on the bus and to guarantee synchronized data sampling.

The CAN interface has been preferred over a wireless solution due to its robust data transmission capability, the guaranteed data rate of 1 Mbit per second and the possibility to directly supply electrical power to the scanners via the flat CAN cable. Thus, no battery is required in the scanner which reduces both, the dimensions and the overall weight of the scanners. All measurement data and configuration commands are sent over the CAN interface. A correctly received command is always acknowledged by the scanner with the transmission of a response message.

Two basic data sampling approaches are supported either autonomous sampling or master sampling. The desired option can be configured and stored in the configuration flash. In auto sampling mode each scanner in the network generates its own sample timing according to a programmed sample rate and transmits the measurement data of each sample to the CAN bus autonomously. The measurement data can be collected on-line or can be stored in the internal flash of the scanner and downloaded off-line after the measurement session.

In master sampling mode the user programmed instrumentation system (SW running on PC or Laptop) acts as sample master and broadcasts each sample start with a SINGLE_SHOT sample command. All scanners in the network receive the sample command at the same time and start the measurement immediately and synchronously. Each scanner writes the measurement data to the CAN bus following a bus collision avoidance protocol. The master collects the response messages of all active scanners in the CAN network, and initiates the next sample according to the desired sample rate. The master sampling mode has the advantage that all scanners connected to the CAN bus are synchronized by the master, even over a long sample period of several hours. The 16 sensors of each scanner are sampled sequentially with an internal scan rate of up to 4 kHz. Hence in master sampling mode all sensors in the CAN network can be sampled nearly synchronous within 4 ms.

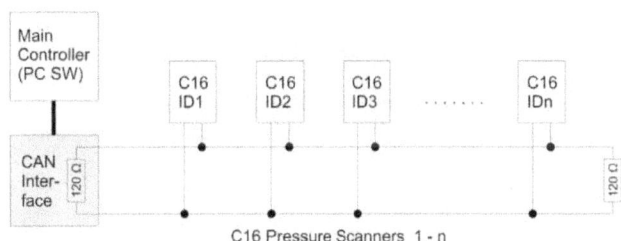

Figure 4 – CAN bus topology

Figure 5 - Pressure scanner electronics with MEMS sensors bonded to FR4 substrate

Pressure strips technology

The pressure strip system is suited for aerodynamics testing on full scale objects in their natural environment but also for models in a wind tunnel (Figure 6). Its main advantage is the light weight and thin, flexible foil appearance which allows non-invasive application to the test surface. The pressure strips are made of thin polymer films and the strip geometry can be customized for nearly seamless fitting to the test object (Figure 7). Tiny micro-channels in the pressure strip propagate the pressure from the tap to the connected pressure scanner. Manufacturing processes have been developed successfully using laser and micro-milling to produce strips with comparatively deep channels. Laser fabrication has the advantage that it can produce channels in soft materials such as silicone or soft PVC, thus increasing the flexibility of the strip significantly without having to reduce the thickness of the strip. On the other hand, channels can be manufactured approximately 5 times faster using micro-milling (Figure 6).

The base material with milled or laser ablated channels is laminated with transparent adhesive tape in order to obtain sealed channels.

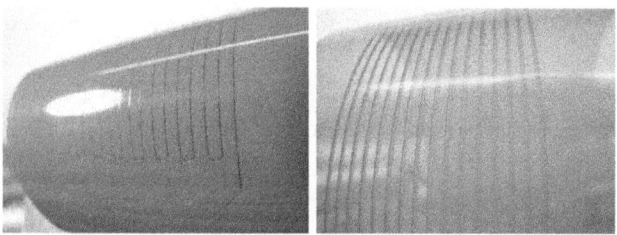

Figure 6 - 400 μm wide and 400 μm deep laser fabricated micro-channels in a silicone strip (left) and micro-milled in a polycarbonate strip (right).

Figure 7 – Pressure strip with 40 taps on three sections tailor-made for the jib of a 1:10 scale model yacht

Pad technology

Pressure pads are a specific form of small pressure strips and provide a simple solution to attach a pressure tube to a very thin structure like a sail or spinnaker. The pads can be individually placed on the test section and connected to the pressure scanner with small plastic tubing (Figure 8). This system has been extensively characterized, in terms of static and dynamic response, which is reported in the following section (Metrological Validation)

The pads provide two pressure taps on one end of the pad and a pressure tube adapter with two metal tubes of 1 mm diameter on the other end. Each of the two taps faces to one side of the sail (leeward or windward). A small hole of 0.8 mm in diameter is made in the sail directly beneath the respective pressure tap to realize a pressure passage to the opposite side of the sail. The pad thickness is between 0.5 and 1 mm and therefore introduces only minor interference to the airflow. The length of the pad depends on the size of the test object. For model sails the length is as short as 30 mm while for real sails the length is up to 150 mm to keep the tube adapter a certain distance out of the air flow of the measurement section (Figure 8). The micro channels are cut in the base layer using the same laser ablation and micro milling techniques as for the pressure strips. The base layer is made of a soft material like PVC soft, silicone or acrylic foam tape and usually has a thickness of 0.3 - 0.5 mm. The cover layer is a transparent adhesive tape of 0.2 mm. The cover layer overlaps the base layer by a few millimeters which results in a smooth transition between the sail and the pad layers after application on the sail.

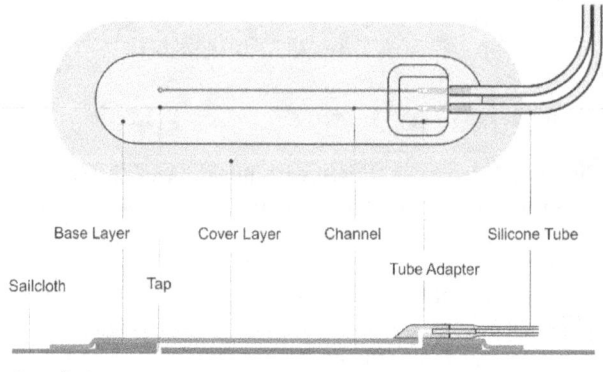

Figure 8 - Conceptual drawing of the pressure pad

METROLOGICAL VALIDATION

Some preliminary tests were performed to verify the measurement quality of the system in terms of static accuracy of the pressure measurements and dynamic response of pressure strips.

Static accuracy

The analysis of the static accuracy of the CSEM pressure measurement system was carried out in the 1m x 1.5m test section of the close circuit wind tunnel of the Aerodynamics Laboratory of Politecnico di Milano, using a constant section NACA 23015 airfoil model. The model has 0.3 m cord length, aspect ratio 3.1 and it is instrumented with pressure taps along the mid-span section. A dedicated pressure strip has been realized spanning from the 25.5%

chord position of the lower surface to the 87.5% chord position of the upper surface of the model (Figure 9).

The strip is provided with a double series of pressure taps (Figure 9): at the same chord position of each pressure tap on the strip another hole through the strip was created to be connected to proper tubes, so that comparative measurements could be taken both for the novel pressure system with strip micro-channels (CSEM) and the consolidated high accuracy pressure scanner system (PSI) (Figure 9). More specifically, the latter relies on an Esterline Pressure Systems DTC ESP miniature pressure scanner with 1 PSI range, controlled by a Chell QUADdaq System.

Figure 9 - NACA23015 airfoil model instrumented with pressure strip for static wind tunnel test

Static measurements were gathered for fixed incidence angles, ranging from -2 degrees to 14 degrees, at respectively 2.5 m/s, 5 m/s, 10 m/s, 15 m/s and 20 m/s wind speed.

For clarity, in Figure 10, only the results for the wing model at 6 degrees of incidence, are reported. The data plotted in the graph are reduced in terms of pressure coefficient, defined as

$$C_p = \frac{p - p_0}{\frac{1}{2}\rho V^2_\infty} \qquad (1)$$

where p is the pressure on pressure tap, p_0 the reference pressure, ρ the air density and V_∞ the wind speed of the incoming inflow.

It can be noticed that there is good agreement between pressure taps data and the pressure strip data, except for a few points near the airfoil leading edge, where the airfoil has strong curvature. In this region the strip installation, even though it was executed with particular accuracy, presents some tiny surface deformations - visible as small air bubbles around the pressure taps next to the leading edge - which are the main source of the differences.

Dynamic response

Further experimental investigation was also carried out to better understand the dynamic capabilities of the pressure strips. A truck hooter has been utilized as pressure wave generator, driven by a signal generator and an amplifier. The pressure measurements were taken by the means of two

CSEM pressure scanners: one with a pressure port connected directly to the pressure wave source by a very short tube (Figure 11), the second with a pressure port connected to the strip channel under test in the same way to be used during the wind tunnel testing, Figure 7.

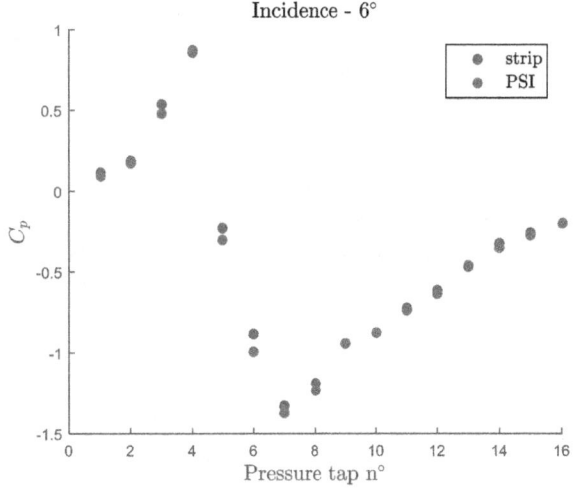

Figure 10 - Pressure distribution along the airfoil for an angle of incidence of 6°

Figure 11 - Test setup for dynamic response evaluation

A special attention was paid to the pressure connections to the source. Two tubes of the same length were adopted to connect the scanner and the strip channel to be tested, Figure 11. In such a way it can be reasonably assumed that the pressure wave measured near the source has the same amplitude and phase of the pressure wave reaching the pressure tap on the strip. The connection and the sealing of the tube on the strip was done by means of modeling clay.

Measurements were carried out on the pressure taps connected with the longest channel of each pressure tap array (bottom, middle, top) both on the mainsail and the jib strip (red circles in Figure 12).

The tests were conducted generating single tone sinusoidal pressure waves and sinusoidal sweeps in the frequency range 0 - 3 Hz the expected frequency range for this phenomenon. The pressure data acquisition was started simultaneously on the two scanners. The choice behind the characterization of the pressure system within this frequency range, is consistent with the interest of investigating the physics of slow varying aerodynamic phenomena connected to the sailing yacht motion, due the combined wind and wave loading. The full-scale cutoff frequency of this range is approximately 2 Hz [10]. Frequencies higher than this range (e.g. turbulence) are not expected to have any relevant influence on the overall dynamics of the boat, in that it represents a mechanical low pass filter.

For the sake of simplicity, only the results obtained for the pressure tap on top array of the mainsail (left) are shown in Figure 13.

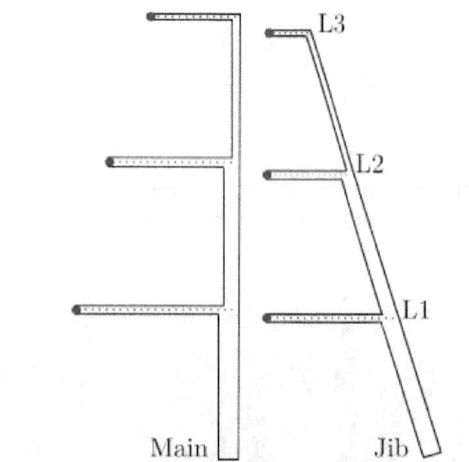

Figure 12 – Positions of the pressure taps tested

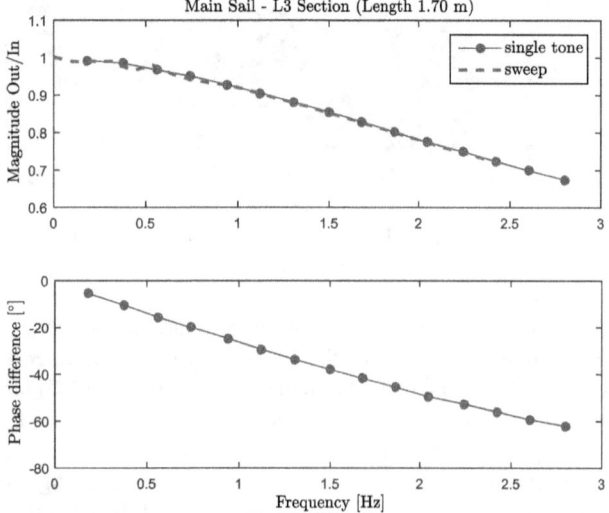

Figure 13 - Dynamic response of the pressure tap on the top array of the mainsail

The obtained results agree with the expected ones for channels with a small section and comparable length. As it can be noticed from Figure 13, the linear trend in the output/input relative phase means a constant shift of the signal in phase (i.e. angular coefficient of the straight line in the frequency graph, Figure 13), so that the time history can be easily and consistently corrected during the post-processing procedures.

WIND TUNNEL TESTS
Test apparatus, program, and procedure

A complete 1:10 scale model of a 48' cruiser-racer, consisting of yacht hull body (above the waterline) with deck, mast, rigging and sails is mounted on a six component balance, which is fitted on the turntable of the wind tunnel (Figure 14). The turntable is automatically operated from the control room enabling a 360° range of headings. This permit to set desired AWA (apparent wind angle) during the tests.

The large size of the low speed test section enables yacht models of quite large size to be used, so that the sails are large enough to be made using normal sail making techniques, the model can be rigged using standard model yacht fittings and small dinghy fittings without the work becoming too small to handle, commercially available model yacht sheet winches can be used and, most importantly, deck layout can be reproduced around the sheet winch, allowing all the sails to be trimmed as in real operating condition. Moreover, the model yacht drum type sheets are operated through a 7 channel proportional radio control system, except that the aerial is replaced by a hard wire link and the usual joystick transmitter is replaced by a console with a 7 multi-turn control knobs that allow winch drum positions to be recorded and re-established if necessary. The sheet trims are controlled by the sail trimmer who operates from the wind tunnel control room.

Figure 14 shows the model mounted in the wind tunnel. Both sides of the main sail and the jib are equipped with tailor-made pressure strips, each providing three test sections and a total of 40 pressure taps. An example of such pressure strip is given in (Figure 7). Wind tunnel pressure set up was realized to measure pressure distributions on both sides of the sails, whereas in the full scale set up only differential pressure could be measured, for the intrinsic difficulty in the definition of a reference pressure signal in the real operating environment: this set up permits to better investigate the flow field around the sail, more in detail than it would be possible for full scale measurements.

A high performance strain gauge dynamic conditioning system is used for balance signal conditioning purposes. The balance is placed inside the yacht hull in such a way that x axis is always aligned with the yacht longitudinal axis, and the model can be heeled with respect to the balance. The wind tunnel is operated at a constant speed after the wind speed profile and optionally wind twist can be properly tuned considering the desired targets, which are previously calculated considering the potential boat performance at different true wind speeds and yacht courses. This allows to reproduce apparent wind speed both in terms of wind magnitude and profile [11]. As previously mentioned, the velocity profile can be simulated by means of independent control of the rotation speed of each fan joined to the traditional spires & roughness technique, while the twist can be simulated by twisting the flexible vanes by different amounts over the height range. The wind tunnel speed is most usually limited by the strength of the model mast and rigging and the power of the sheet winches.

Figure 14 - Yacht model in the boundary layer test section of Politecnico di Milano Wind Tunnel

The model is set at an apparent wind angle and at a fixed heel. An important feature of wind testing procedure is that the model should be easily visible during the tests so that the sail tell-tales can be seen by the sail trimmer. For this purposes some cameras placed in the wind tunnel as well as onboard allow a view similar to the real life situation (Figure 16).

Flying Shape Detection

During the present wind tunnel campaign, a novel sail flying shape detection system, based on Time of Flight technology (TOF), was adopted to perform shape measurements along with pressure and force data, as in the full scale final system [1]; therefore, a master software was programmed to trigger synchronously the acquisition of all devices. A thorough explanation of this TOF novel technology can be found in [12]. Basically, a laser pulse is emitted by the TOF sensor and by measuring the time the pulse takes to hit the target surface and to come back to the receiver, it is possible to estimate the target distance.

Figure 15 - The Flying Shape Detection (TOF) System and its position for wind tunnel tests: one system per sail

The device performs measurements in terms of spatial coordinates of thousands of points belonging to the sail surface without impairing its shape since no contact occurs between object and sensor. These data, coming singularly from two different laser scanners, one for main sail and the other for jib (or gennaker), as in Figure 15, are then processed together to reconstruct the 3D sail surfaces (Figure 17), which can be sectioned in correspondence to the pressure strip heights allowing to compare pressure distribution against sail shape (Figure 21).

Figure 16 - Onboard camera view on jib during testing

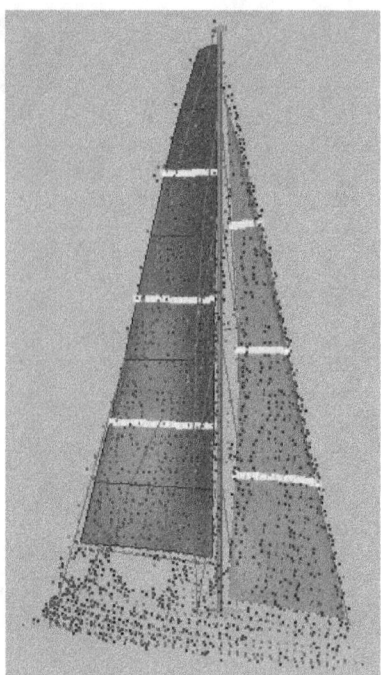

Figure 17 - Example of a point cloud acquired for mainsail (blue) and jib (green): 3D reconstruction sail shapes and highlighted sections in correspondence of the pressure strips

Upwind sails tests and results

The usual way to analyze data of this type of measurements is to compare non-dimensional coefficients [11], [13], allowing to compare the efficiency of sails of different total area at different conditions of dynamic pressure. The first useful parameter to be analyzed is the variation of driving force coefficient Cx versus heeling force coefficient Cy. Figure 18 shows Cx Vs Cy curves for the 6 apparent wind angles (AWA) tested in this campaign. It can be seen that there are some settings at the highest values of heeling force coefficients where the driving force is lower than the maximum value (e.g. below the maximum efficiency curve, isolated points in Figure 18). These non-optimum values were due to an over-sheeting of the sails, such that the mainsail generally had a tight leech and the airflow separated in the head of the sail, [14]. After having maximized the driving force, the sails were adjusted to reduce the heeling force measuring the reduction of the driving force. The reduction in heeling force was achieved by initially easing the main sheet, to twist the mainsail and minimize flow separation, then adjusting the traveller to reduce the angle of attack of the wind on the main. Envelope curves have been drawn through the test points with the greatest driving force at a given heeling force (e.g. depower curves, Figure 18).

The heeling moment is also measured during these wind tunnel tests and it can be used to determine the center of effort position of the rig. For close hauled configuration, the sail plain center of effort height, Ceh, is obtained by dividing the roll moment by the heeling force component in the yacht body reference system and normalized over the height of the

of the mast.

A plot of center of effort height variation with heeling force coefficients for tested apparent wind angles can be seen in Figure 19, in terms of the ratio between center of effort height from the boat deck and the fore-triangle height. As it can be seen, the center of effort height tends to decrease as the heeling force coefficients decrease. This is explained by the way in which the sails are de-powered to reduce Cy: increasing the twist reduces the loading in the head of the sails and then depowering the mainsail leaving the same genoa trim, which has a lower center of effort, tends to reduce it.

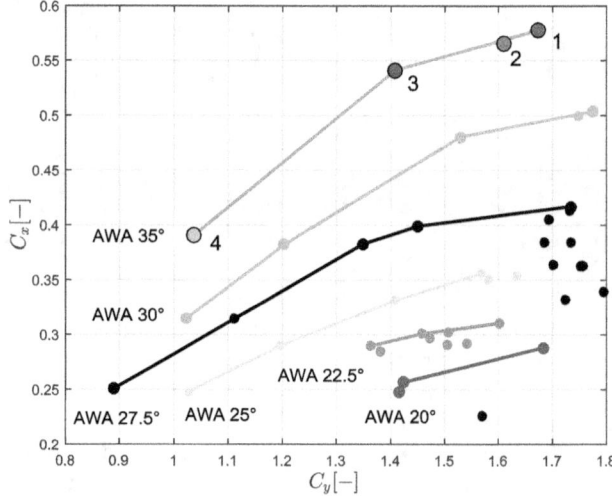

Figure 18 - Driving force coefficient Cx versus lateral force coefficient Cy

Analogously, also the center of effort longitudinal position, Cea, is obtained by dividing the yaw moment by the heeling force component in the yacht body reference system and normalized over the LWL. Its variation with heeling force for all angles can be seen in Figure 20, to be interpreted with reference to the envelope of the points corresponding to maximum driving force at each heeling force are reported (Figure 18).

In Figure 20 Cea is measured from the origin of the balance which is placed behind the mast. As can be seen Cea moves forward as Cy reduces: this is again explained by the way the sails are de-powered, as described above.

As depicted in Figure 18, the points laying on the de-powering curve at maximum apparent wind angle AWA (35°), were chosen for a more thorough investigation in terms of pressure distributions and corresponding flying shapes. More specifically, as it can be noticed in Figure 12, three stations at different height, both on main sail and jib, port/starboard sides, were chosen to put the pressure strips on (Figure 12 and Figure 17). In Figure 12, L1, L2 and L3 stand respectively for the three levels at 25%, 50% and 75% of the sail heights, so the corresponding pressures at the points 1-4 are reported in Figure 28.

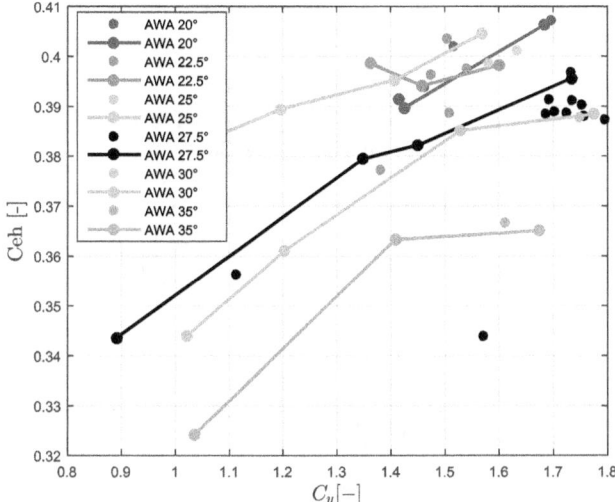

Figure 19 - Centre of effort height vs heeling force coefficient

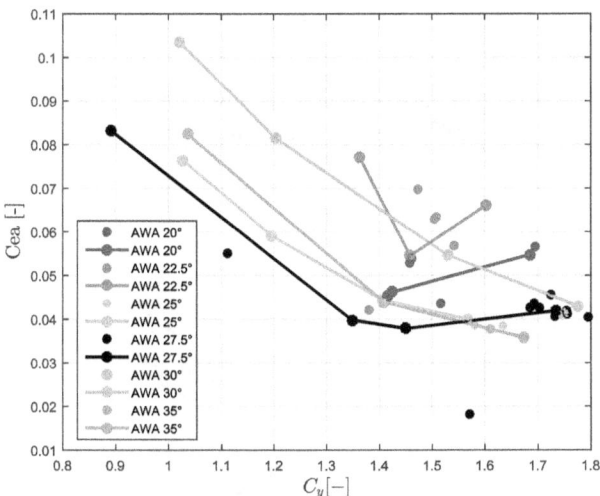

Figure 20 - Centre of effort longitudinal position vs heeling force coefficient

Furthermore, in Figure 28, the pressure channels at the L2 level of the main sail, were doubled to get the same measurements with two different pressure systems to cross-check again, during wind tunnel tests, the reliability of the novel CSEM acquisition system (the empty markers are due to the certified old system, PSI).

Figure 28 suggests a few interesting comments about the aerodynamics behind: analyzing singularly the pressure coefficients of the main sails, especially for L2 and L3, it can be easily seen that the trend is of an increase in the pressure coefficient Cp nearby the main sail's leading edge, where the mast is placed, associated to a separation bubble, which has been documented in the past [14]. Furthermore, moving along the airfoil chord, up to the trailing edge, another greatly negative pressure coefficient is experienced, for points of lower driving force coefficient Cx (i.e. 3-4 of Figure 18), right after the reattachment behind the mast separation, whereas for points with greater efficiency (i.e. 1-

2 of Figure 18) this bubble seems not to be occurring: therefore, after the separation due to the mast, a monotonic trend up to the trailing edge is evident, index of a gradually and efficient reattachment of the flux on the leeward side of the sail. Nevertheless, some considerations can be reported when it comes to analyzing comparatively both jib and main sails at the same time, for example at the level L2 (Figure 28). It is clear how moving to less efficient trims of the sail plain (i.e. 3-4 of Figure 18), which were possible by sheeting differently only the main sail, as thoroughly above explained, has an important influence on the jib sail itself, then modifying the overall driving force coefficient Cx.

More specifically, for worse trimming of the sail, the aerodynamics of the jib is modified in the sense that the suction effect of the main gets weaker (Figure 28, L2 – jib), leading to a different lifting attitude of the jib as well. Therefore, the higher (less negative) pressure coefficient on the leeward side of the jib, visible in the L2 section of Figure 28, can be interpreted as a consequence of this phenomenon.

Moreover, in Figure 29, a comparative plot of two different AWA (i.e. 20° and 35°), both from maximum driving force coefficient configurations (see Figure 18), are reported. Considering again the section L2 both for jib and for main, without loss of generality for the other sectional airfoils, the following explanations can be drawn: the pressure distributions of the jib have basically the same trends and values, since the related shape is due to the maximum power trim; therefore, the main difference can be noticed in the main sail, where the pressure distribution also explains the lower driving force coefficients Cx AWA 20° (Figure 18), associated to a center of effort that is located upward (Figure 19) and slightly forward (Figure 20) with respect to the higher angle of attack (AWA 35°). This trend is also visible in the Figure 21, where it is clear that the driving force, computed as the integral of the pressure distribution on the corresponding section, is mainly given by the jib. In Figure 21, the blue arrows correspond to positive pressure coefficients (windward), whereas red ones to negative (leeward). The pressure measurements of the whole wind tunnel session reported were referenced to the wind tunnel static pressure signal.

Comparing Figure 28 and Figure 29 is also interesting to notice that a less efficient sail plan (e.g. pt.4 of AWA 35°, Figure 28) is aerodynamically kind of equivalent to a sail plain set to a smaller effective apparent wind angle (AWA 20°, Figure 29), which is something that is also known and commonly experienced in sailing. As a matter of fact, great similitude can be found in the results reported in Figure 28 and Figure 29, with the extensively investigated aerodynamics of airfoils with external airfoil flaps [14].

The same considerations can be drawn analyzing pressure distributions along with the sail shapes detected by TOF system. More specifically, in Figure 22 a comparative visualization of the pressure distributions on the effective flying shape, at the corresponding section (L2), is reported for the apparent wind angle AWA 35° and for maximum and minimum efficiency (i.e. pt.1 and 4 of Figure 18).

L2 section - AWA 20°

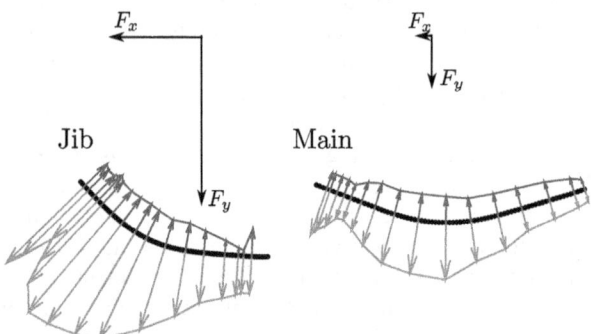

Jib   Main

Figure 21 - Visual pressure distribution on the sail plan based on measurements at section L2, AWA 20°

The pressure distributions are the same that are also reported in Figure 28: it is evident how basically the same jib's shape of the section, which was not modified along the de-power curve of Figure 18, turns out in a different pressure trends due to the different efficiency of the main sail, whose suction effect on the jib is greatly lower in a de-powered regulation. Also the different overall lateral forces Fy depicted in Figure 21, computed from integrating the pressures over the L2 section, confirm the integral measurements reported in Figure 18. Furthermore, in Figure 23, the comparison of different AWA results, both in a maximum power configuration, as reported in Figure 29, for L2 section, are shown in terms pressure/shape visualizations. It is easy to notice that the fairly similar pressure distribution on the jib (Figure 29), due to the maximum power based trimming, is associated to different shape, leading to different driving forces on the jib and comparable lateral forces on the main sail (i.e. Figure 18, for approximately the same Cy, AWA 20° shows a lower driving force coefficient Cx with respect to AWA 35°). Also Figure 24, in which apparent wind directions are highlighted, supports the consideration above reported: more specifically, the less efficient sail plan at AWA 35° shows similarities with lower AWA configurations in terms of main sail pressure distribution. In fact, it is quite evident a reattachment of the flow right behind the mast combined with the co-alignment of the flow with the luff of the mainsail.

It is worth mentioning that these tests represent, according to author's knowledge, one of the first systematic wind tunnel set of measurements of the pressure distributions on flexible sails, instead of rigid, which is a testing condition closer to the real navigation (i.e. full scale), whose measurements (forces, pressures and flying shapes) on SYL, will be taking advantages from the present research.

Furthermore, the rationale behind the way the sails were trimmed, corresponds to a procedure which is consistent to the real sailing navigation. However, a systematic investigation on the aerodynamics of sail plains, relying on this experimental setup, could be based on different depowering approaches, aiming at assessing the effect on

single trim parameters (involving also the jib) on the local and integral efficiency of the sails.

L2 section - AWA 35°

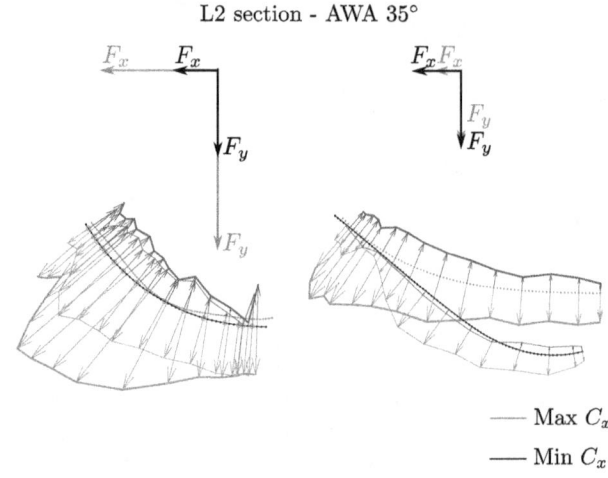

—— Max C_x

—— Min C_x

Figure 22 - Pressure distributions for L2 section for AWA 35°, maximum and minimum power configurations

L2 section

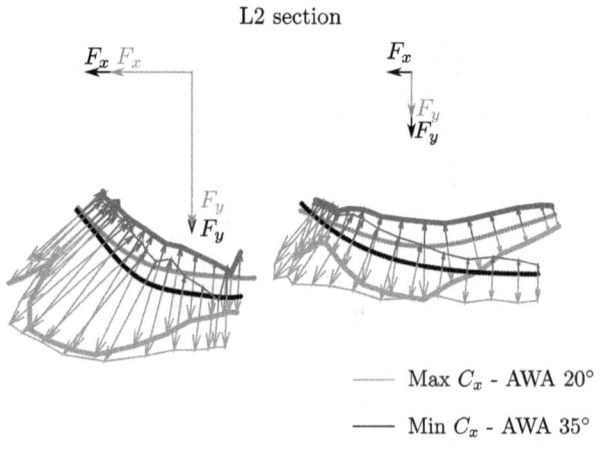

—— Max C_x - AWA 20°

—— Min C_x - AWA 35°

Figure 23 - Pressure distributions for L2 section, maximum power configuration, AWA 20° and 35°

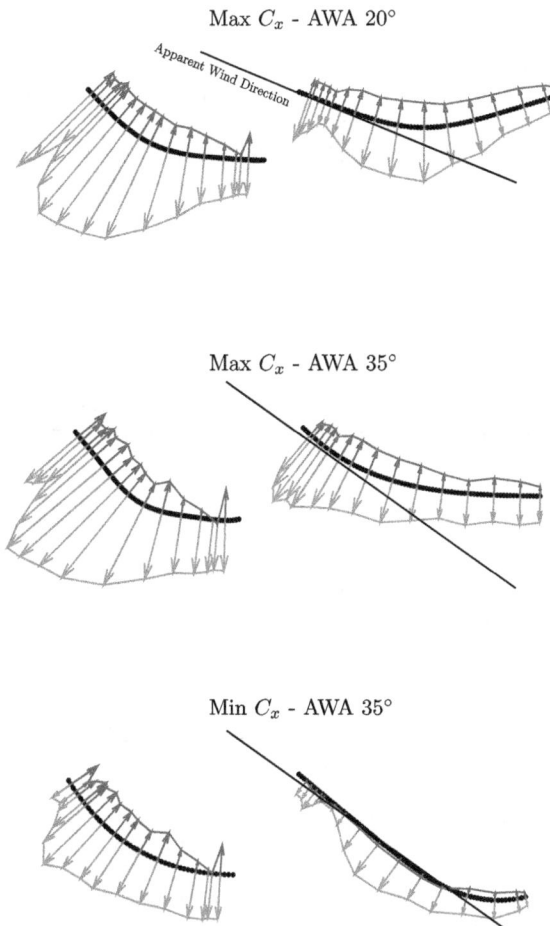

Max C_x - AWA 20°

Apparent Wind Direction

Max C_x - AWA 35°

Min C_x - AWA 35°

Figure 24 - Pressure distributions for various AWA and power configurations

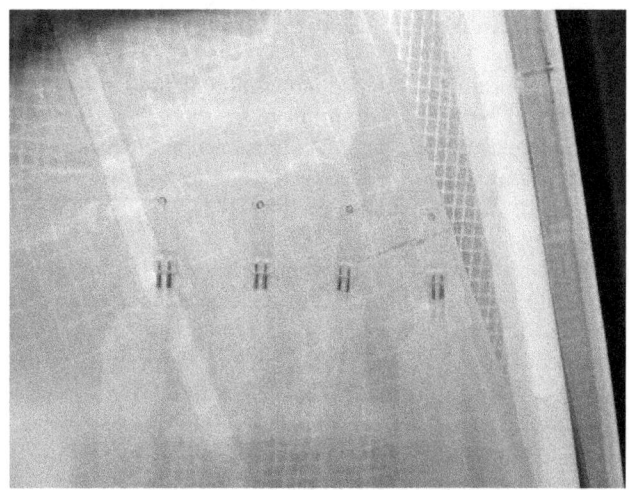

Figure 25 - Example of the full scale pads on gennaker sail for wind tunnel model

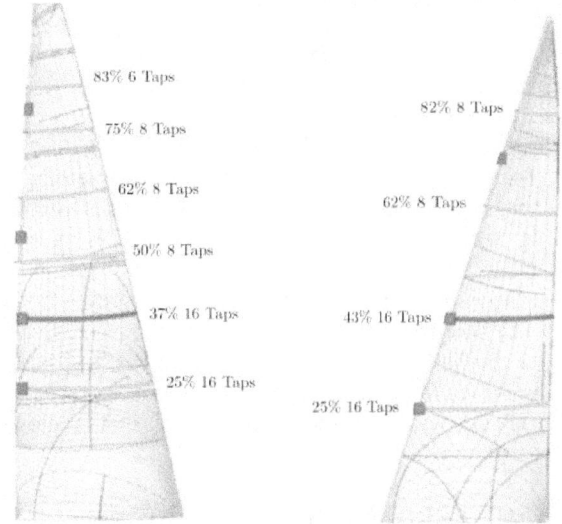

83% 6 Taps
75% 8 Taps
62% 8 Taps
50% 8 Taps
37% 16 Taps
25% 16 Taps

82% 8 Taps
62% 8 Taps
43% 16 Taps
25% 16 Taps

Figure 26 - Full scale mainsail and jib pressure taps layout

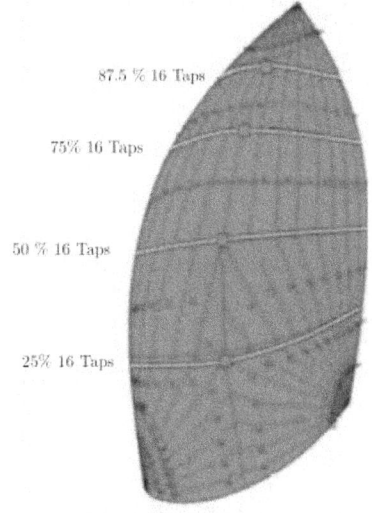

87.5 % 16 Taps
75% 16 Taps
50 % 16 Taps
25% 16 Taps

Figure 27 - Full scale gennaker pressure taps layout

FULL SCALE PRESSURE SYSTEM LAYOUT

On the Sailing Yacht Lab, the pressure distribution on the sails is carried out by means of specifically designed pressure pads which have been designed and produced aiming to provide the differential measurement between the sail leeward and windward side. So that the above reported metrological characterization on the strips and the wind tunnel tests were mainly to assess the capabilities of the measuring system, in advance with respect the final full scale implementation with pads. However, during the wind tunnel tests some tests on pads themselves were conducted just to verify the expected functioning, Figure 25.

With reference to the SYL sail inventory, Figure 26 and Figure 27 show the proposed sail pressure measurement system, the sections considered and a relevant number of pressure taps for the complete sail plane. The full scale pressure system is being finalized at the time of writing this paper.

CONCLUSIONS

A joint project developed among Politecnico di Milano, CSEM and North Sails, aiming at developing a new sail pressure measurement system is presented in the paper. The system was designed for the specific application of full scale measurements on Politecnico di Milano Sailing Yacht Laboratory.

The capabilities of this system were evaluated through a metrological validation of the system alone (static and dynamic tests) and through wind tunnel tests in upwind configuration. The pressure system was integrated in the existent set up, in particular together with the flying shape detection based on Time of Flight (TOF) technology.

Wind tunnel tests allowed both to check the reliability of the new system and to investigate thoroughly upwind soft sails aerodynamics, with the possibility of carrying out regulation as in real navigation. During wind tunnel test session, each side of the sail pressure distributions were measured with reference to a common static pressure, by means of pressure strips, whereas for full scale implementation, pressure pads for differential measurements were developed.

Wind tunnel results give a promising sight of the potentiality of the system described in explaining the dependency of sail plan aerodynamics on the sails trimming, relying on the combined measurements of forces, pressures and flying shapes.

Furthermore, these measurements will represent a great reference database for validation of CFD codes and can be used to complete the interpretation of full scale results.

Further developments in the visualization techniques (PIV) are expected to be combined with this methodology in the near future.

ACKNOWLEDGEMENTS

The authors want to thank the master student Riccardo Romagnoni for his help in performing the calibration tests.

REFERENCES

[1] Fossati F., Bayati I., Orlandini F., Muggiasca S, Vandone A., Mainetti G, Sala R., Bertorello, Begovic E: "A Novel Full Scale Laboratory For Yacht Engineering Research", *Ocean Engineering 104 (2015) 219-237.*

[2] Viola, I.M., Flay, R.G.J., "On-water pressure measurements on a modern asymmetric spinnaker", in *21st HISWA Symposium on Yacht Design and Yacht Construction*, Amsterdam, 2010.

[3] Deparday, J., Bot, P., Hauville, F., Motta, D., Le Pelley, D.J., Flay, R.G.J., Dynamic measurements of pressures, sail shape and forces on a full-scale spinnaker, in 23rd HISWA Symposium on Yacht Design and Yacht Construction, Amsterdam, 2014

[4] Motta, D., Flay, R.G.J., Richards, P., Le Pelley, D.J., Bot, P., Deparday, J., An investigation of the dynamic behavior of asymmetric spinnakers at full-scale, in 5th High Performance Yacht Design, Auckland, 2015

[5] Pot, P., Viola, I.M., Flay, R.G.J., Brett, J.S., Wind-tunnel pressure measurements on model-scale rigid downwind sails, Ocean Engineering, 90, pp. 84-92, 2014

[6] Viola, I.M., Flay, R.G.J., "Sail pressures from full-scale, wind-tunnel and numerical investigations", *Ocean Engineering*, 38 (16), pp.1733–1743, 2011.

[7] Lozej, M., Golob, D., Bokal, D., "Pressure distribution on sail surfaces in real sailing conditions", in *4th High Performance Yacht Design Conference*, Auckland, pp. 242–251, 201, 2012.

[8] Le Pelley, D., Morris, D., Richards, P., "Aerodynamic force deduction on yacht sails using pressure and shape measurements in real time", in *4th High Performance Yacht Design Conference*, Auckland, pp. 28–37, 2012.

[9] Motta, D., Flay, R.G.J., Richards, P.J., Le Pelley, D.J., Deparday, J., Bot, P., "Experimental investigation of asymmetric spinnaker aerodynamics using pressure and sail shape measurements", *Ocean Engineering,* 90, pp. 104-118, 2014.

[10] Fossati, F., Muggiasca, S. Experimental investigation of sail aerodynamic behavior in dynamic conditions (2013) Transactions - Society of Naval Architects and Marine Engineers, 120, pp. 327-367.

[11] Fossati, F., Muggiasca, S., Maria, I., Zasso, A. Wind tunnel techniques for investigation and optimization of sailing yachts aerodynamics (2006) 2nd High Performance Yacht Design Conference 2006, pp. 105-113.

[12] Fossati F., Mainetti G., Sala R., Schito P., Vandone A., " Offwind Sail Flying Shapes Detection, 5th, *High performance Yacht Design Conference*, Auckland, 2015.

[13] Fossati, F., Muggiasca, S., Martina, F. Experimental database of sails performance and flying shapes in upwind conditions, in INNOVSAIL International Conference on Innovation in High Performance Sailing Yachts, Lorient, RINA, pp. 99-114, 2008.

[14] Abbott H., Doenhoff E., Theory of Wing Sections: Including a Summary of Airfoil Data, Dover Publications, Inc. New York, 1959.

[15] Viola I., Flay R., "Pressure Distributions on Sails Investigated Using Three Methods: On-Water Measurements, Wind-Tunnel" *Measurements,and Computational Fluid Dynamics",* 20th Chesapeake Sailing Yacht Symposium, 2011.

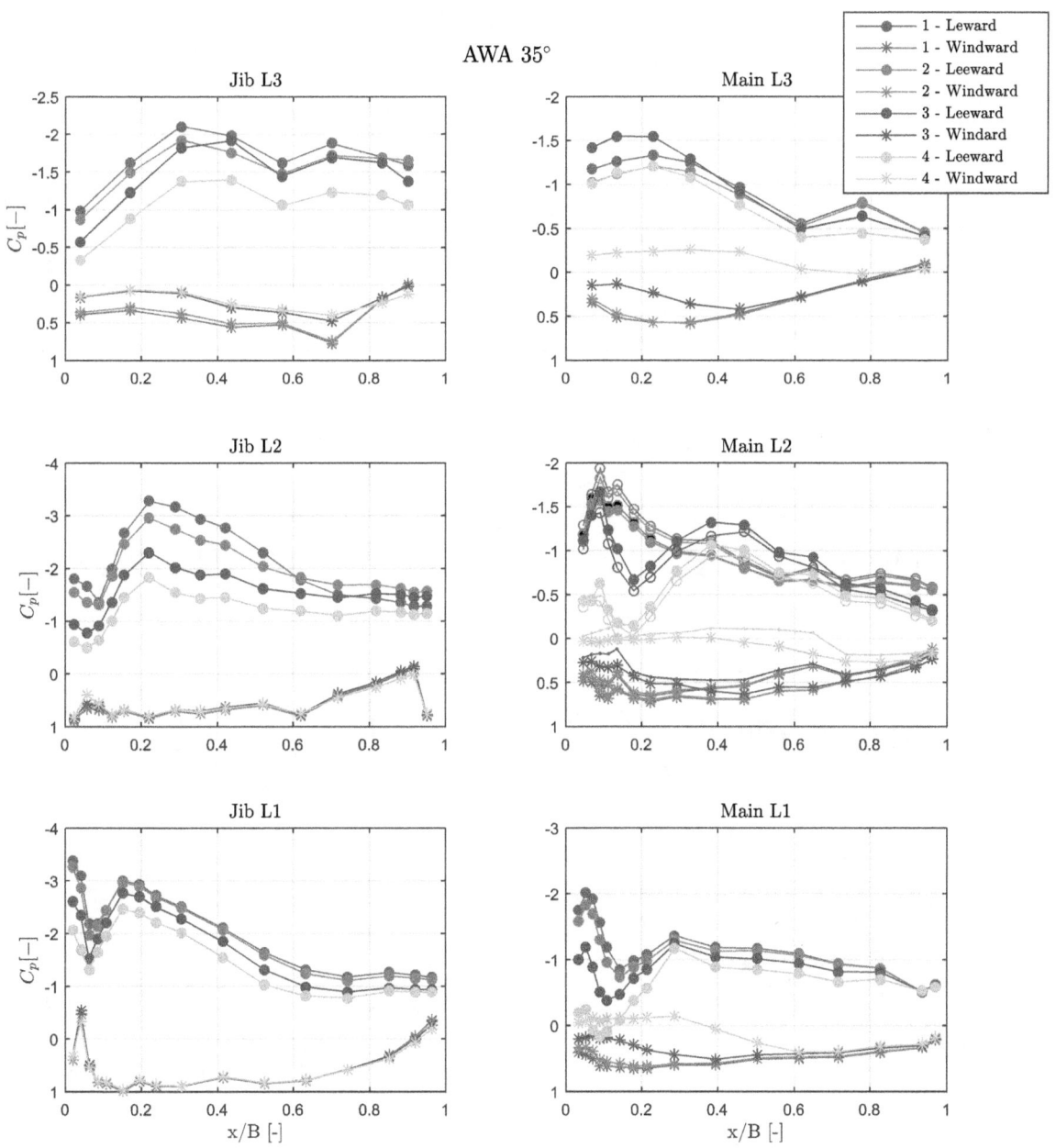

Figure 28 - Pressure distributions on the three stations along the span of the sail plane: points on the maximum power curve for an apparent wind angle AWA of 35°

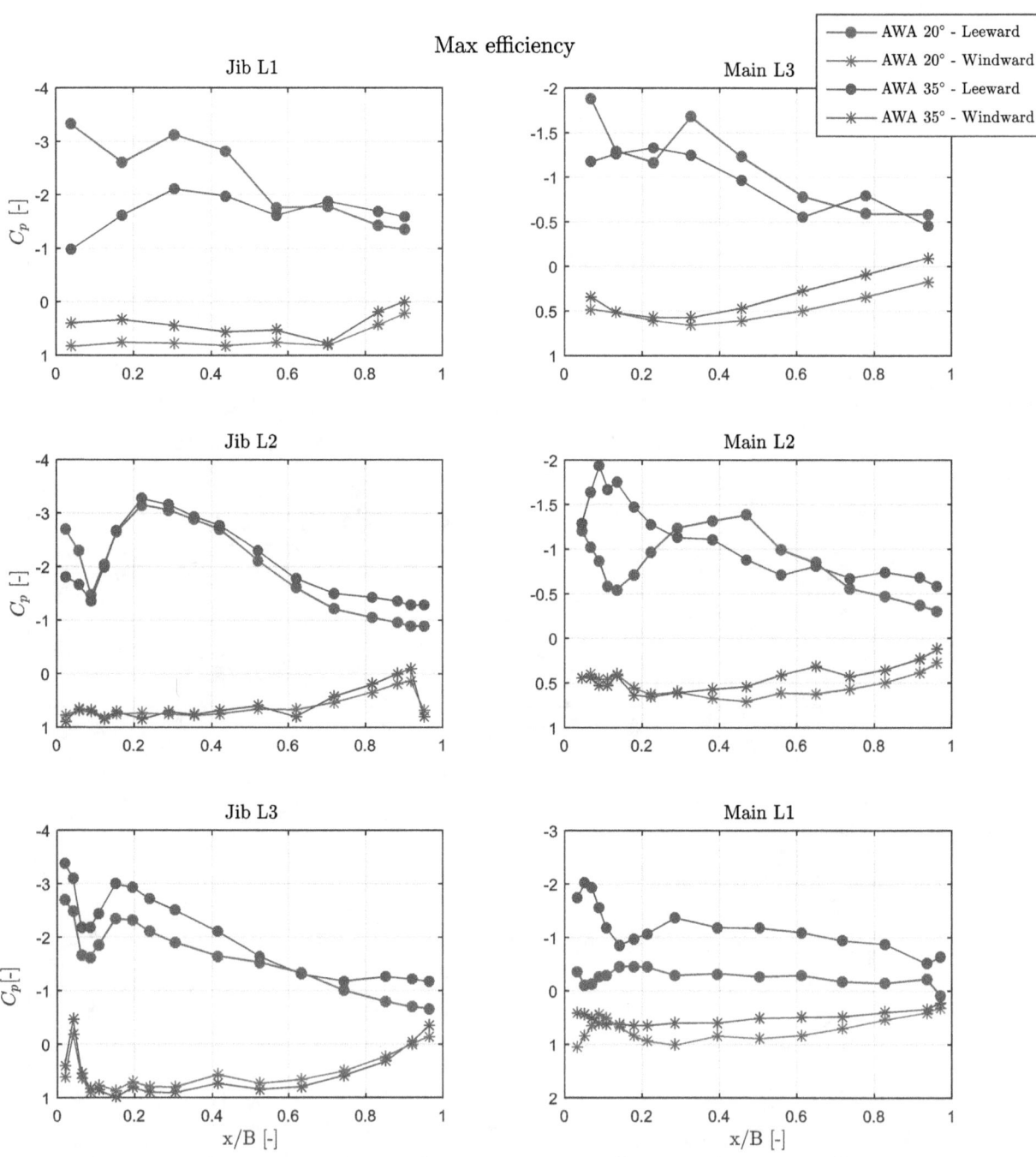

Figure 29 - Comparison of the pressure distribution of two different apparent wind angles AWA (20° and 35°), both considered at the maximum driving force coefficients

Modal Analysis of Pressures on a Full-Scale Spinnaker

Julien Deparday[1], Naval Academy Research Institute, Brest, France
Patrick Bot, Naval Academy Research Institute, Brest, France
Fréderic Hauville, Naval Academy Research Institute, Brest, France
Benoit Augier, Naval Academy Research Institute, Brest, France
Marc Rabaud, Laboratoire FAST, Univ.Paris-Sud, CNRS, Université Paris-Saclay, F-91405, Orsay, France
Dario Motta, University of Auckland, New-Zealand
David Le Pelley, University of Auckland, New-Zealand

ABSTRACT

While sailing offwind, the trimmer typically adjusts the downwind sail "on the verge of luffing", letting occasionally the luff of the sail flapping. Due to the unsteadiness of the spinnaker itself, maintaining the luff on the verge of luffing needs continual adjustments. The propulsive force generated by the offwind sail depends on this trimming and is highly fluctuating. During a flapping sequence, the aerodynamic load can fluctuate by 50% of the average load. On a J/80 class yacht, we simultaneously measured time-resolved pressures on the spinnaker, aerodynamic loads, boat and wind data. Significant spatio-temporal patterns are detected in the pressure distribution. In this paper we present averages and main fluctuations of pressure distributions and of load coefficients for different apparent wind angles as well as a refined analysis of pressure fluctuations, using the Proper Orthogonal Decomposition (POD) method.

POD shows that pressure fluctuations due to luffing of the spinnaker can be well represented by only one proper mode related to a unique spatial pressure pattern and a dynamic behavior evolving with the Apparent Wind Angles. The time evolution of this proper mode is highly correlated with load fluctuations.

Moreover, POD can be employed to filter the measured pressures more efficiently than basic filters. The reconstruction using the first few modes allows to restrict to the most energetic part of the signal and remove insignificant variations and noises. This might be helpful for comparison with other measurements and numerical simulations.

NOTATION

AWA	Apparent Wind Angle
AWS	Apparent Wind Speed
IQR	Inter Quartile Range (Q3-Q1)
Q1	First Quartile
Q3	Third Quartile
POD	Proper Orthogonal Decomposition
C_F	Load Coefficient ($\frac{\text{Load}}{\frac{1}{2}\rho\,\text{S}(\text{AWS})^2}$)
ΔC_P	Differential pressure coefficient ($\frac{P_{leeward}-P_{windward}}{\frac{1}{2}\rho(\text{AWS})^2}$)
ρ	Density of air (1.25 kg/m^3)
S	Sail Area of the asymmetrical Spinnaker (65 m^2)

INTRODUCTION

In research and development in sail aerodynamics, full-scale testing, wind tunnel testing and numerical simulation have always been complementary. Numerical simulation allows to efficiently investigate different designs without the cost of creating sails (Chapin *et al.*, 2011, Durand *et al.*, 2014, Ranzenbach *et al.*, 2013, Viola *et al.*, 2015). Nowadays, advanced computational resources have enhanced numerical simulation and have allowed to couple fluid and structural solvers to create Fluid-Structure Interaction simulations (Chapin *et al.*, 2011, Durand *et al.*, 2014, Ranzenbach *et al.*, 2013, Augier *et al.*, 2014, Lombardi *et al.*, 2012, Renzsch and Graf, 2010, Trimarchi *et al.*, 2013). However wind tunnel testing and full-scale testing are required for comparison and validation (Hansen *et al.*, 2002, Renzsch and Graf, 2013, Viola and Flay, 2011). Wind tunnel testing has the advantage to be in a controlled environment where a balance can be used to measure the forces created by the sails on the boat frame (Campbell, 2014a, Flay, 1996, Graf and Müller, 2009, Zasso *et al.*, 2005). Those results can easily be used to create a Velocity Prediction Program (Campbell, 2014b, Le Pelley and Richards, 2011). Nevertheless with wind tunnel testing, some rules of similitude are violated as the Reynolds number, or the ratio of fabric weight to wind

---
[1]julien.deparday@ecole-navale.fr

pressure or the ratio of membrane stress to wind pressure. Full-Scale testing does not have those issues, and permits to determine yacht performance in real sailing conditions. Those experiments need complex set-up in a harsh environment but actual aerodynamic loads can be assessed in a variety of ways. Sail boat dynamometers (Herman, 1989, Hochkirch and Brandt, 1999, Masuyama, 2014) measured forces from upwind sails transmitted to the boat frame. Fossati *et al.* (2015) created a sail boat dynamometer with the possibility of measuring aerodynamic forces of downwind sails. Augier *et al.* (2012) carried out experiments where loads on the rigging lines and sails were measured. They contributed to a better comprehension on interaction between the wind, the rigging and the sails. Le Pelley *et al.* (2015) measured the forces and the directions on the three corners of spinnakers. Le Pelley *et al.* (2012), Lozej *et al.* (2012), Motta *et al.* (2014), Viola and Flay (2010) measured pressures on sails for upwind and downwind sails.

However downwind sails are more complex to study than upwind sails mainly due to their non-developable 3D shape with highly cambered sections and massively detached flow around a thin and very flexible membrane. Due to the dynamic behavior of this unsteady fluid-structure interaction, the pressures on the sail vary quickly. Even in stable conditions, offwind sails have an inherent unsteadiness. One key feature of spinnaker unsteadiness comes from the flapping at the leading edge, also called luffing. We have previously investigated pressure evolution during luffing (Deparday *et al.*, 2014, Motta *et al.*, 2015). In Deparday *et al.* (2014), we showed an example where flapping of spinnaker creates pressure peaks at the leading edge increasing the aerodynamic force dynamically by 50%. Due to the non-stationarity of the environment while sailing, spatio-temporal pressure data are complex to analyze and therefore to simulate. However significant and different spatio-temporal patterns can be spotted (Motta *et al.*, 2015) and might be produced by different physical causes (Fluid-Structure Interaction, wind variations, boat motions, etc.). In this paper we present an approach to decompose complex pressure evolutions into simpler modes. It would then allow easier analysis and comparison with simulations.

This paper presents results of full-scale experiments of an instrumented J/80 class yacht in offwind conditions were loads, pressures on the spinnaker, boat and wind data were measured. After presenting the experimental apparatus, average and fluctuations of pressures and loads are presented. The next section is the use of the Proper Orthogonal Decomposition (POD) method on pressures to create a simpler model of the complex variations of pressure distribution in time. We show then that the method also helps to highlight the correlation between the main evolution of pressures and the variations of loads.

EXPERIMENTAL SETUP

An instrumented J/80 class sailing yacht, an 8 meter one-design cruiser racer was used during those experiments. A tri-radial asymmetrical spinnaker with a surface of about $65\,m^2$ with a 12 meter long rounded luff was hoisted as well as a mainsail of $17\,m^2$. Boat and wind data, loads on the standing rigging and on the sails were recorded. Moreover pressure taps, developed by the Yacht Research Unit from the University of Auckland were stuck on the spinnaker to acquire the dynamic pressure distribution. They were synchronized with the other data thanks to an acquisition software, RTMaps developed by Intempora which received every signal at their own rate and timestamped them "on the flow". A resampling was applied during the post processing to obtain synchronous data for easier analysis. Figure 1 shows the arrangement of all the sensors set onboard.

This setup for downwind navigation is a further development of the experimental system described in Augier *et al.* (2012) which was used for measurements in upwind navigation.

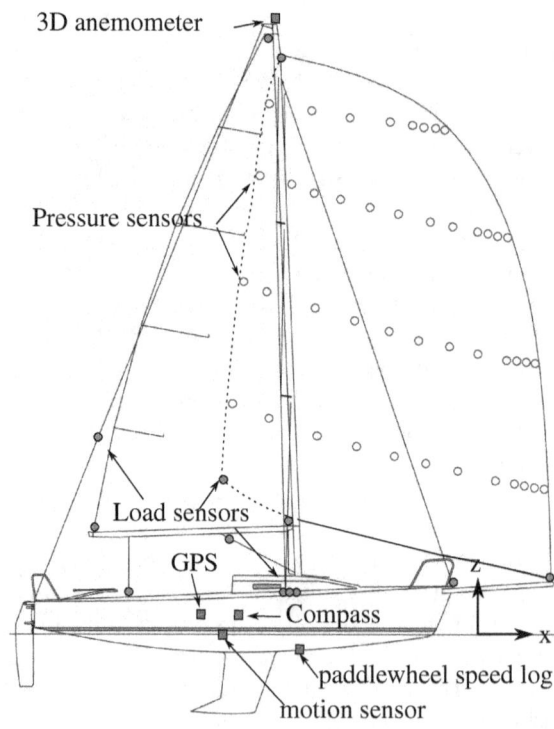

Figure 1: General arrangement of the experimental set-up on the J/80. 16 load sensors (green discs), 44 pressure taps (red circles), and wind and boat sensors (blue squares).

Loads

The standing rigging (shrouds, forestay and backstay) is fitted with custom-made turnbuckles and shackles equipped with strain gages. The running rigging (the corners of the mainsail and of the spinnaker -head, tack and clew-) is equipped with instrumented shackles too. For the standing rigging and the mainsail the sensors are connected to a load acquisition system Spider8 from HBM. Voltages are received from all strain gages and amplified. They are then converted in digital data at a rate of 25 Hz. Thereafter they are transferred to the real-time acquisition software, RTMaps. Due to the high displacements of the spinnaker -in the order of magnitude of 1 to 5 meters for the spinnaker used in those experiments-, the instrumented shackles on the three corners of the spinnaker communicate wirelessly to the acquisition system. The clew sensor is connected via a wire running along the foot of the sail to a small box located near the tack point of the spinnaker (see Figure 2). This box contains two strain gage amplifiers, one for the tack sensor and one for the clew. A microcontroller receives and transmits data at a sampling frequency of 25 Hz to the receiver inside the boat via a wireless and low consumption ZigBee network. Another box is located at the head position for the head instrumented shackle. The delay between the emission and reception of data is insignificant compared to dynamics in sailing.

The errors of measurement are less than 2% of the measurement range (10 000 N for the shrouds, forestay and for the mainsail sheet, 5000 N for the backstay and other instrumented shackles on the mainsail and spinnaker).

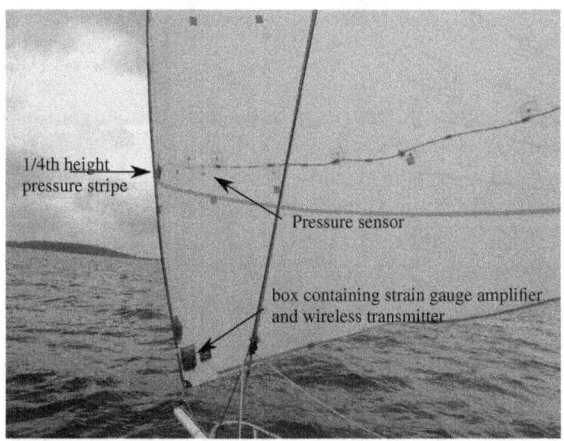

Figure 2: Photograph of the tack of the spinnaker used.

Pressures on spinnaker

On the spinnaker, 44 low range differential pressure sensors (Honeywell XSCL04DC) are located on the surface along 4 horizontal stripes: at 1/4, 1/2, 3/4 and 7/8th height of the spinnaker (see Figures 1 and 2). 12 transducers are used on each of the first 3 stripes and 8 for the top one. There is a higher concentration of pressure taps near the leading edge to be able to record potential leading edge suction peaks. Those sensors measure a difference of pressure between the suction side and the pressure side using the piezoresistive effect. There is no need for a measure of a reference pressure, a complicated task in full-scale experiments. Those sensors are stuck on one side -at the pressure side when sailing on portside tack- and are positioned facing 2mm-diameter holes on the sail to measure the pressure jump across the sail without significant air leak. Punctured light sail cloth patches are applied on the pressure taps to profile the sensors. This custom built pressure system was designed by the Yacht Research Unit at the University of Auckland. These pressure transducers are connected by wires to the receiver inside the boat and thus are synchronized with the other data.

The pressure sensors have a sampling frequency of approximatively 10 Hz with a maximum range of ± 1 kPa and a resolution of 0.5 Pa. The pressure acquisition system is more described in Motta et al. (2014).

Procedure

Sea trials were performed in the bay of Brest, France, offshore Ecole Navale. During those experiments the weather conditions were stable:

- average true wind speed: 6 m/s (12 kn)
- gust: 8 m/s (16 kn)
- wind direction: 270° (westerly wind). Stable. Flat water.

Even in conditions considered as "stable" (with no gust, no wind shift, on flat water and fixed trimming), offwind sails have an inherent unsteadiness, like luffing (flapping at the leading edge). To keep "stable" conditions, a standard procedure must be followed. During those experiments, the controlled inputs were the apparent wind angle and the trim of the spinnaker. Trimmer and helmsman were kept the same for the whole test. The trimmer adjusted the spinnaker at the optimum trim (i.e. on the verge of luffing at the leading edge). The helmsman kept the apparent wind angle as constant as possible.

During the post-processing routine, periods of 5 seconds minimum were labelled "stable" when the standard deviation of the apparent wind angle (AWA) was below 4° and the standard deviation of the apparent wind speed (AWS) was below 10% of the average. Those periods were extended in time as long as those criteria were met. A large range of AWA (between 55° and 140°) is swept by the "stable" periods found, with a certain redundancy for most of the AWA. Each stable period is processed individually.

AVERAGES AND FLUCTUATIONS

Pressures

The average pressure distribution and loads are compared according to the apparent wind angle. A large range of apparent wind angles (AWA) has been met in a rather constant true wind speed (TWS) between 5.8 m/s and 7.1 m/s thus between 11.2 kn to 13.8 kn. The apparent wind angle is measured at the mast head. One should be aware this measure is affected by the twist of the wind, the upwash effect from the sails and the heel.

To display the pressure distribution on the whole sail from discrete measurement points, a linear Radial Basis Function interpolation has been used. On following figures where the pressure distribution is displayed on the spinnaker, blue crosses show where the pressure measurement sensors were located on the sail. Thus pressures at those blue crosses are actual measured values, when the pressure distribution is interpolated between the stripes and pressure taps. With no information on the sail boundaries, values at the top (above 7/8th) and at the bottom (below 1/4th) are extrapolated. The shapes used to display the pressure distributions come from other experiments with the same spinnaker where photogrammetric measurements were carried out to acquire the flying shapes.

Time-Averaged pressure distributions for similar apparent wind angles have good repeatability. Moreover the pressure distribution evolves clearly with the AWA. Figure 3a presents 3 characteristic pressure spatial distributions at 66°, 118° and 140°. It shows the coefficient of the difference of pressure as commonly defined in aerodynamics:

$$\Delta C_P = \frac{P_{\text{leeward}} - P_{\text{windward}}}{\frac{1}{2}\rho(\text{AWS})^2}.$$

At tight angles as AWA 66°, a bulb of high suction is found at the leading edge in the top half of the spinnaker ($\Delta C_P \approx -3$) which produces high aerodynamic force. ΔC_P on the rest of the sail is around -2 increasing to -1/-0.5 at the trailing edge.

At AWA around 110°-120°, the area where the peak of suction occurs is smaller around half of the spinnaker height and the absolute value lower. On the rest of the spinnaker, the pressure coefficient on the spinnaker is rather constant around -1.2 and increasing to -0.5 at the trailing edge.

At AWA 140°, the decrease of suction is even more visible on the whole sail with almost no suction peak at the leading edge and with a reduction of $|\Delta C_P|$ along the flow up to a positive ΔC_P at the trailing edge even on the actual measured points. Positive pressure coefficient means a collapse of the sail at the trailing edge and thus an unstable flying shape. It is consistent with what the authors have noticed during experiments: at large AWA, the spinnaker starts collapsing first at the leech and not at the luff.

While the AWA is increased, not only is a clear decrease of absolute differential pressure coefficient (from -3 down

to 0 about), but also the AWS decreases (from 7 m/s to 3.5 m/s about). So the absolute values of ΔP decrease even more dramatically: At tight AWA, around 65°, the order of magnitude of differential pressure is -40 Pa, and only -4 Pa at large AWA –around 140°-.

Figure 3b shows the standard deviation for the corresponding AWA. Standard deviation on the whole sail is interpolated from the standard deviations calculated on the pressure taps only. Higher standard deviations mean bigger variations of pressure during a "stable" period. Despite a clear difference for the pressure distribution on the whole spinnaker depending on the AWA, pressure variations during the "stable" periods have similar spatial patterns. Strong variations are found at the leading edge, around 1, on the whole height for 66° and 118° while the rest of the spinnaker has a standard deviation of about 0.2. However, while the order of magnitude of standard deviation of ΔC_P is similar for every AWA, the relative variation of pressures compared with the average pressure coefficient varies. Variations are more significant for large AWA (around 120°-140°) than for tight AWA. For tight AWA, the standard deviation is around 30 Pa thus 75% of the average pressure. For large AWA, the standard deviation is around 8 Pa thus 2 times bigger than the average pressure.

Loads

Figure 4 displays the load coefficients on the three corners of the spinnaker according to the apparent wind angle:

$$C_F = \frac{\text{Load}}{\frac{1}{2}\rho\,\text{S}(\text{AWS})^2}$$

with S the sail area of the spinnaker.

In Figure 4, only "stable" periods of 10 seconds minimum are taken. Even though the periods chosen are "stable", loads can vary significantly. Therefore each period for a specific average apparent wind angle is displayed as a box plot. The central red mark is the median, and the edges of the box are the lower and upper quartile. The lower quartile (Q1) splits off the lowest 25% of data from the highest 75%. The upper quartile (Q3) splits off the highest 25% of loads from the lowest 75%. The box represents the interquartile range (IQR = Q3 –Q1). It contains 50% of the loads recorded during one "stable" period. The whiskers show the maximum and minimum loads recorded.

Figure 4 shows also the general trend of the load coefficients on the three corners according to the apparent wind angle. Head and tack have similar evolution with a decrease especially between 110° and 140° respectively from 0.8 to 0.5 and from 0.7 to 0.3. Whereas clew load coefficient is approximatively constant around 0.4.

As explained previously, when the AWA is increased, the AWS decreases. While at clew point, the absolute loads mostly decrease only due to the decrease of the AWS, at

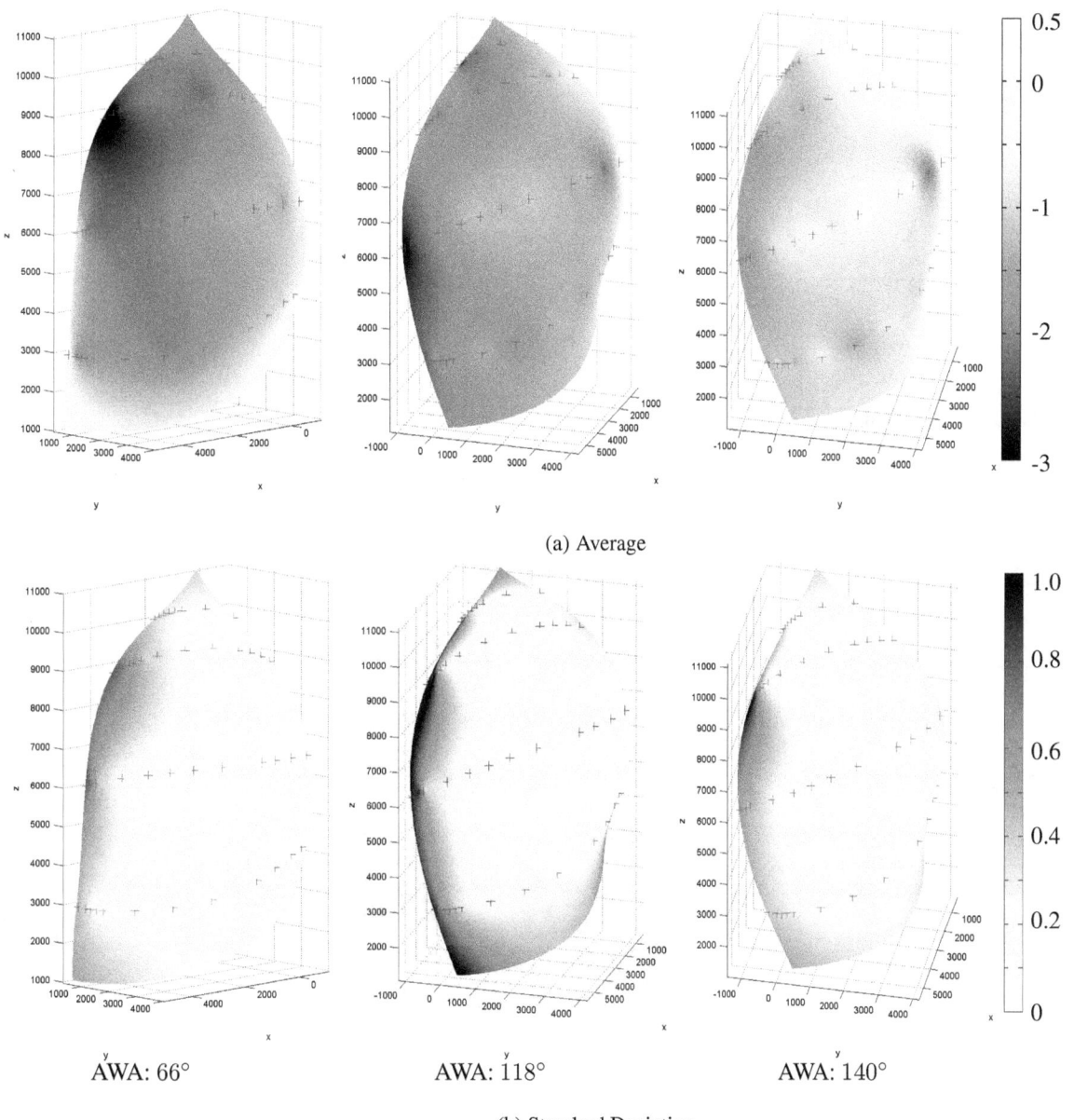

(a) Average

| AWA: 66° | AWA: 118° | AWA: 140° |

(b) Standard Deviation

Figure 3: Pressure distributions, time average pressure coefficient ΔC_P (a) and the fluctuations (b) for 3 typical AWA (66°, 118°, 140°). Blue crosses show the positions of the pressure taps.

tack and head points the absolute loads decrease even more significantly. To confirm this, Figure 5 displays the evolution of load coefficient using the True Wind Speed (TWS) for the non-dimensional coefficient:

$$C_F = \frac{\text{Load}}{\frac{1}{2}\rho \, \text{S(TWS)}^2} \text{ for Figure 5 only.}$$

with the TWS formula from Fossati (2009):

$$TWS = \sqrt{(\text{AWS}\cos(\text{AWA}) - BS)^2 + \overline{(\text{AWS}\,sin(\text{AWA})cos(\text{heel}))^2}}$$

TWS is rather constant for every AWA (around 12 kn). A higher decrease is noticed when the AWA is increased for the head and tack loads than for the clew load which varies only a little.

To analyze variations of load coefficients from an aerodynamic point of view the AWS is used for the non dimensional coefficient C_F. Q3 can be seen as an arbitrary separation between the small variations of loads around the median (inside the IQR) and the peaks of loads (in the top quarter). In Figure 4, the IQR has the same relative range for every AWA, between –10% and +10% of the median for the most loaded corners (head and tack) and between

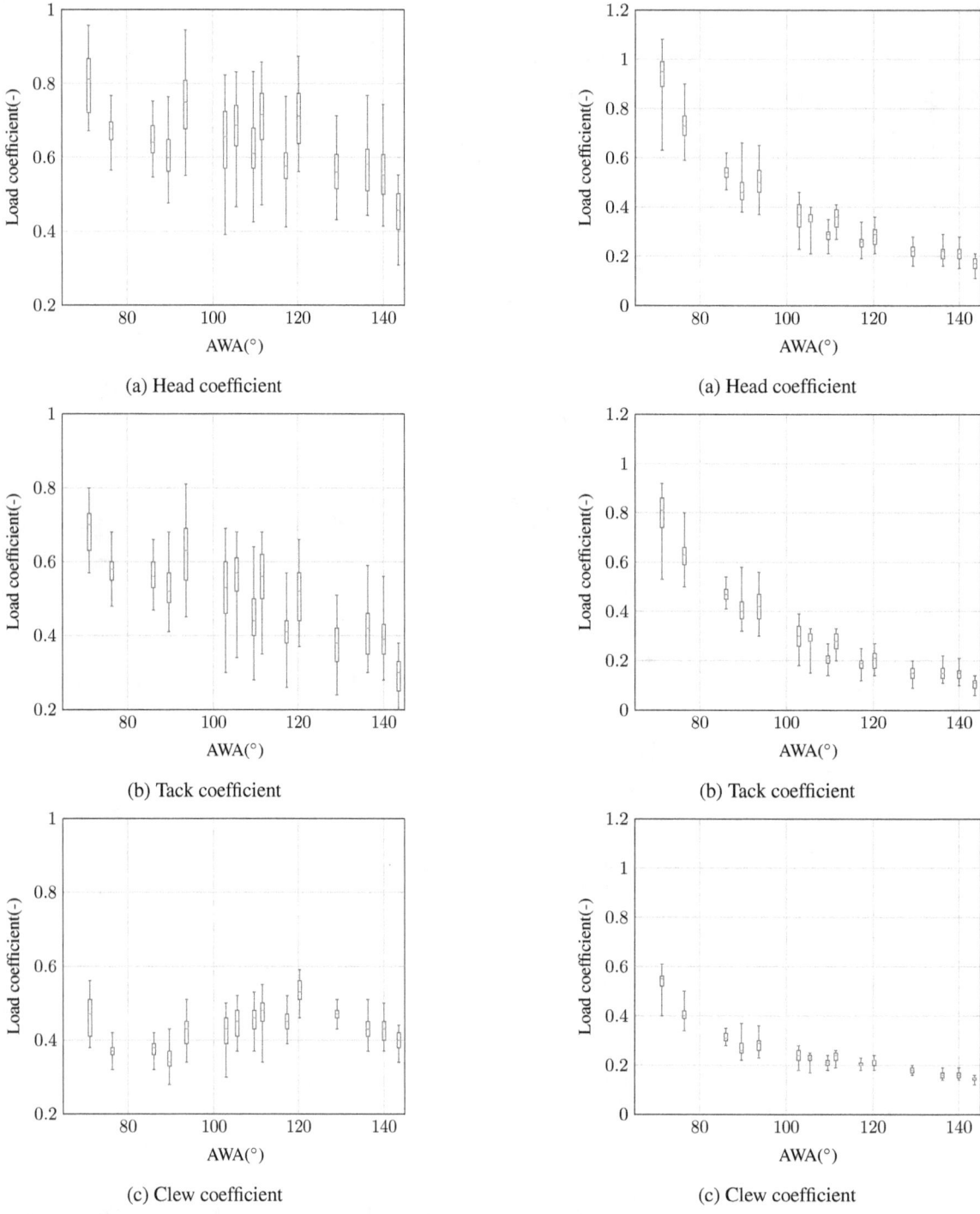

(a) Head coefficient

(b) Tack coefficient

(c) Clew coefficient

Figure 4: Boxplot for load coefficients C_F at the three corners of the spinnaker for different AWA. Central red mark is the median, the box represent the interquartile range between the lower and upper quartiles. It contains 50% of the loads. The upper and lower whiskers indicate the minimum and maximum values.

(a) Head coefficient

(b) Tack coefficient

(c) Clew coefficient

Figure 5: Boxplot for load coefficients C_F with the *TWS* as reference at the three corners of the spinnaker for different AWA. Central red mark is the median, the box represent the interquartile range between the lower and upper quartiles. It contains 50% of the loads. The upper and lower whiskers indicate the minimum and maximum values.

-5% and +5% for the clew. Since the IQR is relatively constant and small, one can conclude that for every AWA, without taking into account peaks of loads, the averaged load coefficient is rather stable and varies slightly.

For the most loaded corners (head and tack) the upper whisker (maximum load) is about 20% higher than the median for tight angles (AWA < 100°) and 30% for large angles (AWA > 120°). For the clew point the upper whisker is always about 15% higher than the median for all AWA. Relative variations of loads seem fairly constant for every AWA, and slightly bigger for large AWA at the head and tack points.

It is interesting to note that most variations of loads are present at the head and tack points, the closest points of the leading edge where the highest variations of pressure occur. While at clew point, the relative variation of loads is smaller.

Those variations of loads and pressures are unsteady even during "stable" periods. However specific patterns might be spotted and might be linked to different causes. The yacht motion and its influence on the apparent wind (pitching and rolling of the boat), gusts (pure aerodynamic cause), vortex shedding, or a change of the spinnaker shape as luffing (unsteady fluid-structure interaction) could make the spinnaker forces vary.

Therefore, we would like to extract patterns in order to decompose complex pressure evolutions into simpler modes. Those pressure modes could help to describe a temporal global behavior in a better way than analyzing each pressure sensor signal, and could be correlated with other recorded data. We decided to use the Proper Orthogonal Decomposition method to characterize the spatial pattern of pressure variations.

DECOMPOSITION INTO MODES

Proper Orthogonal Decomposition method

The Proper Orthogonal Decomposition (POD), is based on the Karhunen-Loeve expansion and also called Principal Component Analysis, PCA. It was first introduced in the context of Fluid Mechanics by (Lumley, 1967). The input data (in our case $\Delta C_P(x,t)$) can be expanded into orthogonal basis functions $\phi_i(x)$ with time coefficient $a_n(t)$:

$$U(x,t) = \sum_n a_n(t) \cdot \phi_n(x).$$

As proper modes are derived from the data itself (data driven decomposition), there is no need of a-priori knowledge or education scheme. Moreover, each basis function has its own amount of fluctuation energy different from each other. These functions are statistically optimal in the least mean-square sense. As a result, fluctuation energy drops down quickly which means a low number of modes is needed in the expansion to reproduce the main variations of the field. POD is a powerful tool for generating lower dimensional models of dynamical systems.

Most of the time, POD is used on the fluctuations of

the input data only. After subtracting the average component (seen as the zeroth mode) from the data, a matrix U is created as a set of N observations (commonly called snapshots) of M records. Each column contains all fluctuating input data (M values) from a specific snapshot and each row contains all snapshots (N snapshots) from a specific measurement point.

$$\mathbf{U} = \begin{bmatrix} u_{11} & u_{12} & \cdots & u_{1N} \\ u_{21} & u_{22} & \cdots & u_{2N} \\ \vdots & \ddots & \ddots & \vdots \\ u_{M1} & u_{M2} & \cdots & u_{MN} \end{bmatrix}$$

Then the auto covariance matrix C (MxM) is calculated as:

$$\mathbf{C} = \mathbf{U} * \mathbf{U}^T$$

because $M >> N$. We have M= 44 measurement points and $N \approx 20000$. However in fluid mechanics, it is common to have $N >> M$ when using PIV or CFD results for example. For those cases the so-called "Snapshot POD" introduced first by (Sirovich, 1987) is used. For our experiments, the "Direct POD" has been applied. The corresponding eigenvalue problem of the auto covariance matrix is solved:

$$\mathbf{C} * \mathbf{\Phi} = \lambda * \mathbf{\Phi}$$

The eigenvectors $\Phi(i)$ are the POD modes. POD modes are sorted in descending order according to their corresponding eigenvalue $\lambda(i)$ which represent their energy. The POD mode with the highest corresponding eigenvalue is mode 1. The expansion coefficient (or mode time coefficient) is calculated as follows:

$$\mathbf{a} = \mathbf{U}^T * \mathbf{\Phi}$$

POD results

Following results presented here are for a "stable" period with an average AWA of 69°, but is representative to what we have observed for different periods at different AWA. This point will be discussed further in the article. Figure 6 shows the energy distribution for each POD mode. The first mode contains almost 45% of the fluctuation energy. Mode 2 and 3 represent only 15% each. And most of the time other modes have less than 5% of the fluctuation energy. It is clear that the first mode is dominant compared to the others.

The pressure distribution evolution can be simplified by taking only the first terms of the expansion. The reconstruction using the first modes allows to only keeping the most energetic part of the signal and removing insignificant variations and noises. The precision of the reconstruction has been calculated according to the number of modes. With mode 0 (the average) and mode 1, 85% of the signal is already reconstructed. With 3 modes, the error of reconstruction of the pressure signals is 10%. About 10 modes are

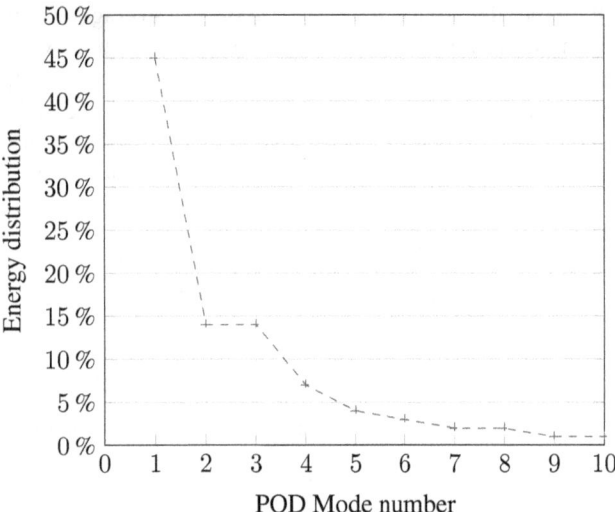

Figure 6: Energy distribution of the fluctuations for the first 10 modes for AWA 69°.

required to achieve a reconstruction with less than 5% of difference.

Figure 7 presents the first 3 modes. The scale value is arbitrary. To represent a fluctuation of Cp, they must be multiplied by the corresponding mode time coefficient depending on the time - which can be positive or negative - (presented in Figure 9). Mode 1 has a bulb of pressure on the top half of the spinnaker at the leading edge and a smaller bulb of opposite sign on the bottom half height of the spinnaker. Mode 2 is similar with the standard deviation pattern presented before. Mode 3 and further modes display smaller coherent patterns and may change with the period used.

Table 1: Energy distribution of the fluctuations for the first 10 modes for different AWA.

Modes	66 deg	98 deg	120 deg	140 deg
1	43%	46%	45%	45%
2	21%	16%	19%	15%
3	8%	9%	10%	11%
4	7%	6%	6%	7%
5	4%	5%	3%	5%
6	3%	4%	3%	3%
7	2%	3%	2%	2%
8	2%	2%	2%	2%
9	2%	2%	2%	2%
10	1%	1%	1%	1%

Table 1 shows the energy distribution for different "stable" periods at different AWA. The ratio of energy of each mode number is rather constant for every AWA. Moreover each mode number has a similar pattern of pressure distribution, even though it happens that mode 2 and mode 3 are inverted in a few cases.

To compare modes, the maximum value of time coef-

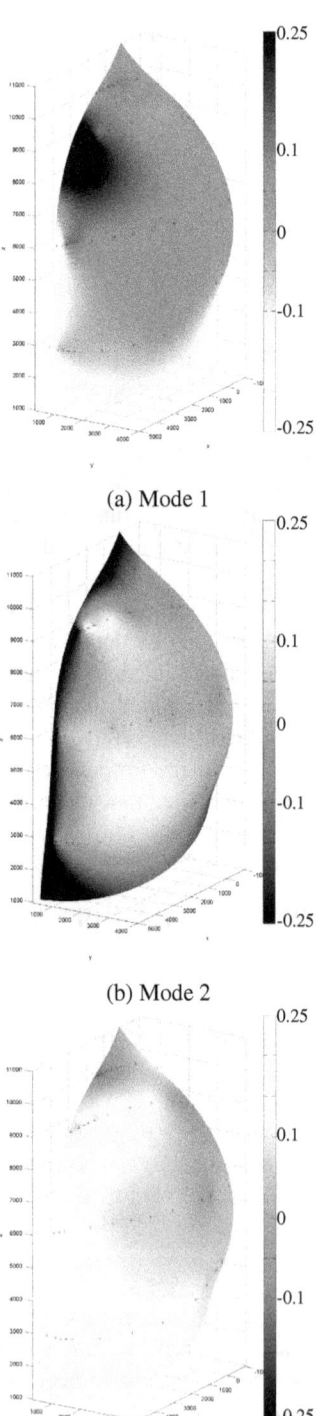

(a) Mode 1

(b) Mode 2

(c) Mode 3

Figure 7: First 3 spatial orthogonal modes from POD method for AWA 69°. Mode 1 is the most energetic mode.

ficient is taken to be multiplied by the spatial mode: $\max(a_1(t)) \cdot \phi_1(x)$. Maximum value of time coefficient is used since we want to analyze and compare dominant variations. Moreover, by mathematical definition: $\forall n, \overline{a_n(t)} = 0$ and thus is irrelevant to be used for comparison. Maximum values of mode 1 for different AWA are presented in Figure 8 at 4 different stripes where the pressures are measured (1/4, 1/2, 3/4, 7/8th height of the spinnaker). Differential pressure coefficients have comparable shapes for every AWA, except that $|\Delta C_P|$ on the bottom half is slightly smaller for deeper AWA. The bulb of suction at the leading edge at 7/8 and 3/4 of the spinnaker height is always present and a smaller bulb of positive ΔC_P at 1/2 and 1/4 height is also spotted. Even if the POD method is a data driven decomposition (i.e. modes are derived from the data itself), there is a good repeatability of POD modes for "stable" periods when the spinnaker has a fixed trim. Moreover mode 1, which plays an important role in the fluctuation of pressures, could be defined as a unique mode whatever the AWA.

POD modes evolve in time. When the time coefficient of a corresponding mode is at an extremum, the corresponding mode is then preponderant. Analyzing time coefficients would then help to link pressure variations with other recorded data.

Figure 9 shows the evolution of the time coefficient for the first 3 modes. Amplitudes of mode 1 are bigger than the other modes as expected due to the larger energy it possesses. Mode 1 and mode 2 are slightly correlated; mode 2 is shifted of a quarter of a pseudo-period. It means when mode 1 is maximal, mode 2 is null. A typical pseudo period for mode 1 stands out for this AWA 69°. Furthermore, for different "stable" periods, at different AWA -not displayed here-, similar variations of the temporal coefficient of mode 1 are detected. However the dynamics change with the AWA. The pseudo-period is measured, and the corresponding frequency is displayed in Table 2 with the corresponding average AWS of the "stable" period. The reduced frequency is calculated as follows:

$$f_r = \frac{f_s \cdot \sqrt{S}}{AWS}$$

with S the sail area, thus $\sqrt{S} = 8.3$ m.
When the AWA is increased, the typical pseudo frequency of the time coefficient of mode 1 is reduced by a factor of 2.5. The AWS decreases at a similar rate. A ratio of 2.1 is found between AWA 57° and 140°. Therefore the reduced frequency is nearly constant with a small decrease. Figure 10 displays the pseudo-frequency of the time coefficient of mode 1 as a function of the AWS. There is a linear dependence of the pseudo-frequencies with the AWS. It demonstrates that mode 1 is mostly driven by aerodynamic phenomena as expected, and not by mechanical resonance of the rigging or of the membrane of the sail.

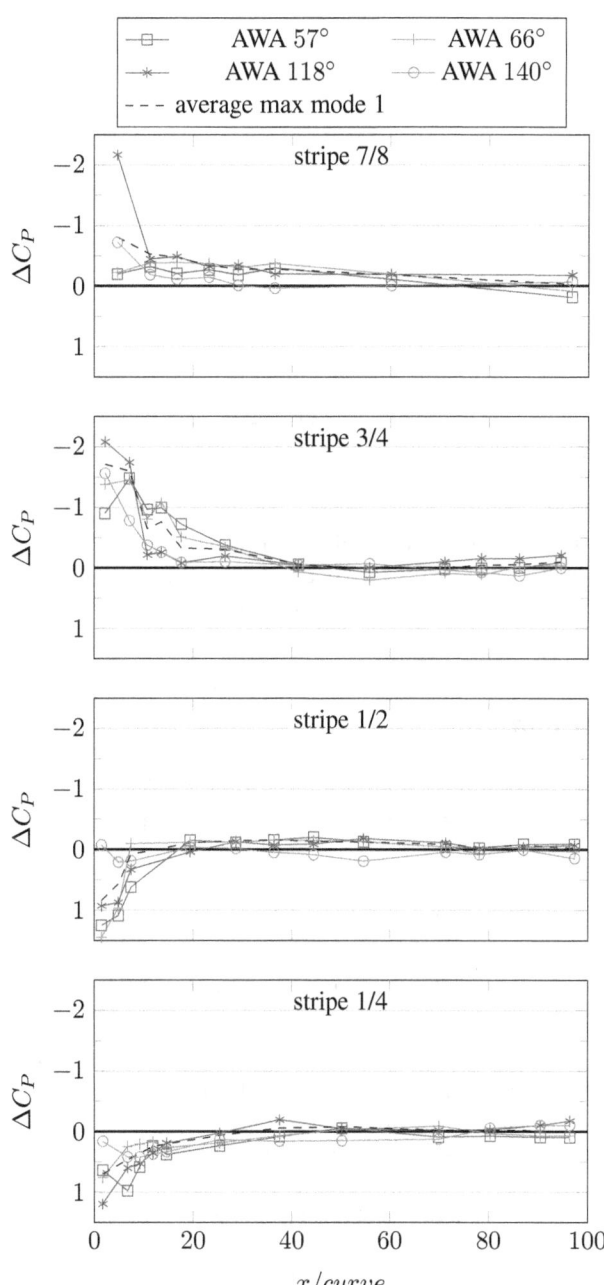

Figure 8: Mode 1 for maximum time coefficient $max(a_1(t) \cdot \phi_1(x)$ for different AWA.

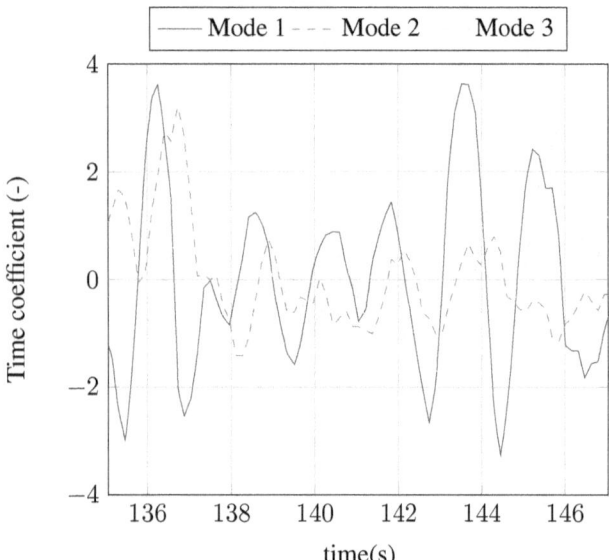

Figure 9: Time coefficient $a_n(t)$ for the first three modes for AWA 69°.

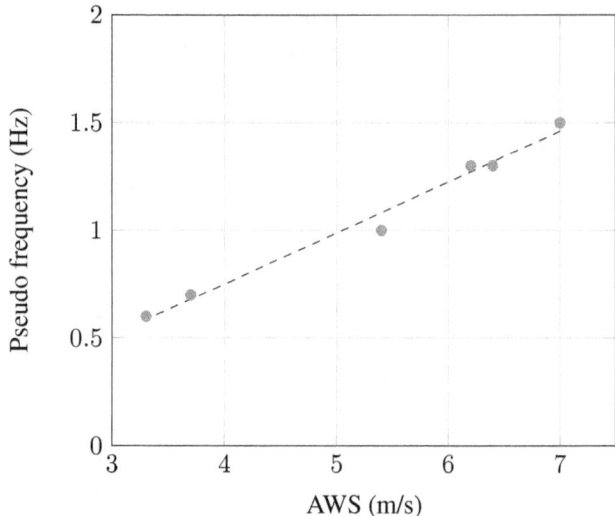

Figure 10: Pseudo-frequency of the time coefficient of the first POD mode (red dots) for different AWA as a function of the AWS. A linear interpolations fits the experimental data.

In conclusion, the spatial pattern of mode 1 does not change with the AWA, but only the temporal dynamics with an increase of the pseudo-period when the AWA increases.

Table 2: Pseudo-frequency and reduced frequency for different AWA.

AWA(°)	f_s(Hz)	AWS(m/s)	f_r(-)
57	1.5	7.0	1.78
66	1.3	6.4	1.69
69	1.3	6.2	1.73
98	1	5.4	1.53
118	0.7	3.7	1.57
140	0.6	3.3	1.52

Table 3 presents the cross-correlation of all data measured for the specific period presented in (Deparday *et al.*, 2014) where loads and flapping were strongly correlated. The normalized cross-correlation is calculated with the time coefficients of the first three modes. Cross correlation between two signals X and Y is defined as follow:

$$C_{xy}(\tau) = \mathcal{E}[(X(t_2) - \mu_{X(t_2)}) - (Y(t_1) - \mu_{Y(t_1)})]$$

where $\mathcal{E}[\cdot]$ is the expected value operator, $\tau = t_2 - t_1$ is the shift applied between two signals. μ_X and μ_{UY} are the mean functions.

The cross correlation matrix is calculated to determine the correlations of every signal with each other. The values are between 0 when not correlated at all -in white in the table- and 1 when signals have the same dynamics -in red in the table. Colors in Table 3 highlight the correlations between experimental data. The diagonal represents the auto-correlation of every signal.

A very strong correlation is present between the loads except with the forestay and the shroud D1 leeward as they are not loaded and slack. Loads are also slightly correlated with the yaw. The shift between the two signals is about 1 second in advance for the loads. A peak of aerodynamic loads might modify the aerodynamic center of effort and thus change the equilibrium of the sailing yacht and make the course vary.

The time coefficient of mode 1 is also well correlated with the loads (around 0.9). There is no delay between mode 1 and the spinnaker aerodynamic loads. Spinnaker aerodynamic loads are 0.1 s in advance of the standing rigging loads. In this case the pressure evolution is instantaneously transmitted to the corners of the spinnaker which then transmit this increase of loads to the shrouds and backstay. Mode 1 is a good parameter to define peaks of loads due to flapping of the spinnaker.

Here mode 2 is reasonably correlated with mode 1 with a coefficient of 0.72. The shift between the two modes is 0.4 s. As explained previously with Figure 9, mode 2 is shifted of a quarter of a pseudo-period. Thus the pseudo-frequency for this specific period should be around $T = 4 \cdot 0.4\,\text{s} = 1.6\,\text{s}$. This pseudo-period corresponds to what was presented in Deparday *et al.* (2014).

In conclusion, mode 1 describes well the flapping of the luff. It can be represented as a unique spatial pattern for every AWA and thus is a good indicator of the dynamics of the spinnaker. Its temporal pseudo-period slows down as the AWS decreases when the AWA is increased.

Table 3: Cross-correlation between different signals recorded during experiments, and first 3 modes of the POD. Correlation values (and their corresponding colors) vary between 0 (in white) meaning no correlation and 1 (in red), same dynamics.

	forestay	backstay	V1 windw'd	D1 windw'd	D1 leew'd	Head	Tack	Clew	mode 1	mode 2	mode 3	roll	pitch	yaw	AWA	AWS	BS
forestay	1.00	0.58	0.62	0.63	0.78	0.59	0.59	0.59	0.65	0.62	0.59	0.60	0.69	0.51	0.73	0.59	0.40
backstay	0.58	1.00	0.88	0.92	0.62	0.96	0.95	0.93	0.91	0.66	0.45	0.58	0.61	0.75	0.58	0.49	0.57
V1 windw'd	0.62	0.88	1.00	0.99	0.79	0.94	0.94	0.94	0.85	0.66	0.41	0.48	0.61	0.68	0.63	0.50	0.44
D1 windw'd	0.63	0.92	0.99	1.00	0.79	0.95	0.95	0.95	0.88	0.68	0.41	0.54	0.61	0.71	0.63	0.54	0.46
D1 leew'd	0.78	0.62	0.79	0.79	1.00	0.69	0.70	0.76	0.66	0.53	0.52	0.49	0.52	0.65	0.60	0.54	0.44
Head	0.59	0.96	0.94	0.95	0.69	1.00	1.00	0.98	0.90	0.63	0.47	0.58	0.57	0.71	0.63	0.53	0.57
Tack	0.59	0.95	0.94	0.95	0.70	1.00	1.00	0.98	0.88	0.61	0.45	0.54	0.55	0.71	0.62	0.51	0.58
Clew	0.59	0.93	0.94	0.95	0.76	0.98	0.98	1.00	0.85	0.65	0.44	0.63	0.58	0.70	0.67	0.61	0.60
mode 1	0.65	0.91	0.85	0.88	0.66	0.90	0.88	0.85	1.00	0.72	0.43	0.52	0.63	0.74	0.54	0.47	0.38
mode 2	0.62	0.66	0.66	0.68	0.53	0.63	0.61	0.65	0.72	1.00	0.46	0.64	0.84	0.60	0.68	0.48	0.29
mode 3	0.59	0.45	0.41	0.41	0.52	0.47	0.45	0.44	0.43	0.46	1.00	0.66	0.41	0.45	0.56	0.75	0.51
roll	0.60	0.58	0.48	0.54	0.49	0.58	0.54	0.63	0.52	0.64	0.66	1.00	0.83	0.83	0.77	0.88	0.47
pitch	0.69	0.61	0.61	0.61	0.52	0.57	0.55	0.58	0.63	0.84	0.41	0.83	1.00	0.76	0.85	0.68	0.28
yaw	0.51	0.75	0.68	0.71	0.65	0.71	0.71	0.70	0.74	0.60	0.45	0.83	0.76	1.00	0.66	0.56	0.46
AWA	0.73	0.58	0.63	0.63	0.60	0.63	0.62	0.67	0.54	0.68	0.56	0.77	0.85	0.66	1.00	0.58	0.52
AWS	0.59	0.49	0.50	0.54	0.54	0.53	0.51	0.61	0.47	0.48	0.75	0.88	0.68	0.56	0.58	1.00	0.50
BS	0.40	0.57	0.44	0.46	0.44	0.57	0.58	0.60	0.38	0.29	0.51	0.47	0.28	0.46	0.52	0.50	1.00

CONCLUSIONS

Full-Scale experiments were carried out on an instrumented J/80 sailing yacht where loads, pressure distribution on the spinnaker boat and wind data were measured at different downwind angles.

We show pressure coefficients and load coefficients decrease when the AWA is increased, whereas variations of load and pressure coefficients are mainly constant. Therefore variations relative to the average loads or pressures are bigger for larger AWA. Moreover we found that most of the pressure and load variations are mainly at the luff for every AWA.

A POD analysis has been used on pressure signals in order to identify the most energetic patterns. The first mode of the POD method is uniquely identified for every AWA, and is well associated with the flapping of the luff producing correlated variations of loads on the 3 corners of the spinnaker. It has a unique spatial mode of pressure fluctuations for all AWA. Only the temporal behavior differs with the AWA. An identified typical pseudo-period of flapping has been calculated thanks to this first mode. It shows a linear decrease of the pseudo-frequency with the AWS. Flapping of the luff might be the result of an aerodynamic phenomenon.

Moreover POD enables to characterize a global unsteady behavior instead of analyzing all local pressure time series

This paper presented a way to characterize the pressure evolution due to flapping. It permits to show to sail designers where the highest variations of pressure occur when flapping. Moreover thanks to the POD method, the first modes allow to reconstruct a signal with the main variations (i.e. with the most energetic part) and remove noises. Therefore comparison with other measurements or numerical simulations is simplified.

ACKNOWLEDGEMENTS

This project has received funding from the European Union's Seventh Programme for research, technological development and demonstration under grant agreement No PIRSES-GA-2012-318924, and from the Royal Society of New Zealand for the UK-France-NZ collaboration project SAILING FLUIDS.

References

CHAPIN, V.G., DE CARLAN, N. and HEPPEL, P., *A Multidisciplinary Computational Framework for Sailing Yacht Rig Design & Optimization through Viscous FSI*, 20th

Chesapeake Sailing Yacht Symposium, March, (1–17), Annapolis (2011).

DURAND, M., LEROYER, A., LOTHODÉ, C., HAUVILLE, F., VISONNEAU, M., FLOCH, R. and GUILLAUME, L., *FSI investigation on stability of downwind sails with an automatic dynamic trimming*, Ocean Engineering, **90**, (2014), 129–139.

RANZENBACH, R., ARMITAGE, D. and CARRAU, A., *Mainsail Planform Optimization for IRC 52 Using Fluid Structure Interaction*, 21st Chesapeake Sailing Yacht Symposium, March, (50–58), Annapolis (2013).

VIOLA, I.M., BIANCOLINI, M.E., SACHER, M. and CELLA, U., *A CFD-based wing sail optimisation method coupled to a VPP*, 5th High Performance Yacht Design Conference, (1–7), Auckland (2015).

AUGIER, B., HAUVILLE, F., BOT, P., AUBIN, N. and DU-RAND, M., *Numerical study of a flexible sail plan submitted to pitching: Hysteresis phenomenon and effect of rig adjustments*, Ocean Engineering, **90**, (2014), 119–128.

LOMBARDI, M., CREMONESI, M., GIAMPIERI, A. and PAROLINI, N., *A strongly coupled fluid-structure inter-action model for wind-sail simulation*, 4th High Perfor-mance Yacht Design Conference, (212–221), Auckland (2012).

RENZSCH, H. and GRAF, K., *Fluid Structure Interaction Simulation of Spinnakers–Towards Simulation Driven Sail Design*, 21st HISWA Symposium on Yacht Design and Yacht Construction (2010).

TRIMARCHI, D., VIDRASCU, M., TAUNTON, D., TURNOCK, S. and CHAPELLE, D., *Wrinkle development analysis in thin sail-like structures using MITC shell fi-nite elements*, Finite Elements in Analysis and Design, **64**, (2013), 48–64.

HANSEN, H., JACKSON, P. and HOCHKIRCH, K., *Com-parison of Wind Tunnel and Full-Scale Aerodynamic Sail Force measurements*, High Performance Yacht Design Conference, Auckland (2002).

RENZSCH, H. and GRAF, K., *An experimental validation case for fluid-structure-interaction simulations of down-wind sails*, 21st Chesapeake Sailing Yacht Symposium, March, (59–66), Annapolis (2013).

VIOLA, I.M. and FLAY, R.G., *Sail pressures from full-scale, wind-tunnel and numerical investigations*, Ocean Engineering, 38(16), (2011), 1733–1743.

CAMPBELL, I.M.C., *A comparison of downwind sail coef-ficients from tests in different wind tunnels*, Ocean Engi-neering, **90**, (2014a), 62–71.

FLAY, R.G., *A twisted flow wind tunnel for testing yacht sails*, Journal of Wind Engineering and Industrial Aero-dynamics, **63**(1-3), (1996), 171–182.

GRAF, K. and MÜLLER, O., *Photogrammetric Investiga-tion of the Flying Shape of Spinnakers in a Twisted Flow Wind Tunnel*, 19th Cheasapeake Sailing yacht Sympo-sium, March, Annapolis (2009).

ZASSO, A., FOSSATI, F. and VIOLA, I.M., *Twisted Flow Wind Tunnel Design for Yacht Aerodynamic Studies*, EACWE4 - The 4th European & African Conference on Wind Engineering, (1–14), Prague (2005).

CAMPBELL, I.M.C., *Comparison of downwind sailing per-formance predicted from wind tunnel tests with full-scale trials from America's Cup class yachts*, 23rd HISWA Symposium on Yacht Design and Yacht Construction (2014b).

LE PELLEY, D. and RICHARDS, P., *Effective Wind Tunnel Tesing of Yacht Sails Using a Real-Time Velocity Predic-tion Program*, 20th Cheasapeake Sailing Yacht Sympo-sium, Annapolis (2011).

HERMAN, J.S., *A sail force dynamometer: design, imple-mentation and data handling*, Ph.D. thesis, Massachus-sets Institute of Technology (1989).

HOCHKIRCH, K. and BRANDT, H., *Fullscale hydrody-namic force measurement on the Berlin sailing dy-namometer*, 14th Chesapeake Sailing Yacht Symposium, (33–44), Annapolis (1999).

MASUYAMA, Y., *The work achieved with the sail dy-namometer boat "Fujin", and the role of full scale tests as the bridge between model tests and CFD*, Ocean Engi-neering, **90**, (2014), 72–83.

FOSSATI, F., BAYATI, I., ORLANDINI, F., MUG-GIASCA, S., VANDONE, A., MAINETTI, G., SALA, R., BERTORELLO, C. and BEGOVIC, E., *A novel full scale laboratory for yacht engineering research*, Ocean Engi-neering, **104**, (2015), 219–237.

AUGIER, B., BOT, P., HAUVILLE, F. and DURAND, M., *Experimental validation of unsteady models for fluid structure interaction: Application to yacht sails and rigs*, Journal of Wind Engineering and Industrial Aerodynam-ics, **101**, (2012), 53–66.

LE PELLEY, D.J., RICHARDS, P.J. and BERTHIER, A., *De-velopment of a directional load cell to measure flying sail aerodynamic loads*, 5th High Performance Yacht Design Conference, (66–75), Auckland (2015).

LE PELLEY, D., MORRIS, D. and RICHARDS, P., *Aerody-namic force deduction on yacht sails using pressure and shape measurements in real time*, 4th High Performance Yacht Design Conference, (28–37), Auckland (2012).

109

LOZEJ, M., GOLOB, D. and BOKAL, D., *Pressure distribution on sail surfaces in real sailing conditions*, 4th High Performance Yacht Design Conference, (242–251), Auckland (2012).

MOTTA, D., FLAY, R.G., RICHARDS, P.J., LE PELLEY, D.J., DEPARDAY, J. and BOT, P., *Experimental investigation of asymmetric spinnaker aerodynamics using pressure and sail shape measurements*, Ocean Engineering, **90**, (2014), 104–118.

VIOLA, I.M. and FLAY, R.G., *On-water pressure measurements on a modern asymmetric spinnaker*, 21st HISWA Symposium on Yacht Design and Yacht Construction, November, Amsterdam (2010).

DEPARDAY, J., BOT, P., HAUVILLE, F., MOTTA, D., LE PELLEY, D.J. and FLAY, R.G.J., *Dynamic measurements of pressures, sail shape and forces on a full-scale spinnaker*, 23rd HISWA Symposium on Yacht Design and Yacht Construction, Amsterdam (2014).

MOTTA, D., FLAY, R., RICHARDS, P., PELLEY, D.L., BOT, P. and DEPARDAY, J., *An investigation of the dynamic behaviour of asymmetric spinnakers at full-scale*, 5th High Performance Yacht Design Conference, (76–85), Auckland (2015).

FOSSATI, F., Aero-hydrodynamics and the performance of sailing yachts, International Marine / Mc Graw Hill (2009).

LUMLEY, J., *The structure of inhomogeneous turbulence*, A. Yaglom and V. Tatarski, eds., Atmospheric Turbulence and Wave Propagation, (166–178), Nauka, Moscow (1967).

SIROVICH, L., *Turbulence and the dynamics of coherent structures part i: coherent structures*, Quarterly of Applied Mathematics, **XLV**(3), (1987), 561–571.

Wind tunnel investigation of dynamic trimming on upwind sail aerodynamics

Aubin N., Naval academy research Institut - IRENAV, France [1]

Augier B., Naval academy research Institut - IRENAV, France

Bot P., Naval academy research Institut - IRENAV, France

Hauville F., Naval academy research Institut - IRENAV, France

Sacher M., Naval academy research Institut - IRENAV, France

Flay R. G. J., Yacht Research Unit, Department of Mechanical Engineering, The University of Auckland, New Zealand

ABSTRACT

An experiment was performed in the Yacht Research Unit's Twisted Flow Wind Tunnel (University of Auckland) to test the effect of dynamic trimming on three IMOCA 60 inspired mainsail models in an upwind ($AWA = 60°$) unheeled configuration. This study presents dynamic fluid structure interaction results in well controlled conditions (wind, sheet length) with a dynamic trimming system. Trimming oscillations are done around an optimum value of CF_{obj} previously found with a steady trim. Different oscillation amplitudes and frequencies of trimming are investigated. Measurements are done with a 6 component force balance and a load sensor giving access to the unsteady mainsail sheet load. The driving CF_x and optimization target CF_{obj} coefficient first decrease at low reduced frequency f_r for quasi-steady state then increase, becoming higher than the steady state situation. The driving force CF_x and the optimization target coefficient CF_{obj} show an optimum for the three different design sail shapes located at $f_r = 0.255$. This optimum is linked to the power transmitted to the rig and sail system by the trimming device. The effect of the camber of the design shape is also investigated. The flat mainsail design benefits more than the other mainsail designs from the dynamic trimming compared to their respective steady situation. This study presents dynamic results that cannot be accurately predicted with a steady approach. These results are therefore valuable for future FSI numerical tools validations in unsteady conditions.

NOTATION

FSI	Fluid-structure interaction
VSPARS	Visual Sail Position And Rig Shape
YRU	Yacht Research Unit
A	Dynamic trimming amplitude (mm)
AWS	Apparent wind speed (m s^{-1})
AWA	Apparent wind angle (°)
c	Reference chord (m)
CF_i	Force coefficient in the i axis direction (-)
CF_{obj}	Optimization target coefficient (-)
CF_{sheet}	Force coefficient in the mainsail sheet (-)
f	Input frequency (Hz)
f_r	Reduced frequency (-)
F_i	Force in i axis direction (N)
F_{sheet}	Force in the main sail sheet (N)
h	Mainsail luff length (m)
L_{car}	Car traveller line length (mm)
L_{sheet}	Mainsail sheet length (mm)
$MSmax$	Mainsail with maximum camber for the design shape
$MSstd$	Mainsail with standard camber for the design shape
$MSflat$	Mainsail with zero camber for the design shape
P	Mechanical power from the sheet (mW)
q	Dynamic pressure (Pa)
S	Sail mould area (m^2)
T	Time period of oscillation (s)
U_{ref}	Reference wind velocity (m s^{-1})
ρ	Density of air (kg m^{-3})

INTRODUCTION

A challenging task in yacht design modeling and simulation is the analysis of dynamic effects in the Fluid Structure Interaction (FSI) of the yacht sails and rig. The dynamic behavior can be caused by the sea state or the wind, but can

---

[1] nicolas.aubin@ecole-navale.fr

also be caused by the action of the crew while trimming. Literature has pointed out the difficulty of considering the realistic sailing environment of a yacht (Charvet *et al.*, 1996, Marchaj, 1996, Garrett, 1996). Recent studies have underlined the importance of considering the dynamic behavior: forced pitching motion in the wind tunnel (Fossati and Muggiasca, 2012), 2D simplified pitching (Gerhardt *et al.*, 2011), interaction of yacht sails in unsteady conditions (Gerhardt, 2010), full scale experiments and simulations (Augier *et al.*, 2012, 2013, 2014), and downwind sails (Collie and Gerritsen, 2006, Deparday *et al.*, 2014). Downwind sail design is where the gain from a dynamic aero-elastic analysis seems to be potentially the greatest due to the large motion and the induced large load variation. The main findings of these different studies are the same, i.e. the aerodynamics can be predicted more accurately with an unsteady approach.

To account for this dynamic behavior, several Dynamic Velocity Prediction Programs (DVPPs) have been developed (Masuyama *et al.*, 1993, Masuyama and Fukasawa, 1997, Richardt *et al.*, 2005, Keuning *et al.*, 2005) which need models of dynamic aerodynamic and hydrodynamic forces. While the dynamic effects on hydrodynamic forces have been studied extensively, the unsteady aerodynamic behavior of sails has received much less attention. (Schoop and Bessert, 2001) first developed an unsteady aeroelastic model in potential flow dedicated to flexible membranes but neglected the inertia. In a quasi-static approach, a first step is to add the velocity induced by the yacht's motion to the steady apparent wind to build an instantaneous apparent wind (Richardt *et al.*, 2005, Keuning *et al.*, 2005) and to consider the aerodynamic forces corresponding to this instantaneous apparent wind using force models obtained in the steady state.

Recently, advanced computational resources have enhanced numerical simulations and have allowed coupling of fluid and structural solvers dedicated to yacht sails (Renzsh and Graf, 2010, Chapin and Heppel, 2010, Trimarchi *et al.*, 2013, Ranzenbach *et al.*, 2013). In past years, IRE-Nav and the K-Epsilon company have developed numerical tools dedicated to the simulation of the dynamic behavior of yacht sails. The FSI potential model ARAVANTI has been validated by full scale measurements (Augier *et al.*, 2012) and enables numerical studies of a yacht pitching in a head swell (Augier *et al.*, 2013, 2014), showing a clear break with the quasi-static approach. The recent RANS FSI coupling ARA-FINE™/Marine (Durand *et al.*, 2014) is required to simulate cases with strong separation for downwind simulations, but it is very time and CPU consuming.

Even though some advanced models are now available for sail aerodynamics, there is a real need for detailed validation of numerical simulations in order to provide reliable design tools for the sailing industry. Controlled experiments are also a great opportunity to understand the physics of FSI of yacht sails. Unfortunately, realistic and reliable experimental data is scarce and the validation of models in real conditions is difficult (Augier *et al.*, 2012, Fossati *et al.*, 2015). In

this context, wind tunnel testing and full-scale testing are required for comparison and validation (Flay, 1996, Renzsch and Graf, 2013, Le Pelley *et al.*, 2002). Wind tunnel testing has the advantage of being in a controlled environment where a balance can be used to measure the forces created by the sails on the boat frame (Viola and Flay, 2010, Fossati, 2010, Fossati and Muggiasca, 2009, 2010, Wright *et al.*, 2010). Pressure and flying shape measurements can also be performed in wind tunnels (Lasher and Richards, 2007, Graf and Müller, 2009, Viola and Flay, 2011, Viola *et al.*, 2013). In a recent study, (Gerhardt *et al.*, 2011) developed an analytical model to predict the unsteady aerodynamics of interacting yacht sails in 2D potential flow, and performed 2D wind tunnel oscillation tests with a motion range typical of a 82-foot (25m) racing yacht (1992 International America's Cup Class). Recently (Fossati and Muggiasca, 2012, 2009, 2010, 2011) studied the aerodynamics of model-scale rigid sails in a wind tunnel, and showed that pitching motion has a strong and non-trivial effect on aerodynamic forces.

A dedicated experiment has been developed in the Yacht Research Unit Twisted Flow Wind Tunnel, University of Auckland, to study the aerodynamics of dynamic trimming. The model was simplified to a simple model-sized IMOCA 60 mainsail and a mast with no shrouds. We measured the effect of dynamic trimming on the forces (F_x, F_y) with the balance and the load in the sheet (F_{sheet}) for a given incoming wind ($U_{ref} = 3.5 \,\mathrm{m\,s^{-1}}$ at $1.5\,\mathrm{m}$ height at model-scale location in an empty wind tunnel configuration) for 3 different sail design shapes.

In the first part of the paper, we describe the experimental set up and we define the optimum trimming. In the second part, the results are presented for different trimming oscillation amplitudes and frequencies and for different sail design shapes. Finally the influence of these different parameters on the global performance of the rig is discussed.

EXPERIMENTAL SETUP

Experiments were performed thanks to the Sailing Fluids collaboration program in the Twisted Flow Wind Tunnel of the Yacht Research Unit of the University of Auckland described in (Flay, 1996).

An 1/13th scale IMOCA 60 foot design mainsail was designed and built by INCIDENCE SAILS, using SAIL-PACK software developed by BSG DEVELOPPEMENTS, for these experiments. A system of three stepper motors and a control card was used in order to modify the main sheet length L_{sheet} and main car position L_{car} (see Fig. 1). Therefore, the mainsail trimming was imposed remotely without any human contact in the wind tunnel. The uncertainty of imposed trimming was estimated to be $\pm 2\,\mathrm{mm}$ through repeated measurements. Fig. 2 shows the $2.2\,\mathrm{m}$ long mast with the scaled mainsail in the $7.2\,\mathrm{m}$ wide by $3.5\,\mathrm{m}$ tall open jet test section of the YRU wind tunnel. Sail geometry is defined in Fig. 1. The rig is composed of a single $14\,\mathrm{mm}$ circular section carbon mast without spread-

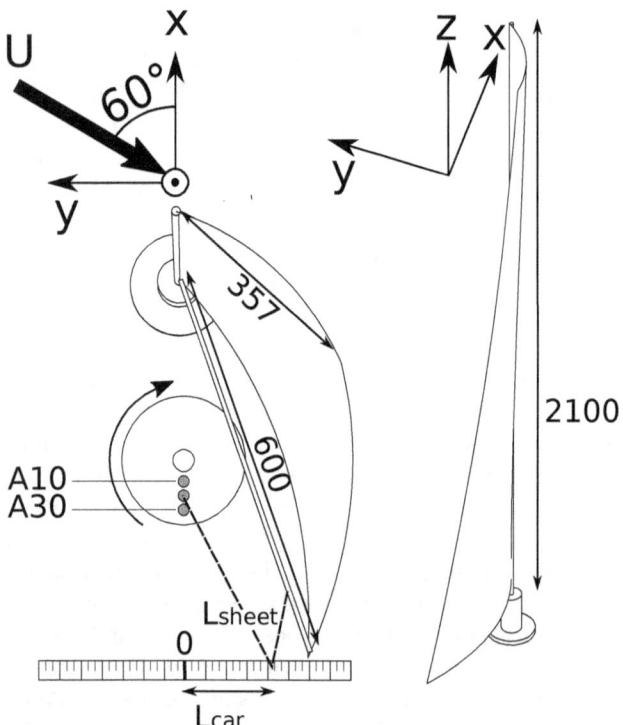

Figure 1: Experimental set up for dynamic trimming. Dimensions are in mm.

Figure 2: Model mainsail in YRU Twisted Flow Wind Tunnel, University of Auckland

ers, backstay or forestay. The objective is to create a simple bench experiment to study the aerodynamic effect of the dynamic trimming and to validate trimming optimization methods. The experiment includes Fluid Structure Interaction with mast deformation for use in numerical model comparisons. A six-component force balance located under the floor of the wind tunnel measures aerodynamic forces. The X-direction is aligned with the model longitudinal direction forward (driving force), the Y-direction is perpendicular positive port-side and measures the side force and the Z-direction is vertical as shown in Fig. 1. The balance precision was verified by calibration testing and the uncertainty on X, Y and Z axis are ± 0.09 N, ± 0.11 N and ± 0.27 N respectively. A load sensor of 50 N range measures the sheet load with a precision of ± 0.02 N. The flying shape is measured with five orange stripes (see Fig.2) through the VS-PARS acquisition system (Le Pelley and Modral, 2008). The sampling frequency of the system measurement is 200 Hz and every run is recorded over 30 s.

The velocity profile follows the empty wind tunnel boundary layer profile and is not twisted (no vanes in the flow). The apparent wind speed (AWS) is $U_{ref} = 3.5 \, \mathrm{m\,s^{-1}} \pm 0.15 \, \mathrm{m\,s^{-1}}$ - measured at 1.5 m high at the model-scale location in an empty configuration- and an apparent wind angle (AWA) set to $60° \pm 2°$.

A Pitot tube in the wind tunnel roof, was used to measure the dynamic pressure during each run. The mean value $\overline{q(t)}$ calculated for each test was used for the normalization of

equations in order to correct for the possible fluctuations in the wind tunnel flow speed.

Optimum trimming

Different sail design shapes were tested. Three sails, made from the same sail cloth were designed with different cambers:

- MSstd = camber of the full scale sail (9.19% at the reference stripe)

- MSflat = no camber

- MSmax = more camber than MSstd (11.67% at the reference stripe)

A first test was performed in order to determine the best trim for the studied $AWA = 60°$. The model was placed on the balance and the sail was statically trimmed to the optimum $CF_{obj} = CF_x - 0.1|CF_y|$. This optimization target takes into account the contribution of the side force on the aerodynamic force and can be found in the design process of sailing yacht to consider the penalty due to the added hydrodynamic drag and leeway. For more details on the optimization function readers should refer to (Sacher *et al.*, 2015). Three stepper motors were used as winches to trim the sail: two motors used to trim the traveller position L_{car} and one centered motor used to trim the main sheet length L_{sheet}. Here we were looking for the best 2 trimming parameters (L_{sheet}, L_{car}). Optimum trimming was extracted from the test using the algorithm described in (Sacher *et al.*,

2016) and used as the reference for the dynamic trimming described in the following sections.

Dynamic trimming

The dynamic trimming consists of an oscillation in the sheet length L_{sheet} around the optimum trimming length obtained previously. The dynamic trimming was done with a fixed traveller position L_{car} (obtained from the optimum trimming) and the instantaneous sheet length $L_{sheet}(t)$ could be calculated from the controlled and recorded angular position of the rotating plate (see Fig. 3 and 1). $L_{sheet}(t)$, the instantaneous length of the sheet, is a function of A the amplitude of variation in mm, f the frequency of oscillation (rotation frequency of the stepper motor controlling the rotating plate) in Hz and the model-scale configuration geometry. The frequency f and amplitude A of oscillation were controlled by the rotating plate placed at the center-line of the boat as illustrated in Fig. 3. The sheet was connected to a pin fixed on the plate. The amplitude of oscillation depends on radial position of the pin. $A = 10 \, \text{mm}$ stands for an eccentric of $10 \, \text{mm}$ and corresponds to a peak to peak amplitude of motion of 20 mm on L_{sheet} (oscillation of L_{sheet} of $\pm 10 \, \text{mm}$).

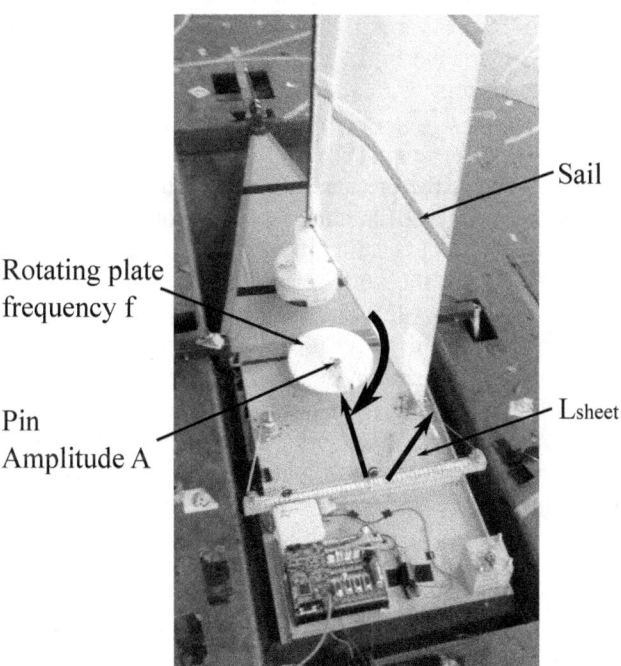

Figure 3: Experimental set up for dynamic trimming: rotating disk (photograph taken without wind)

RESULTS

We examine here the influence of the dynamic trimming on the aerodynamic forces of the sail. Three different sails were tested for 3 amplitudes of oscillation of 10 mm, 20 mm and 30 mm and 7 ordered frequencies from 0 Hz to 3 Hz. From these frequencies, non-dimensional reduced frequencies f_r are defined in the post processing parameters.

Post processing parameters

In this study we define the reduced frequency $f_r = f.c/U_{ref}$, with f the frequency of oscillation in Hz, c the reference chord length $c = S/h = 0.475 \, \text{m}$ and $U_{ref} = 3.5 \, \text{m s}^{-1}$ the reference flow speed. The reduced frequency is a non-dimensional indicator defined as the ratio of the oscillating motion to the reference convection time, from 0 to 0.38.

In the following part, aerodynamic forces and sheet loads are normalized as:

- the instantaneous aerodynamic driving force is defined using $CF_x(t) = \frac{F_x(t)}{q(t)S}$

- its mean value presented in our study $CF_x = \overline{CF_x(t)} = \frac{\overline{F_x(t)}}{q(t)S}$

- equivalent definition is used for the side force coefficient $CF_y = \overline{CF_y(t)} = \frac{\overline{F_y(t)}}{q(t)S}$

- equivalent definition is used for the sheet load coefficient $CF_{sheet} = \overline{CF_{sheet}(t)} = \frac{\overline{F_{sheet}(t)}}{q(t)S}$

- $q(t) = \frac{1}{2}\rho U(t)^2$ is the dynamic pressure measured during the run by the pitot tube.

Forces were averaged over an integer number of period of oscillation regardless of the reduced frequency in order to compare relevant mean values. Time series were filtered with a low pass filter frequencies defined as a Savitzky-Golay filter of order 1 of span 21 samples (Schafer, 2011).

Effect of the reduced frequency f_r

We focus here on the effect of the reduced frequency f_r on the forces for the case of the standard mainsail (MSstd) for an oscillation amplitude $A = 20 \, \text{mm}$ round the optimum L_{sheet}. Coefficients were averaged over the maximum number of integer oscillation periods found in the 30 s recording. Results are presented in Fig. 4. Measurements were doubled and showed good repeatability. Up and down triangles represent the maximum amplitude i.e. the maximum and minimum value of the time series.

For the first oscillation frequency studied, $f_r < 0.02$, the force coefficients decrease compared to the static situation $f_r = 0$ values. The oscillation is very slow and could be considered as quasi-steady. This quasi-steady oscillation around the optimum L_{sheet} degrades the performance because the sail is trimmed at a non-optimum point most of the time. For $f_r > 0.02$, dynamic trimming increases the mean force coefficient, which reaches a maximum around $f_r = 0.255$.

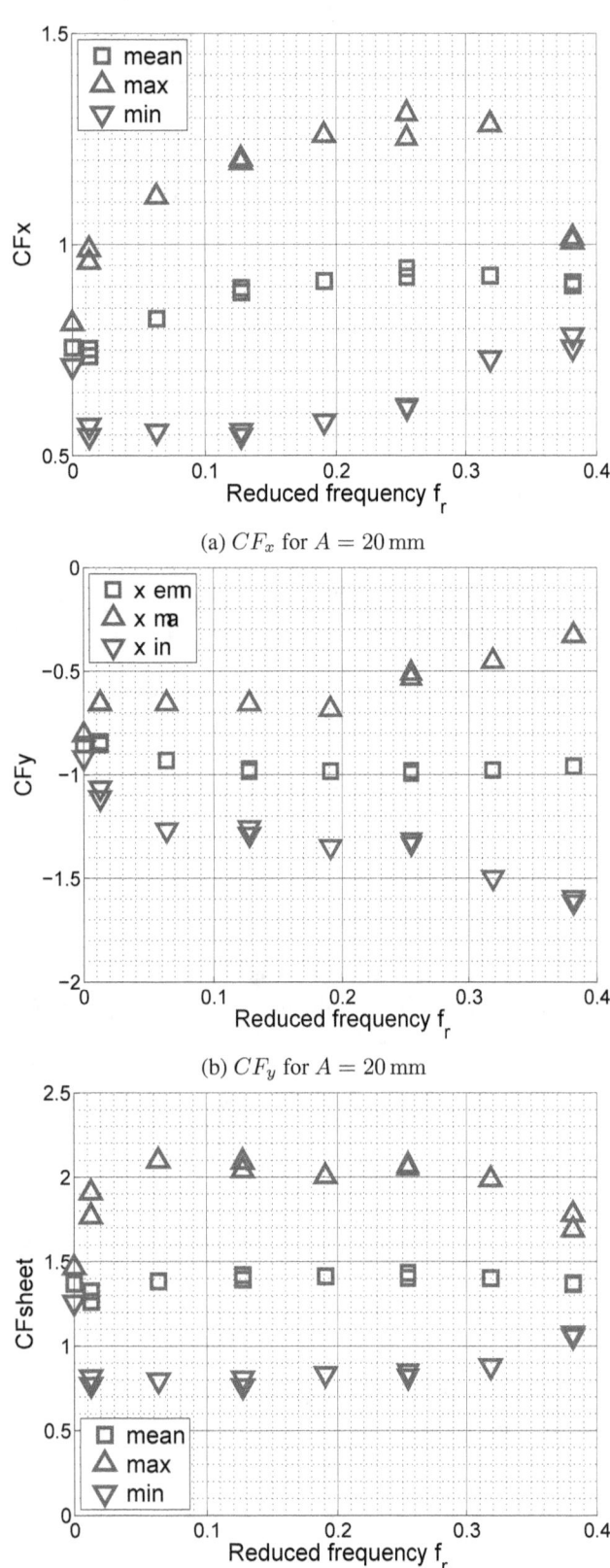

(a) CF_x for $A = 20\,\mathrm{mm}$

(b) CF_y for $A = 20\,\mathrm{mm}$

(c) CF_{sheet} for $A = 20\,\mathrm{mm}$

Figure 4: Effect of f_r at amplitude $A = 20\,\mathrm{mm}$ on CF_x (fig 4a), CF_y (fig 4b) and CF_{sheet} (fig 4c) for the standard mainsail. Up and down triangles represent the maximum and minimu amplitude of the time series respectively

The aerodynamic forces seem to benefit from an unsteady propulsion phenomenon due to the flapping of the sail. This unsteady propulsion is maximized for a defined range of frequencies and its effect decreases above $f_r = 0.255$. Amplitudes of variation of the force coefficients, illustrated by the triangles in Fig. 4, increase significantly until $f_r = 0.255$ and collapse dramatically at higher frequencies for CF_x and CF_{sheet}. In the case of CF_y, the amplitude of variation keeps increasing with the frequency of oscillation. The results show the effect of dynamic trimming compared to the steady trimming maximizes CF_x at a specific range of reduced frequency around 0.255.

Figs. 5 and 6 present the temporal evolution of the load in the sheet and the driving coefficient with the sheet length. This type of Lissajou representation was first proposed for a sailing yacht study by Fossati and Muggiasca (2009, 2010, 2011) in wind tunnel testing and was then used by (Augier et al., 2013, 2014) in simulations and full scale measurements. For more clarity, signals are represented for only 12 s. We present 4 of the 8 studied frequencies, but the trends are identical. The top graph illustrates the static case. The number of cycles represented increases with the frequency f_r. All the curves have been centered around their respective optimimum L_{sheet}, which are slightly different for the different design shapes. $L_{sheet} = 0\,\mathrm{mm}$ is set at the static optimum trim $L_{sheet\ static}$ for the optimum of the optimization target CF_{obj}.

CF_{sheet} vs L_{sheet} describes a loop which witnesses a hysteresis phenomenon (Fig. 5). In this case, the area inside the loop is the mechanical work exchanged with the rig system from the trimming stepper motor. The counter-clockwise sense of rotation, indicated by the arrow on the figure, shows that the work is negative, i.e. given to the system. This confirms that the sail and rig system are forced by the motion of the sheet for the whole range of studied frequencies. The area in the loop increases slightly until $fr = 0.255$ where it reaches a maximum. The loop collapses at $fr = 0.38$. The work exchanged with the rig system is a maximum at $fr = 0.255$ which corresponds to the optimum CF_x observed in Fig. 4a.

CF_x vs L_{sheet} describes a loop as well (Fig. 6). One should realise that the area inside the loops is not actual physical work however it follows the same trend as the work energy from F_x along the x-direction. It is very interesting to observe that the sense of rotation switches for the different frequencies. For $f_r = 0.013$ and 0.38, the system dissipates energy as it turns counter-clockwise. The system gains energy from the oscillation at $f_r = 0.255$ (clockwise rotation). The $f_r = 0.127$ case is a transition where the loop describes a figure 8 shape.

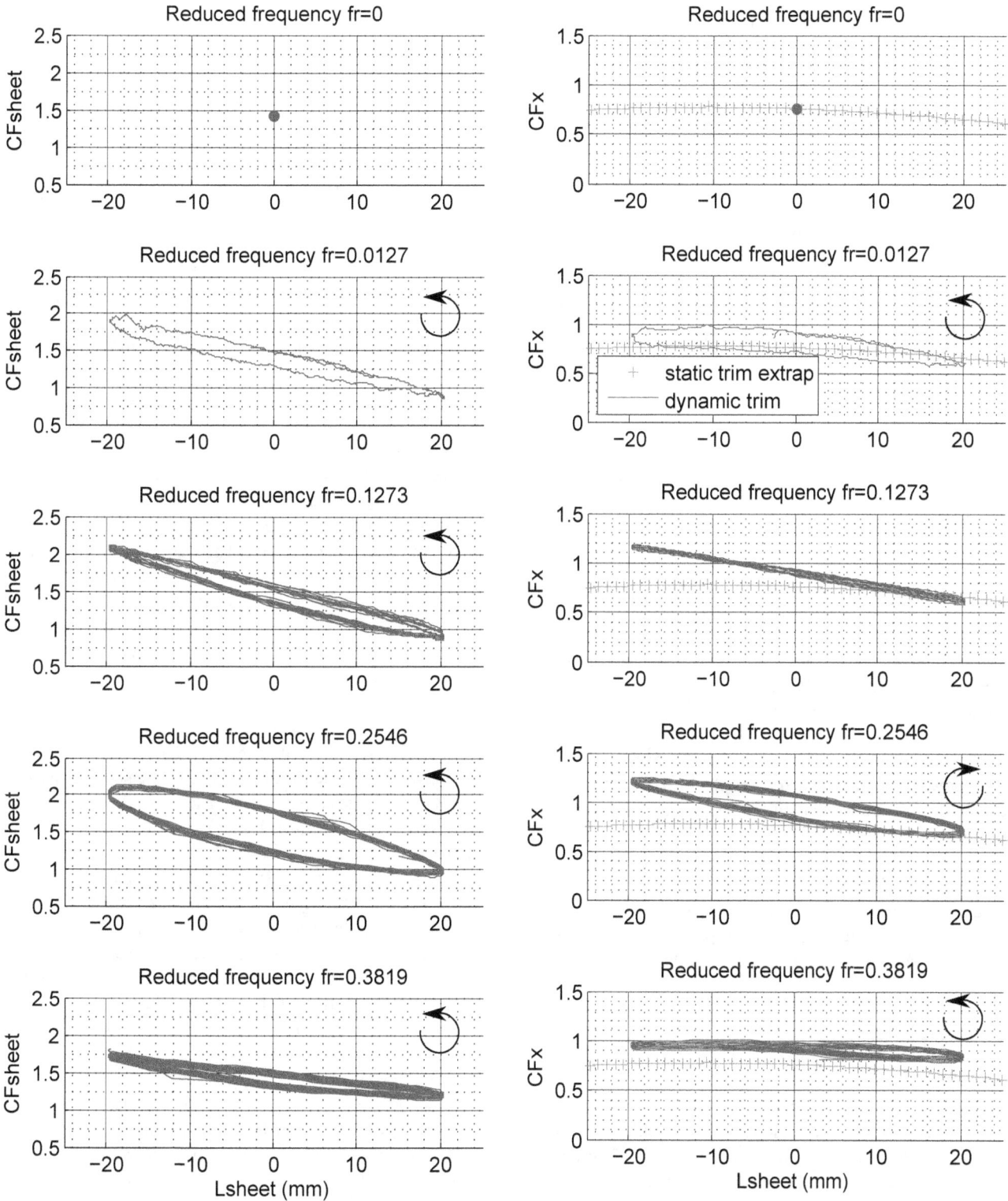

Figure 5: Evolution of CF_{sheet} with L_{sheet} at amplitude $A = 20\,\mathrm{mm}$ for the standard mainsail for different reduced frequencies. Signals are presented for $12\,\mathrm{s}$. The steady part was done without load sensor, so no steady load sheet is available for this configuration.

Figure 6: Evolution of CF_x with L_{sheet} at amplitude $A = 20\,\mathrm{mm}$ for the standard mainsail for different reduced frequencies. Signals are presented for $12\,\mathrm{s}$. Red crosses represent the steady state extrapolated from the 2D optimization part data.

Power is calculated at each reduced frequency and presented in Fig. 7. Power is proportional to the area in the loop illustrated in Fig. 5 and is defined as:

$$P = \frac{\overline{q(t)}S}{T} \oint_{one\ loop} CF_{sheet}(L_{sheet})dL_{sheet}$$

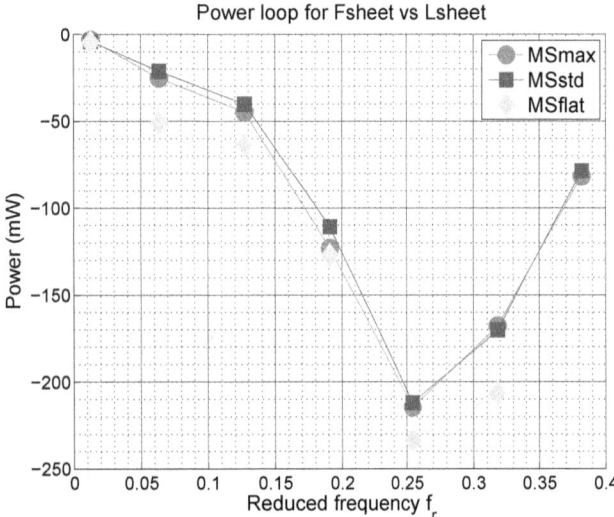

Figure 7: Power given to the system by the sheet oscillation forcing at $A = 20$ mm. Power is proportional to the area in the loop represented in Fig. 5

The power value shown on Fig. 7 is the averaged value of each power value calculated using the previous equation on each entire period oscillation loop. The exchange of energy of the forcing is related to the maximum of force obtained at $f_r = 0.255$ and the sudden collapse of the amplitude of variation of CF_x and CF_{sheet} at $f_r = 0.38$. Nevertheless a part of the trend observed in the force coefficient needs to be explained by studying the other parameters of the other forcing parameters such as the sail camber i.e. the design shape.

Effect of the design shape

The effects of the reduced frequency f_r on the forces are presented for the 3 design shapes and the 3 amplitudes of oscillation. For each sail, the trimming oscillation is done around its specific optimum. $L_{sheet\ Static}$ are different for each case. Again, coefficients are averaged over the maximum number of full oscillation periods found in the 30 s recording. Results are presented in Figs. 8, 9 and 10, CF_x, CF_{sheet} and CF_{obj} respectively. Oscillation amplitudes (maximum and minimum value) of force coefficients are not displayed for clarity but trends are identical to those described in the previous section. Due to the parameters of the optimum trimming (L_{sheet}, L_{car}) for the flat mainsail design shape, high frequency oscillations could not be explored at $A = 30$ mm because the forcing was too strong.

The general trends described in the previous section are identical for the 3 studied sail design shape and the different amplitudes of oscillation. The tendencies observed at $A = 20$ mm are amplified at greater amplitude $A = 30$ mm and slightly minimized at $A = 10$ mm.

It is interesting to notice that the effect of the dynamic trimming is greater for the flat mainsail design MSflat. The CF_x coefficient are nearly identical for the two cambered sails for $A = 10$ mm and $A = 20$ mm whereas the static performances are significantly worse. It seems that the dynamic behavior due to flapping catches/compensates for the defect of flat mainsail design MSflat in static conditions. The unsteady propulsion phenomenon is high enough to compensate for the poor aerodynamic performance of the flat sail in a steady trimming. The oscillation needs a minimum of amplitude of $A > 10$ mm to have a significant effect on the MSflat. However, the optimum of MSflat is reached for a specific frequency $f_r = 0.255$ and decreases rapidly around this value, unlike the other sails MSmax and MSstd where the range of optimal frequencies is wider.

The load in the sheet in static situations i.e. $f_r = 0$ is linked to the camber (Fig. 9). The static CF_{sheet} is greater for the maximum camber mainsail MSmax and it is identical for the two other sails. Variations in the load in the sheet CF_{sheet} for different frequencies are consistent with the effect of dynamic trimming observed on MSflat. The trends are identical with CF_x. At low oscillation amplitudes, the sheet tension increases significantly for the flat sail until $f_r = 0.32$, when the CF_{sheet} reaches a maximum and decreases slightly after $f_r = 0.13$ for the other sails. For $A = 20$ mm, the maximum load in the sheet is reached at lower frequencies but a greater load is still necessary to make the flat sail oscillate. It seems that at these amplitudes, the energy brought to the system by the forced oscillation is greater in the case of MSflat, which explains the important gain on the aerodynamic coefficients observed in Figs. 5 and 6. The differences between the sails are smoothed at $A = 30$ mm in CF_{sheet}, as illustrated by the energy brought to the system in CF_x and CF_y. Energy brought to the system by the oscillation of the sheet is illustrated in Fig. 7. The power exchanged is a maximum at $f_r = 0.255$.

The CF_{obj} evolution shown in Fig. 10 depends on both the camber of the sail and the amplitude of oscillation. For the low oscillation $A = 10$ mm in Fig. 10a, the optimization target shows a maximum for the flat mainsail design contrary to the standard and maximum camber designs which present a plateau from $f_r = 0.255$. This plateau disappears for higher amplitude oscillations and all the curves present a maximum. The maximum camber design presents either the best optimization value or is fairly close to the maximum optimization target value regarding the different oscilation frequencies and amplitudes. For this $AWA = 60°$ this trend confirms the sailors' knowledge causing them to try to increase the camber by easing the outhaul of the sail to improve their performance in a situation such as a dogleg while sailing perpendicular to the true wind direction.

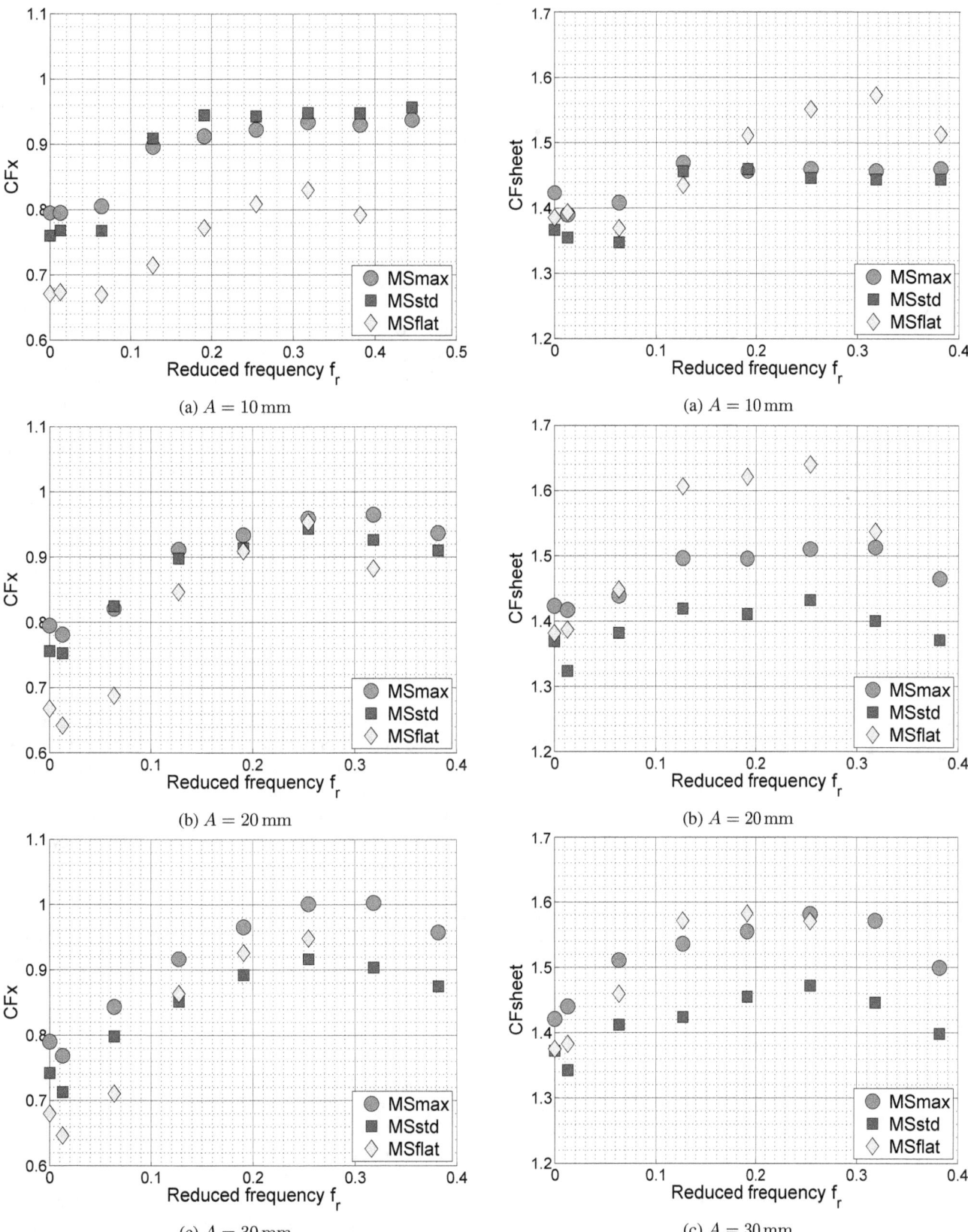

Figure 8: Effect of f_r on CF_x for the 3 design shape at amplitude (a) $A = 10\,\text{mm}$, (b) $A = 20\,\text{mm}$ and (c) $A = 30\,\text{mm}$.

Figure 9: Effect of f_r on CF_{sheet} for the 3 design shape at amplitude (a) $A = 10\,\text{mm}$, (b) $A = 20\,\text{mm}$ and (c) $A = 30\,\text{mm}$.

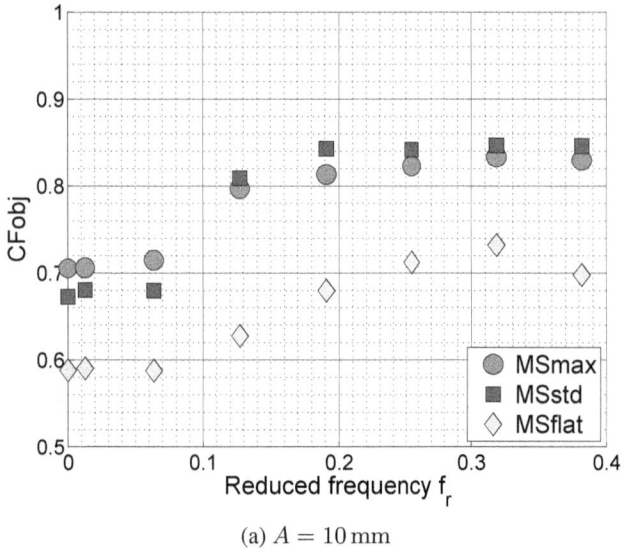

(a) $A = 10\,\text{mm}$

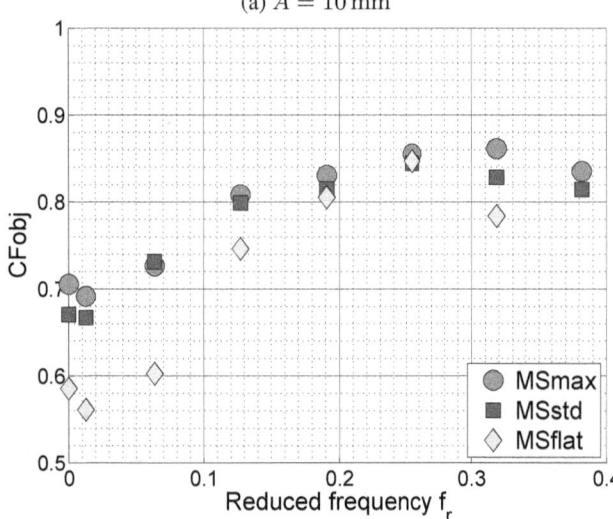

(b) $A = 20\,\text{mm}$

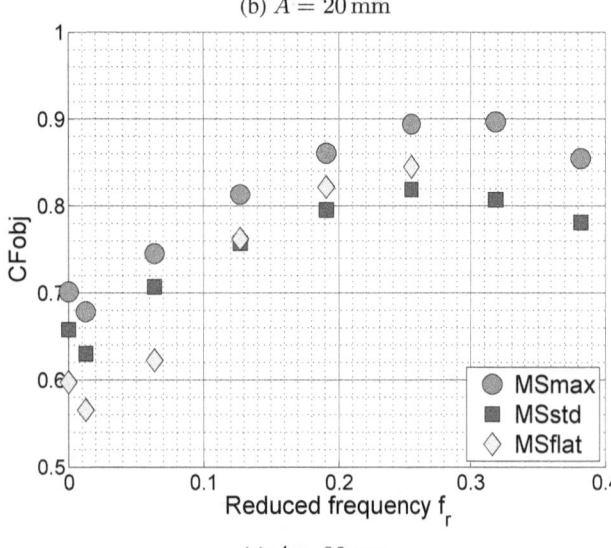

(c) $A = 30\,\text{mm}$

Figure 10: Effect of f_r on CF_{obj} for the 3 design shapes at amplitude (a) $A = 10\,\text{mm}$, (b) $A = 20\,\text{mm}$ and (c) $A = 30\,\text{mm}$.

DISCUSSION

Dynamic contributions to aerodynamic forces can be decomposed into three components. The first one is due to the change in circulation around the profile. At this $AWA = 60°$, the more camber, the more lift in static conditions. The second component is due to the unsteady propulsion caused by the forced oscillation. At a certain range of frequencies studied, flapping might produce vortices structures beneficial to the aerodynamic force produced by the sail. Vortices structures are linked to the sail area, the frequency and the amplitude of oscillation but do not depend on the sail profile. This flapping effect on the flat sail MSflat which suffers from a poor static aerodynamic contribution is then much more significant but on a narrow range of frequencies. The third component is the energy transferred to the system by the oscillation forcing. As mentioned in the section describing the effect of the reduced frequency f_r, mechanical work is given to the system by the forcing in the sheet (Fig. 7). This work is dissipated at most of the frequencies but is beneficial to the thrust at $f_r = 0.255$ (Fig. 8). In the case of a dynamic trimming, the aerodynamic force is composed of the three components with different effects depending on the frequencies and amplitudes of oscillation. It seems that the forcing at $f_r = 0.255$ benefits from all three components, the reason for the local optimum at that frequency.

CONCLUSIONS

An innovative oscillating trimming experiment has been developed in the Twisted Flow Wind Tunnel at the Yacht Research Unit, University of Auckland. The oscillating trimming effect has been studied on different design shapes of IMOCA 60 type mainsails at $AWA = 60°$ with different input parameters: amplitude and reduced frequency. The dynamic oscillations clearly show that quasi-static measurements are not relevant for predicting aerodynamic forces even at quite low reduced frequencies. These results support previous findings that static or quasi-static approaches are not sufficient to capture the complexities of dynamic effects, even for the simplified oscillating trimming simulation.

The sheet load measurement enabled us to calculate the mechanical power transmitted from the trimming device to the entire rig and sail system and could be correlated with the aerodynamic force evolutions of the different sails.

The dynamic effect showed that there was an optimum reduced frequency $f_r = 0.255$ that improved the performance function for the different sails related to a maximum power transmitted to the rig and sails by the sheet. The three different model sails presented the same trends, but the dynamic improvement was more significant for the flat sail: up to an increase of 40% of its CF_{obj} at $f_r = 0.255$ and $A = 30\,\text{mm}$ compared to the steady case.

Oscillations around optimum static trim have also been performed for $AWA = 25°$ and $AWA = 40°$ and will be compared to this paper's results in a future publication. Flying shape and rig part tracking analysis will also be performed. Further work will be done using those data for comparison and validation of unsteady numerical simulation tools.

ACKNOWLEDGEMENTS

This project has received funding from the European Unions Seventh Program for research, technological development and demonstration under grant agreement N° PIRSES-GA-2012-318924 and from the Royal Society of New Zealand for the UK-France-NZ collaboration project, SAILING FLUIDS (see www.sailingfluids.org). This work was supported by the French Naval Academy, Brest Métropole Océane, Région Bretagne and the Marie Curie European Unions Seventh Framework Programme (FP7/2007-2013) under REA grant agreement n°PCOFUND-GA-2013-609102 (PRESTIGE-Campus France). This work was supported by the "Laboratoire d'Excellence" LabexMER (ANR-10-LABX-19) and co-funded by a grant from the French government under the program "Investissements d'Avenir". The authors are grateful to K-Epsilon and VS-PARS company for their continuous collaboration and to Ronan Floch from Incidence Sails for designing and manufacturing the model-scale sails. The authors would like to thank the SEFER services for providing the electronical actuator and remote control parts, Mr David Le Pelley, wind tunnel manager, and Dr Nick Velychko for their help, guidance, their wise advice and comments.

REFERENCES

CHARVET, T., HAUVILLE, F. and HUBERSON, S., *Numerical simulation of the flow over sails in real sailing conditions*, Journal of Wind Engineering and Industrial Aerodynamics, **63**(1-3), (1996), 111 – 129.

MARCHAJ, C., Sail performance: techniques to maximize sail power, International Marine/Ragged Mountain Press (1996).

GARRETT, R., The symmetry of sailing: the physics of sailing for yachtsmen, Sheridan House, Inc. (1996).

FOSSATI, F. and MUGGIASCA, S., *An experimental investigation of unsteady sail aerodynamics including sail flexibility*, 4th High Performance Yacht Design Conference, Auckland, New Zeeland (2012).

GERHARDT, F., FLAY, R.G.J. and RICHARDS, P.J., *Unsteady aerodynamics of two interacting yacht sails in two-dimensional potential flow*, Journal of Fluid Mechanics, **668**(1), (2011), 551–581.

GERHARDT, F.C., *Unsteady Aerodynamics of Upwind-Sailing and Tacking*, Ph.D. thesis, The University of Auckland (2010).

AUGIER, B., BOT, P., HAUVILLE, F. and DURAND, M., *Experimental validation of unsteady models for fluid structure interaction: Application to yacht sails and rigs*, Journal of Wind Engineering and Industrial Aerodynamics, **101**, (2012), 53–66.

AUGIER, B., BOT, P., HAUVILLE, F. and DURAND, M., *Dynamic Behaviour of a Flexible Yacht Sail Plan*, Ocean Engineering, **66**, (2013), 32–43.

AUGIER, B., HAUVILLE, F., BOT, P., AUBIN, N. and DURAND, M., *Numerical study of a flexible sail plan submitted to pitching: Hysteresis phenomenon and effect of rig adjustments*, Ocean Engineering, **90**, (2014), 119–128.

COLLIE, S. and GERRITSEN, M., *The challenging turbulent flows past downwind yacht sails and practical application of CFD to them*, 2nd High Performance Yacht Design Conference, Auckland, New-Zealand (2006).

DEPARDAY, J., BOT, P., HAUVILLE, F., MOTTA, D., LE PELLEY, D.J. and FLAY, R.G.J., *Dynamic measurements of pressures, sail shape and forces on a full-scale spinnaker*, 23rd HISWA Symposium on Yacht Design and Yacht Construction, Amsterdam (2014).

MASUYAMA, Y., TAHARA, Y., FUKASAWA, T. and MAEDA, N., *Dynamic performance of sailing cruiser by a full scale sea reality*, The 11th Chesapeake Sailing Yacht Symposium, Annapolis, USA (1993).

MASUYAMA, Y. and FUKASAWA, T., *Full scale measurement of sail force and the validation of numerical calculation method*, The 13th Chesapeake Sailing Yacht Symposium, Annapolis, USA (1997).

RICHARDT, T., HARRIES, S. and HOCHKIRCH, K., *Maneuvering simulations for ships and sailing yachts using FRIENDSHIP-Equilibrium as an open modular workbench*, International Euro-Conference on Computer Applications and Information Technology in the Maritime Industries (2005).

KEUNING, J., VERMEULEN, K. and DE RIDDER, E., *A generic mathematical model for the manoeuvring and tacking of a sailing yacht*, The 17th Chesapeake Sailing Yacht Symposium, (143–163), Annapolis, USA (2005).

SCHOOP, H. and BESSERT, N., *Instationary aeroelastic computation of yacht sails*, International Journal for Numerical Methods in Engineering, **52**(8), (2001), 787–803.

RENZSH, H. and GRAF, K., *Fluid Structure Interaction simulation of spinnakers - Getting closer to reality*, 2nd International Conference on Innovation in High Performance Sailing Yachts, Lorient, France (2010).

CHAPIN, V. and HEPPEL, P., *Performance optimization of interacting sails through fluid structure coupling*, 2nd International Conference on Innovation in High Performance Sailing Yachts, Lorient, France (2010).

TRIMARCHI, D., VIDRASCU, M., TAUNTON, D., TURNOCK, S. and CHAPELLE, D., *Wrinkle development analysis in thin sail-like structures using MITC shell finite elements*, Finite Elements in Analysis and Design, **64**, (2013), 48–64.

RANZENBACH, R., ARMITAGE, D. and CARRAU, A., *Mainsail Planform Optimization for IRC 52 Using Fluid Structure Interaction*, 21st Chesapeake Sailing Yacht Symposium, March, (50–58), Annapolis (2013).

DURAND, M., LEROYER, A., LOTHODÉ, C., HAUVILLE, F., VISONNEAU, M., FLOCH, R. and GUILLAUME, L., *FSI investigation on stability of downwind sails with an automatic dynamic trimming*, Ocean Engineering, **90**, (2014), 129–139.

FOSSATI, F., BAYATI, I., ORLANDINI, F., MUG-GIASCA, S., VANDONE, A., MAINETTI, G., SALA, R., BERTORELLO, C. and BEGOVIC, E., *A novel full scale laboratory for yacht engineering research*, Ocean Engineering, **104**, (2015), 219–237.

FLAY, R.G.J., *A twisted flow wind tunnel for testing yacht sails*, Journal of Wind Engineering and Industrial Aerodynamics, **63**(1-3), (1996), 171–182.

RENZSCH, H. and GRAF, K., *An experimental validation case for fluid-structure-interaction simulations of downwind sails*, 21st Chesapeake Sailing Yacht Symposium, March, (59–66), Annapolis (2013).

LE PELLEY, D.J., EKBLOM, P. and FLAY, R.G.J., *Wind tunnel testing of downwind sails*, 1st High Performance Yacht Design Conference, (66–75), Auckland (2002).

VIOLA, I.M. and FLAY, R.G.J., *On-water pressure measurements on a modern asymmetric spinnaker*, 21st HISWA Symposium on Yacht Design and Yacht Construction, November, Amsterdam (2010).

FOSSATI, F., Aero-Hydrodynamics and the Performance of Sailing Yachts: The Science Behind Sailing Yachts and Their Design, Adlard Coles Nautical (2010).

FOSSATI, F. and MUGGIASCA, S., *Sails Aerodynamic Behavior in dynamic condition*, The 19th Chesapeake Sailing Yacht Symposium, Annapolis, USA (2009).

FOSSATI, F. and MUGGIASCA, S., *Numerical modelling of sail aerodynamic behavior in dynamic conditions*, 2nd International Conference on Innovation in High Performance Sailing Yachts, Lorient, France (2010).

WRIGHT, A.M., CLAUGHTON, A.R., PATON, J. and LEWIS, R., *Off-wind sail performance prediction and optimisation*, The Second International Conference on Innovation in High Performance Sailing Yachts, Lorient, France (2010).

LASHER, W. and RICHARDS, P., *Validation of Reynolds-averaged NavierStokes simulations for international Americas Cup class spinnaker force coefficients in an atmospheric boundary layer.*, Journal of Ship Reseasrch, **51 (1)**, (2007), 2238.

GRAF, K. and MÜLLER, O., *Photogrammetric Investigation of the Flying Shape of Spinnakers in a Twisted Flow Wind Tunnel*, 19th Chesapeake Sailing Yacht Symposium, March, Annapolis (2009).

VIOLA, I.M. and FLAY, R.G.J., *Sail pressures from full-scale, wind-tunnel and numerical investigations*, Ocean Engineering, **38**(16), (2011), 1733–1743.

VIOLA, I., BOT, P. and RIOTTE, M., *Upwind sail aerodynamics: A RANS numerical investigation validated with wind tunnel pressure measurements*, International Journal of Heat and Fluid Flow, **39**, (2013), 90–101.

FOSSATI, F. and MUGGIASCA, S., *Experimental investigation of sail aerodynamic behavior in dynamic conditions*, Journal of Sailboat Technology, **2**, (2011), 1–41.

LE PELLEY, D. and MODRAL, O., *VSPARS: A combined sail and rig recognition system using imaging techniques*, 3rd High Performance Yacht Design Conference, **14**, (2008), 57–66.

SACHER, M., HAUVILLE, F., BOT, P. and DURAND, M., *Sail trimming FSI simulation - comparison of viscous and inviscid flow models to optimise upwind sails trim*, 5th High Performance Yacht Design Conference, (217–228), Auckland, New-Zealand (2015).

SACHER, M., HAUVILLE, F., DUVIGNEAU, R., LE MATRE, O., AUBIN, N. and DURAND, M., *Experimental and numerical trimming optimizations for a mainsail in upwind conditions*, The 22nd Chesapeake Sailing Yacht symposium, Annapolis, Maryland (2016).

SCHAFER, R., *What Is a Savitzky-Golay Filter?*, IEEE Signal Processing Magazine, **28**(4), (2011), 111–117.

Towards a New Mathematical Model for Investigating Course Stability and Maneuvering Motions of Sailing Yachts

Manolis Angelou, National Technical University of Athens (NTUA)
Kostas J. Spyrou, National Technical University of Athens (NTUA)

ABSTRACT

In order to create capability for analyzing course instabilities of sailing yachts in waves, the authors are at an advanced stage of development of a mathematical model comprised of two major components: an aerodynamic, focused on the calculation of the forces on the sails, taking into account the variation of their shape under wind flow; and a hydrodynamic one, handling the motion of the hull with its appendages in water.

Regarding the first part, sails provide the aerodynamic force necessary for propulsion. But being very thin, they have their shape adapted according to the locally developing pressures. Thus, the flying shape of a sail in real sailing conditions differs from its design shape and it is basically unknown. The authors have tackled the fluid-structure interaction problem of the sails using a 3d approach where the aerodynamic component of the model involves the application of the steady form of the Lifting Surface Theory, in order to obtain the force and moment coefficients, while the deformed shape of each sail is obtained using a relatively simple Shell Finite Element formulation. The hydrodynamic part consists of modeling hull reaction, hydrostatic and wave forces.

A Potential Flow Boundary Element Method is used to calculate the Side Forces and Added Mass of the hull and its appendages. The Side Forces are then incorporated into an approximation method to calculate Hull Reaction terms. The calculation of resistance is performed using a formulation available in the literature. The wave excitation is limited to the calculation of Froude - Krylov forces.

NOTATION

A	Cross sectional area (m²)
A_{ij}	Matrix of influence coefficients
A_R	Aspect ratio
B	Beam (m)
B.E.M.	Boundary Element Method
C_D	Drag Coefficient
C_L	Lift Coefficient
C_F	Force Coefficient
CSYS	Chesapeake Sailing Yacht Symposium
D	Drag force (N)
DoF	Degrees of Freedom
F.E.M.	Finite Element Method
F_i	Force or moment due to "i" excitation
F_N	Froude Number
g	Gravity acceleration (m/sec²)
h	Wave steepness ratio
I, J_{Jib}	Principal dimensions of Jib sail
I_X	Moment of Inertia around X axis (kg m²)
I_Y	Moment of Inertia around Y axis (kg m²)
I_Z	Moment of Inertia around Z axis (kg m²)
k	Wave number (rad/m)
L.S.T.	Lifting Surface Theory
L_i	Lift force of "i" component (N)
L_{WL}	Length of waterline (m)
m	Yacht mass (kg)
m_X	Surge added mass (kg)
m_{YY}	Sway added mass (kg)
$\vec{n}$	Normal vector
$\vec{n}_P$	Panel normal vector (Body Fixed System)
p	Roll velocity (rad/sec)
P, E_{Main}	Principal dimensions of Main sail
q	Pitch velocity (rad/sec)
$\vec{q}_\Gamma$	Vortex induced velocity
r	Yaw velocity (rad/sec)
$\vec{r}_C$	Distance to a point "C" (m)
$\vec{r}_P$	Panel moment vector arm (Body Fixed System)
S_i	Surface of "i" component (m²)
T	Draft (m)
u	Surge velocity (m/sec)
v	Sway velocity (m/sec)
V.L.M.	Vortex Lattice Method

V_{AW}	Apparent Wind speed (m/sec)
V_B	Yacht Speed (m/sec) (m/sec)
V_T	Transverse velocity (m/sec)
Vt	Tangential velocity (m/sec)
V_{TW}	True Wind speed (m/sec)
$\nabla\Phi$	Potential velocity (m/sec)
w	Heave velocity (m/sec)
x_{CG}	Longitudinal center of gravity (m)
x_{CEFF}	Longitudinal center of effort (m)
y_{CEFF}	Transverse center of effort (m)
z_{CG}	Vertical center of gravity (m)
z_{CEFF}	Vertical center of effort (m)
z_P	Vertical distance of panel center to mean water level (m)
α	Angle of attack (rad)
α_R	Effective rudder angle of attack (rad)
α_{TW}	True Wind angle (rad)
α_{AW}	Apparent wind angle (rad)
β	Drift angle (rad)
Γ	Vortex strength (1/sec)
δ	Rudder deflection delta angle (rad)
ζ	Local water surface elevation (m)
θ	Pitch angle (rad)
λ	Wave length (m)
μ	Ship-wave incident angle (rad)
μ_{YY}	Sway sectional added mass (kg)
ρ	Density of water (kg/m³)
σ	Source strength
φ	Roll angle (rad)
φ_0	Incident wave potential
φ_P	Perturbation potential
ψ	Yaw angle (rad)
ω	Ship-wave frequency of encounter

INTRODUCTION

In the scientific literature of sailing yacht modeling of behavior, one comes across a number of significant studies on the detailed calculation of fluid dynamic and structural behavior of certain yacht components (e.g. Jones, 2001, Kagemoto, 2000). On the other hand, the course stability of sailing yachts is a topic that has not attracted much attention, although, historically, several records have existed describing to broaching-to incidents of ships with sails (Spyrou, 2010). The present study is a step towards setting up a framework for the systematic study of the course stability of sailing yachts operating in wind and waves. A mathematical model is under development (described next), consisted of: an aerodynamic component, addressing the forces on the sails and the variation of their

shape due to wind flow; and a hydrodynamic, modeling the hull and its appendages.

MATHEMATICAL MODELLING

Equations of Motions and Coordinate Systems

The model is intended for performing course stability analysis and, as a matter of fact, both upwind and downwind cases should be under consideration. The model is built for simulating ship motions in 6 degrees of freedom. Three different coordinate systems are used: an earth-fixed, non-rotating, coordinate system (x_O, y_O, z_O); a wave fixed system that travels with the wave celerity (x_W, y_W, z_W) and a body fixed (x, y, z) system with its origin located on the midship point where the centerplane and the waterplane intersect (Figure 1).

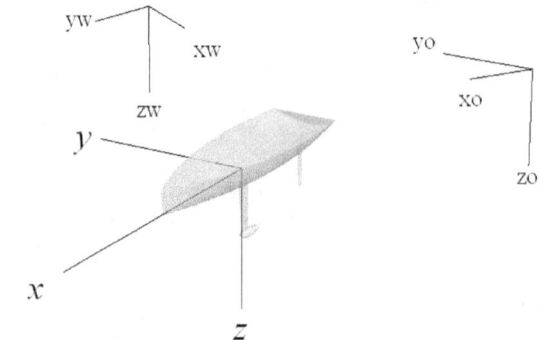

Figure 1 – Coordinate Systems.

The systems are in accordance with the right-hand rule where 'x' axis points positive forward, having on its right the positive 'y' axis, while positive 'z' axis points downwards.

Assuming the hull as a rigid body, the general form of the equations of motions for the 6 DoF, as they accrue from application of Newton's law, is as follows (SNAME 1950):

$$m\left[\dot{u} + qw - rv - x_{CG}\left(q^2 + r^2\right) + z_{CG}\left(pr + \dot{q}\right)\right] = X - mg\sin\theta \tag{1.1}$$

$$m\left[\dot{v} + ru - pw + z_{CG}\left(qr - \dot{p}\right) + x_{CG}\left(qp + \dot{r}\right)\right] = Y + mg\sin\varphi\cos\theta \tag{1.2}$$

$$m\left[\dot{w} + pv - qu - z_{CG}\left(p^2 + q^2\right) + x_{CG}\left(rp - \dot{q}\right)\right] = Z + mg\cos\varphi\cos\theta \tag{1.3}$$

$$I_X\dot{p} - mz_{CG}\left(\dot{v} + ru - pw\right) - mx_{CG}z_{CG}\left(\dot{r} + pq\right) = K \tag{1.4}$$

$$I_Y \dot{q} + (I_X - I_Z) rp + m z_{CG} (\dot{u} + qw - rv)$$

$$- m x_{CG} (\dot{w} + pv - qu) + m x_{CG} z_{CG} (p^2 - r^2) \quad (1.5)$$

$$= M - m x_{cg} g \cos\varphi \cos\theta$$

$$I_Z \dot{r} + (I_Y - I_X) pq + m x_{cg} (\dot{v} + ru - pw)$$
$$+ m z_{CG} x_{CG} (rq - \dot{p}) = N \qquad (1.6)$$

The right-hand-side of the above equations, containing forces and moments, can be expanded in modular form for each equation mode i as

$$F_i = F_{HS} + F_{HR} + F_R + F_S + F_W \qquad (1.7)$$

where the subscripts indicate force contribution from *HydroStatic*, *Hull Reaction*, *Rudder*, *Wave* and *Sails* respectively, in accordance with the excitation being of hydrodynamic or aerodynamic origin.

SAILS MODEL

Sail excitation is calculated allowing for the deformation of sail surface due to pressure's variation, using as input parameter the relative-to-the-sail(s) wind direction. Modeling effort is hence split towards creating an aerodynamic and a structural module.

Whilst we are already into trying transient methods for both upwind and downwind scenarios, we are not in a position yet to report verified results. On the other hand, the integration of the aerodynamic and structural models with a hydrodynamic model of vessel's movement in waves is the focus of the current paper. Hence, only simple steady methods of sail modeling will be found in this paper.

Upwind Case

The small thickness of the sails makes them ideal for being modeled with a potential flow method, such as the one using the Lifting Surface Theory (L.S.T.) which is usually applied through a numerical scheme based on the Vortex Lattice Method (V.L.M.). While the lifting surface bears minimal computational cost, it requires that the flow always remains attached to the surface, thus restraining L.S.T.'s applicability to a relatively small range of fluid inflow angles. The structural part is treated using simple flat shell elements. The sail system is considered to consist of a main and jib sail.

Upon the calculation of the apparent wind angle and velocity the forces and moments for each sail expressed in the body fixed system are (Figure 2):

$$X_S = L_S \cdot \sin(a_{AW}) - D_S \cdot \cos(a_{AW}) \qquad (1.8)$$

$$Y_S = L_S \cdot \cos(a_{AW}) + D_S \cdot \sin(a_{AW}) \qquad (1.9)$$

$$K_S = -Y_S \cdot z_{CEFF} \qquad (1.10)$$

$$N_S = Y_S \cdot x_{CEFF} + X_S \cdot y_{CEFF} \qquad (1.11)$$

where

$$L_S = \frac{1}{2} \rho C_L V^2_{aAW} S \qquad (1.12)$$

$$D_S = \frac{1}{2} \rho C_D V^2_{aAW} S \qquad (1.13)$$

are the lift and drag forces for the sails respectively and their calculation is discussed in the next section.

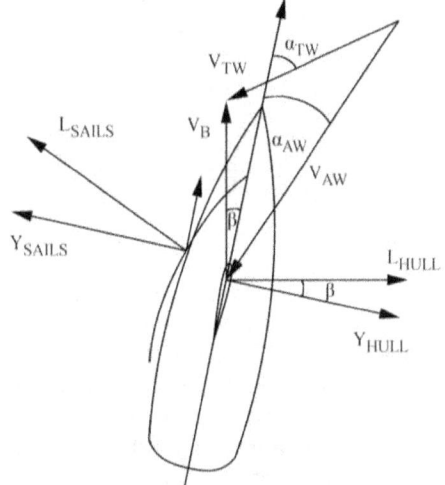

Figure 2 – Horizontal Plane Upwind Yacht Dynamics.

Aerodynamic Part

The L.S.T. is actually a linearized Boundary Element Method (B.E.M.) offering a formulation that allows the effects of camber and thickness to be decoupled, and the no-entrance boundary condition to be transferred to the mean camber line of a surface.

The Vortex Lattice Method is a numerical scheme that discretizes a surface into a series of rectangular panels, allowing so the real flow to be approximated by placing a series of discrete lines of vorticity instead of a continuous distribution along the field. In terms of the Lifting Surface Theory, a system of constant strength horseshoe vortices is placed on every panel.

The vortex induced velocity is evaluated using the Biot-Savart Type Integral for the induced velocity of a vortex segment of length and constant strength dl and Γ to any point C at a distance $|\vec{r}_C|$

$$\vec{q}_\Gamma = \frac{\Gamma}{4\pi} \int \frac{dl \times \vec{r}_C}{|\vec{r}_C|^3} \qquad (1.14)$$

The vortex consists of a parallel to the leading panel edge filament and two trailing filaments that run towards the trailing edge where the Kutta condition is satisfied and further to the wake. As the wake is a free shear layer, it can only carry vorticity. The flow being steady, it allows a first approximation for the free vortex sheet, with the vortex lines aligned with the initially undisturbed flow (frozen wake propagating aft on the direction of the free stream).

The induced velocity of every horseshoe vortex j to every point i is represented by a matrix of influence coefficients and the no-entrance boundary condition $\{\nabla\Phi \cdot \vec{n} = 0\}$ reduces to a system of linear equations $A_{ij}\Gamma_j = \overrightarrow{V_\infty}_i \cdot \vec{n}_i$. Once the coefficient matrix has been calculated, the linear system is solved for the strength of the horseshoe vortices and subsequently the total velocities are calculated. The wake vortex lines are then rotated in order to be aligned with the local total velocity vectors and to apply the force free condition, providing so the roll-up of the wake. Finally, the lift forces on the sail are calculated using the Kutta - Jukowski theorem

$$\vec{L} = \rho \cdot \vec{V}_t \times \vec{\Gamma} \qquad (1.15)$$

where $\vec{V}_t$ is the tangential velocity. The forces are then directed as input to the structural part, to calculate the nodal displacements. The deformed surface is returned to the Lifting Surface module and the procedure commences iteratively until the aerodynamic forces have converged. The number of necessary iterations for convergence is increasing in respect of the relative wind inflow angle, and reaches 25-30 iterations for large angles. An example of converged shapes appears in Figure 3.

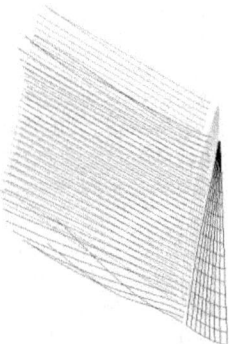

Figure 3 – Two sails system – flying shapes and wakes.

Aeroelastic Part (static aeroelasticity)

The variation of the sail shape is tracked using flat triangular shell elements that consist of a membrane (Constant Strain Triangles) and a bending part (Discrete Kirchhoff Triangles). The elements are thus developed by superimposing the stiffness of the membrane and the bending components.

The membrane contribution to the shell element is provided by the Constant Strain Triangle under plane stress conditions. As in Zienkiewicz (2000a and 2000b), the element is considered to be a flat triangle with three nodes and two degrees of freedom per node, i.e., in the two planar directions.

For the plate bending part the adapted formulation is the one of Batoz (1980) and consider the Discrete Kirchhoff Triangle, an element with three nodes and three degrees of freedom per node; that being the normal to the triangle displacement and the rotations around the x and y axis respectively.

The authors acknowledge that the structural formulation implemented so far is simplistic; the solution is linear, as membrane and bending contributions are solved independently.

Additionally, the current method is appropriate only for small displacements. However as most non-linear models known to the authors tackle the large displacement behavior implicitly (considering a single time step, the exciting load is applied incrementally in sub-steps), the simulations are carried out in the same context.

Results

Lift and drag coefficients for a system of two sails, a main and a jib, are calculated for a small range of incoming flow angles (0° - 25°) (Figure 4a and 4b). The sails are assumed to be made of Kevlar and exhibit isotropic behavior, while their principal dimensions are:

$$I_{jib} = 13.0 \text{ m} , J_{jib} = 5.0 \text{ m} , \text{ for the Jib}$$

$$P_{main} = 15.0 \text{ m} , E_{main} = 6.0 \text{ m} , \text{ for the Main.}$$

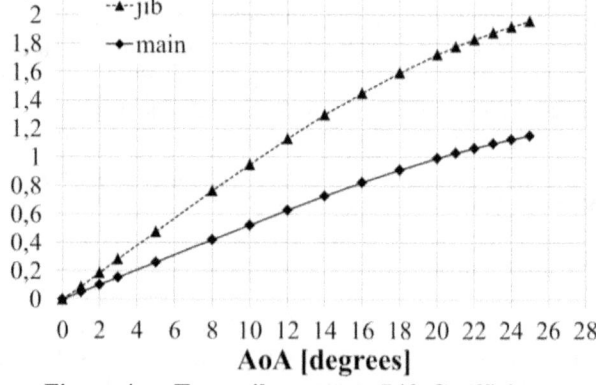

Figure 4a – Two sails system – Lift Coefficients.

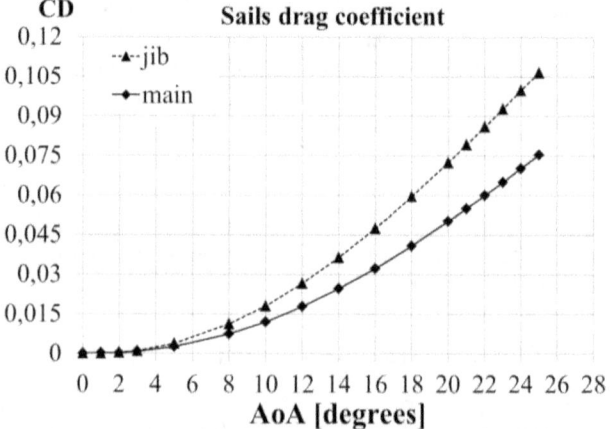

Figure 4b – Two sails system – Drag Coefficients.

The angle of attack was considered as in the case of airfoils, i.e. the relative angle between the incoming flow and a line segment that connects the leading and trailing edge of the camber line.

Downwind Case

To examine numerically the behavior of a sail in a wider operational range where drag effects become dominant, the use of viscous flows methods such as RANSE or LES solvers is unavoidable.

In the context of the authors' progress so far, in a recent paper (Angelou and Spyrou, 2015) the sails model was expanded from the outskirts of upwind sailing, to beam and fully downwind cases. The method was a pseudo-3d approach, based on the evaluation of the flow field characteristics around certain cross sections of the sails.

The 2-d fluid domain of each section was obtained by solving the system of non-conservative Vorticity Transport and Stream Function Equations using the Finite Volume method, while the deformed shape of each section was obtained using a finite element formulation for flexure elements.

Apart from ignoring any 3-dimensional cross flow effects, the method also suffered from considerable numerical diffusion making its' applicability questionable. Moreover the induced computational cost prohibited the simultaneous solution of the method along with the hydrodynamic model; thus it was abandoned. However as the long term objective of this study is directional stability analysis using 6 degrees of freedom while taking into account the instant position and shape of the sail(s), this method proved useful as an intermediate step towards a Lagrangian "free" vorticity formulation, where remeshing of the domain and the large computational cost would be avoided.

In the Lagrangian approach a B.E.M. solves for the surface vorticity distribution around the sail. The vorticity is then shed to the domain by assigning its strength to newly created numerical quantities (vortex blobs) that are allowed to convect and diffuse according to the Vorticity Transport Equation. The numerical solution of the latter is dictated by the viscous split technique where the convection and diffusion terms are treated separately.

Convection was treated in a time marching scheme with the addition of mollifier terms to regularize the vortex kernels and desingularize the velocity near the vortex core, while diffusion was modeled implicitly; vorticity values were not actually changed and the diffusion effects were introduced by small perturbations of the translational displacements of the blobs by application of the random walk method. At the time this paper is written, the aforementioned method is still under development and, so far, only moderate inflow angles have been treated successfully (Figure 5).

Figure 5 – Numerical Vortex blobs evolution behind a sail chord line.

Since there are no result yet available to the authors for purely downwind cases, the driving coefficients were adopted from available literature. Lasher et al (2003), measured lift and drag coefficients for a series of investigated spinnakers (Figure 6).

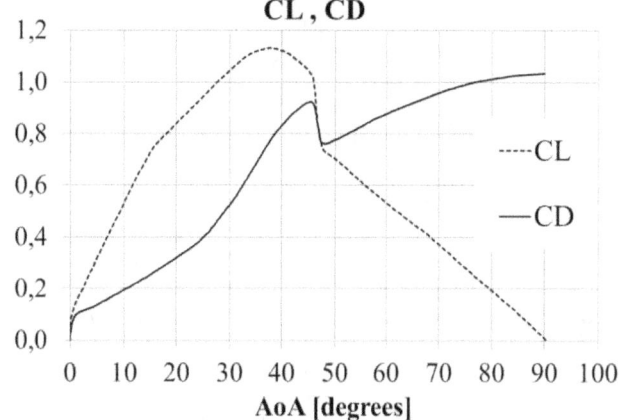

Figure 6 – Lift and Drag coefficients for a spinnaker. Curves are recreated based on data from Lasher (2003).

The sail forces and moments for the downwind case, where the yacht is sailing equipped with a spinnaker sail are:

$$X_S = L_S \cdot \sin(a_{AW}) + D_S \cdot \cos(a_{AW}) \quad (1.16)$$

$$Y_S = D_S \cdot \sin(a_{AW}) + L_S \cdot \cos(a_{AW}) \quad (1.17)$$

$$K_S = -Y_S \cdot z_{CEFF} \quad (1.18)$$

$$N_S = Y_S \cdot x_{CEFF} + X_S \cdot y_{CEFF} \quad (1.19)$$

HULL MODEL

Hydrostatic Forces and Moments

In order to calculate the hydrostatic pressure terms, the hull is discretized into rectangular panels (Figure 7) each of which has its submergence evaluated during every time step. Numerical integration of equations (1.20) and (1.21) for the static pressure sustained by the submerged panels, either on still water or on wavy conditions, provides the instant values of hydrostatic forces and moments.

$$\vec{F}_{HS} = -\rho g \iint\limits_{S} (z_p + \zeta) \cdot \vec{n}_p \, dS \qquad (1.20)$$

$$\vec{M}_{HS} = -\rho g \iint\limits_{S} (z_p + \zeta) \cdot (\vec{r}_p \times \vec{n}_p) \, dS \qquad (1.21)$$

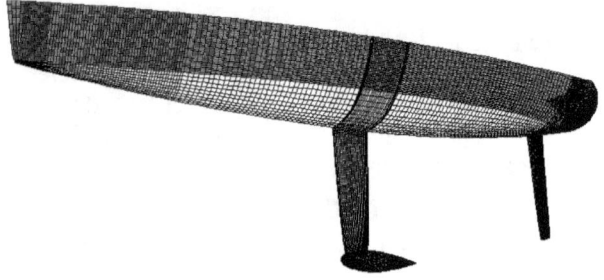

Figure 7 – Hull Paneling.

Hull Reaction Forces and Moments

Following the approach of Oltmann and Sharma (1985), the terms involved with the hydrodynamic reaction of the hull and the appendages are distinguished to *Ideal Fluid*, *Cross-Flow* and *Lifting* effects.

$$\vec{F}_{HR} = \vec{F}_{IF} + \vec{F}_{CF} + \vec{F}_{L} \qquad (1.22)$$

The first two contributions are treated by a strip theory method, in contrast to the third which is obtained from a 3-dimensional method.

Ideal fluid effects

The potential flow effects on the hull are obtained as in Yuanxie (1986), where the acceleration and velocity hydrodynamic derivatives are based on the calculation of sectional added mass, and treated for the influence of the instantaneous submerged section under wave conditions in a similar manner to Tigkas and Spyrou (2012).

In this paper the sway added mass of each section is calculated by a simple B.E.M. under the assumption of a pure drift motion of zero frequency.

Considering a body moving inside a large volume of inviscid, incompressible and irrotational fluid where the fluid velocity can be described by the gradient of a scalar harmonic potential Φ, then the added mass tensor for the body in the aforementioned domain, is given by Newman (1977):

$$m_{ij} = \rho \iint\limits_{S_B} \left(\varphi_i \frac{\partial \varphi_j}{\partial n} \right) dS \qquad (1.23)$$

or

$$m_{ij} = \rho \int\limits_{X_B} \mu_{ij} \, dx \quad , \quad \mu_{ij} = \rho \int\limits_{Y_B} \varphi_i \frac{\partial \varphi_j}{\partial n} \, ds \qquad (1.24)$$

Then, the problem reduces into finding the added mass

μ_{ij} for 2-dimensional sections of the body by employing a representation for the associated potential. As in the numerical application of the L.S.T., this can be achieved by distributing a series of singularities, in this case source and sinks $\{\sigma\}$, on each section and reducing to a system of linear equations $A_{ij}\sigma_j = \vec{V}_{\infty i} \cdot \vec{n}_i$ by enforcing the no-entrance boundary condition. The induced potential and the corresponding perturbation velocities induced from these panel singularities on a random field point P are given by Katz and Plotkin (2001):

$$\varphi_P = -\frac{1}{4\pi} \int\limits_{L_{per}} \left(\sigma/r\right) ds, \nabla \varphi_P = -\frac{1}{4\pi} \int\limits_{L_{per}} \nabla \left(\sigma/r\right) ds \quad (1.25)$$

where, solving for the source strengths and substituting the latter in equation (1.25a), provides the potential distribution on the sectional body and the accompanying added mass.

The method is verified (Table 1) for the calculation of the added mass of an ellipsoid body of semi-axis α and β, using the exact solutions provided by Newman (1977).

$m_{yy} = \pi \cdot \rho \cdot \beta^2$ $\alpha = 1, \beta = 0.5$	Exact Value	Current Method - Number of panels		
		100	200	500
Calculated	805.03	834.77	819.93	805.83
Error		3.69%	1.85%	0.01%

Table 1 – Sectional Added Mass Verification

Even though application of the method employing a crude panel density (of 100 panels) is characterized by an error of acceptable magnitude, the number of the panels during simulations has been kept relatively high (around 500), since the computational burden was insignificant.

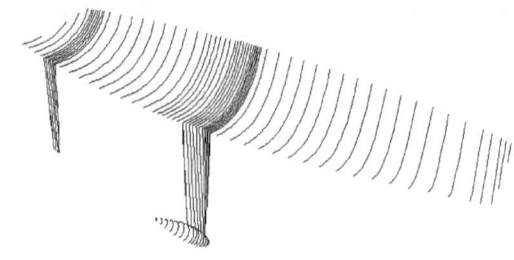

Figure 8 - Perspective view of hull port sections.

In terms of a yacht, the added mass is evaluated in certain positions by defining "y-z" plane sections along the hull length (Figure 8). The density of sectioning was greater in the area of the keel and the rudder, where geometry changes are abrupt.

At any instant in time during yacht's movement, the instant wetted part of each section is defined and mirrored across its waterline (Figure 9). The derived sectional sway added mass is halved, in order to be used in the succeeding calculations.

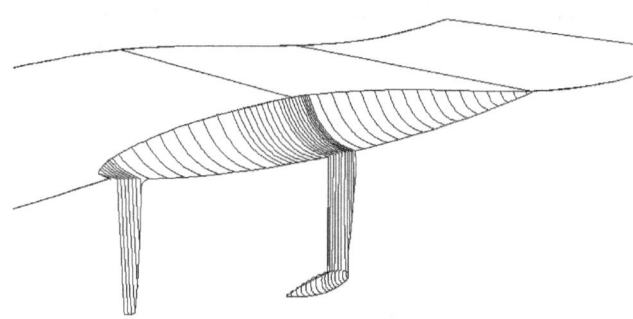

Figure 10a – Wetted hull sections for a following harmonic wave (λ=2L_{WL}, h=1/50) - hull on wave crest

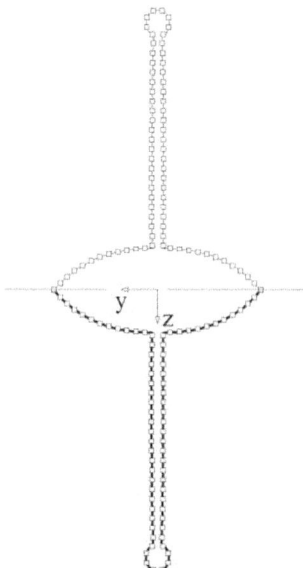

Figure 9 – Canoe-body and keel section (bold line) and mirrored configuration (crude panel density for imaging purpose) – view from bow

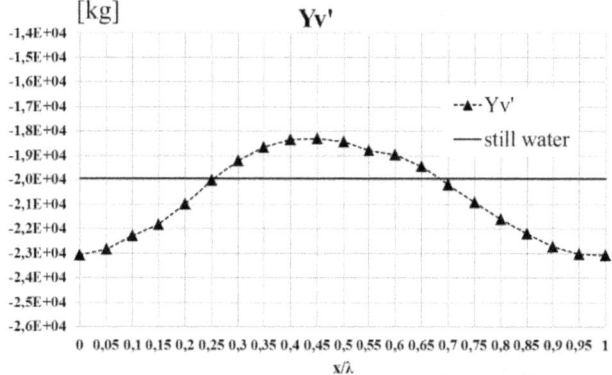

Figure 10b – Variation of $Y_{\dot{v}}$ on wave position.

Following Yuanxie (1986) the acceleration terms of the linear hydrodynamic derivatives on sway and yaw modes can be calculated by integrating sectional added mass dependent kernels.

$$Y_{\dot{v}} = m_{yy} = -\int_{-L/2}^{L/2} \mu_{yy}\, dx \qquad (1.26)$$

$$N_{\dot{r}} = -\int_{-L/2}^{L/2} x^2 \cdot \mu_{yy}\, dx \qquad (1.27)$$

$$Y_{\dot{r}} = N_{\dot{v}} = -\int_{-L/2}^{L/2} x \cdot \mu_{yy}\, dx \qquad (1.28)$$

The variation of the acceleration derivatives due to purely following sea waves of harmonic type, characterized by a length of $\lambda = 2.0 \cdot L_{WL}$ and a steepness of $h = 1/50$, is depicted in Figures 10(a-d) for different relative hull-wave positions. The vertical axis corresponds to the respective force or moment, while the horizontal axis represents the relative position of the hull on the wave length.

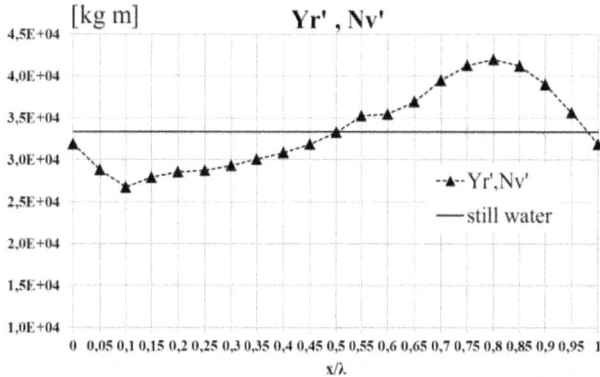

Figure 10c – Variation of $Y_{\dot{r}}$, $N_{\dot{v}}$ on wave position.

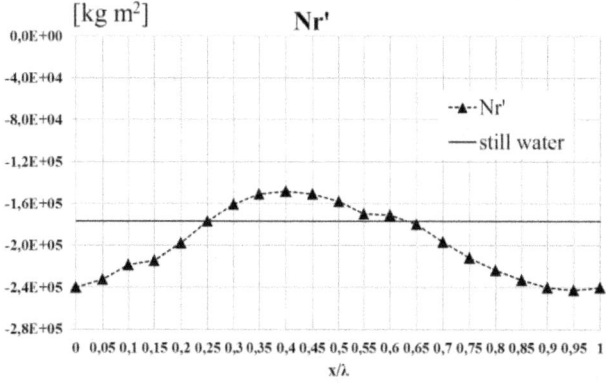

Figure 10d – Variation of $N_{\dot{r}}$ on wave position.

Moreover, assuming that surge and sway velocities as also yaw acceleration, are independent of the longitudinal position of the hull section, the velocity related terms are calculated from

$$Y_v = \int_{-L/2}^{L/2} \frac{\partial \{\mu_{yy}(x)\}}{\partial x}\, dx \qquad (1.29)$$

$$N_v = \int_{-L/2}^{L/2} x \frac{\partial \{\mu_{yy}(x)\}}{\partial x}\, dx \qquad (1.30)$$

$$Y_r = -\int_{-L/2}^{L/2} \left[x \frac{\partial \{\mu_{yy}(x)\}}{\partial x} + \mu_{yy}(x) \right] dx \qquad (1.31)$$

$$-m_x \cdot v_0$$

$$N_r = -\int_{-L/2}^{L/2} x \left[x \frac{\partial \{\mu_{yy}(x)\}}{\partial x} + \mu_{yy}(x) \right] dx \qquad (1.32)$$

And their variation for the same incident wave in this example is on Figures 11(a-d).

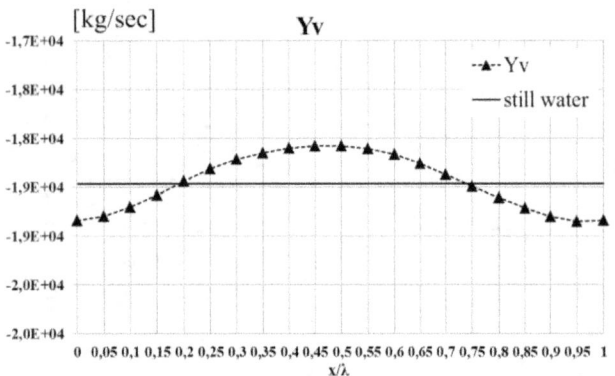

Figure 11a – Variation of Y_v on wave position.

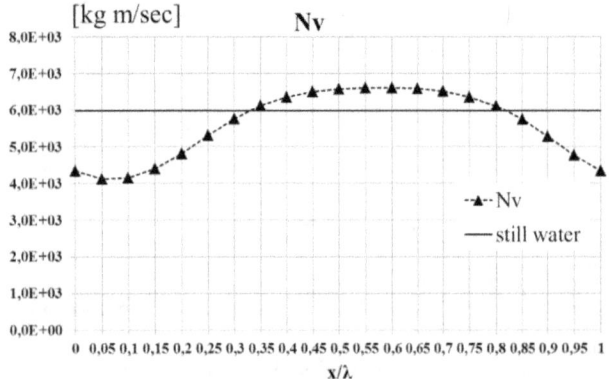

Figure 11b – Variation of N_v on wave position.

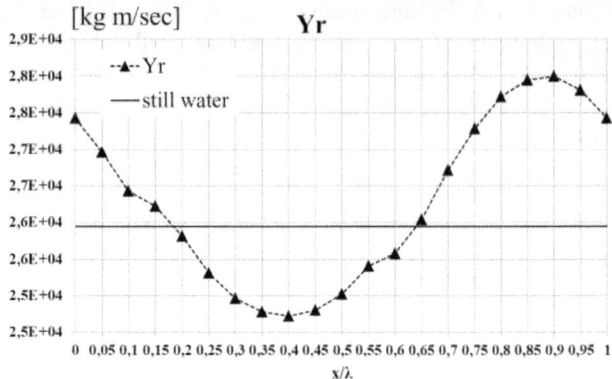

Figure 11c – Variation of Y_r on wave position.

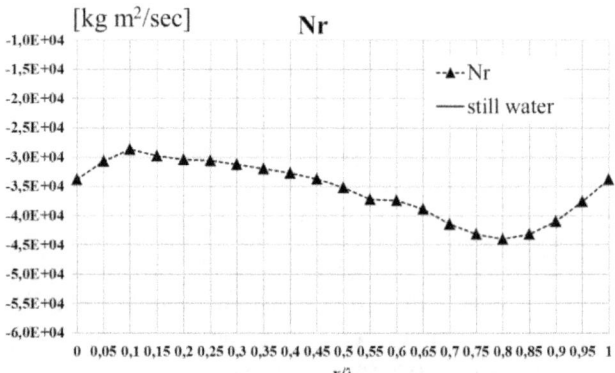

Figure 11d – Variation of N_r on wave position.

Cross-Flow effects

The cross-flow terms are the outcome of local transverse velocity on the hull, as the latter commences a turn or while sustaining a considerable drift angle. If $v_T = (v + r \cdot x)$ is the transverse velocity at the position of a section, then a hull of length L sustains, due to cross flow effects, a sway force and a yaw moment equal to:

$$Y_{CF} = -\frac{\rho}{2} \int_{-L/2}^{L/2} \{A(x,T) \cdot C_D(x,T,v_T) v_T |v_T| \cos\varphi\} dx \qquad (1.33)$$

$$N_{CF} = -\frac{\rho}{2} \int_{-L/2}^{L/2} \{A(x,T) \cdot C_D(x,T,v_T) v_T |v_T| \cdot x \cos\varphi\} dx \quad (1.34)$$

where $A(x,T)$ is the cross sectional area of a section at longitudinal position x and instant draft T , $C_D(x,T,v)$ is the cross-flow drag coefficient for that section as a function also of the transverse velocity defining laminar or turbulent flow behavior.

This coefficient can be approximated from published in the literature results that concern equivalent ellipsoidal or triangular two-dimensional shapes, mainly from Hoerner (1965). However, extracting values, for quantitative use, from an old, low resolution, printed source, entails the possibility of large inaccuracy in the prediction of C_D. This comes in contrast to the 2nd optional method of using state-of-the-art viscous flow CFD solvers that can provide C_D values, in principle with good accuracy.

It should be noted that this coefficient involves information about the viscous-pressure drag of the section. However the potential influence of the cross-flow drag has already been implemented inside the velocity derivatives of the *ideal-fluid effects* sub-model. This introduces a problem of finding a representation regarding the purely viscous portion of C_D, as adopting values for it from either proposed methods will lead to loss of the wave-variation character of the velocity derivatives.

That being, the choice was made to use commercial RANSE software to calculate the total drag coefficient of a series of double bodies of sections of representative shape and for a range of drafts and then subtract the pressure component. The obtained purely viscous based terms for each section and instant position were fitted for a range of velocities using functions of $f(v) = a \cdot v^b$ form (Figure 12).

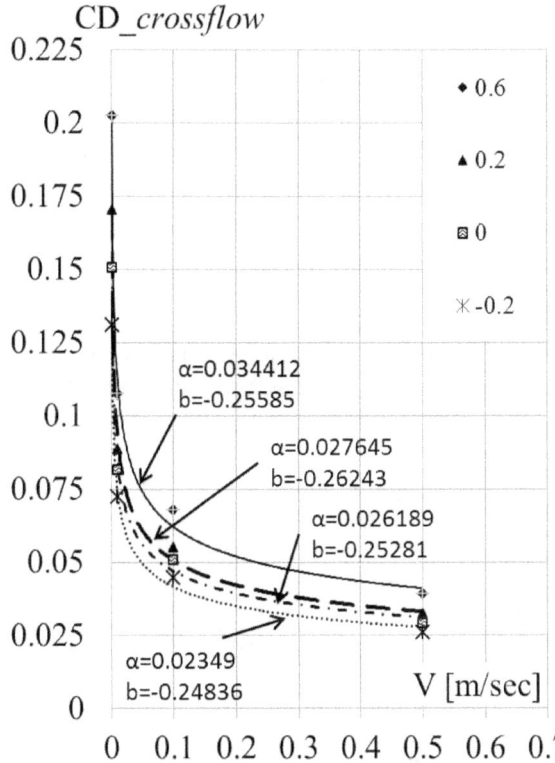

Figure 12 – Cross-flow drag coefficient vs. velocity and fitting functions for a section with keel, with parameter section's draft. Markers and numbers in legend represent emergence (negative) or submergence (positive) deflection from design draft.

The calculations were performed in ANSYS Fluent, incorporating the Spallart – Almaras turbulence model, and in an effort to create meshes that would allow wall y^+

values to retain a magnitude of less than unity during the solution. It should be mentioned that there were cases where this was very difficult to achieve and the drag might have been underestimated.

This raises a question regarding the accuracy of the method. However, the sections resemble bluff bodies, where the pressure drag is expected to dominate in magnitude, over its viscous counterpart. Indeed, the overall influence of the developed viscous forces was of very small order and loses in accuracy may be considered acceptable.

Moreover calculating the viscous force for a very large set of combinations of draft and heel angles for every section, would involve a huge computational load of pre-runs in order to equip the model.

Lifting effects
The keel and the rudder of the hull, due to their slenderness can have their lift and lift-induced drag calculated using the horseshoe vortex lifting surface method, as was the case of sails (respective curves on figures 13 and 14).

In the case of the canoe body of the hull though, the method falls short in applicability, as thickness effects cannot be disregarded. Its' shape being far from slender, makes the treatment even with the thickness corrected version of the lifting surface method unsuitable.

Application of other 3-dimensional methods for the direct or indirect representation of the potential of thick lifting bodies, such as variations of Hess-Smith or Morino methods, are questionable in this case, as there is no specific trailing edge for the Kutta condition to be applied and blunt aft shapes are prone to occurrences of flow separation. Moreover the points of separation are basically unknown.

The canoe body of the hull can be approximated as a very low aspect ratio $\{A_R\}$ airfoil with a round planform. This configuration is characterized by a strong dependence to non-linear lift and according to Hoerner (1985) it may be approximated by (Figure 13)

$$C_L = 0.5\,\pi\,\sin a \cos a (A_R + a) \qquad (1.35)$$

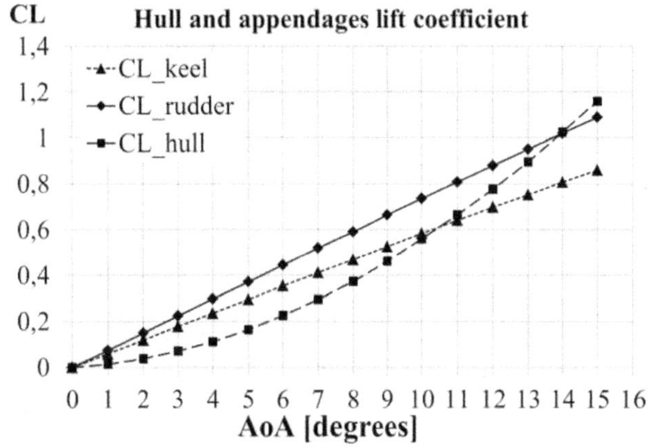

CL   **Hull and appendages lift coefficient**

Figure 13 - Lift coefficient for underwater surfaces.

In the same context, the lift induced drag is approximated by Hoerner (1965):

$$C_{DL} = C_{L\_linear} \tan(0.5a) + \Delta C_{L\_non-linear} \tan a$$
$$= 0.5\pi \sin a\, A_R \tan(0.5a) + k \sin^2 a \tan a \quad (1.36)$$

where $k \approx 1.25$ and $A_R \approx 0.045$ for the wetted part of the canoe body (Figure 14).

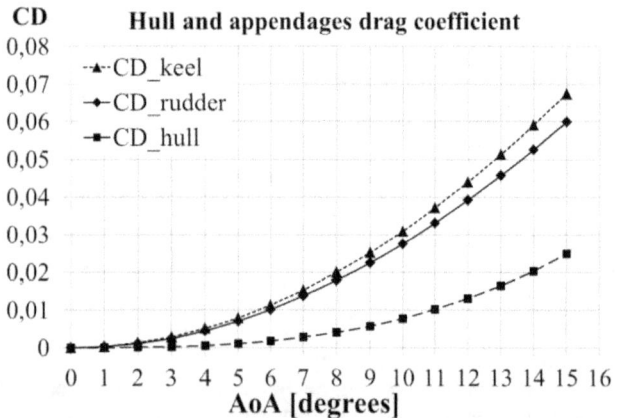

Figure 14 - Drag coefficient for underwater surfaces.

The lift and drag forces of the keel, are functions of drift angle (β) and their contribution on the equations of motions is obtained as:

$$L_{KEEL} = \frac{1}{2}\rho C_{L\_K}(\beta) V_B^2 S_K \quad (1.37)$$

$$D_{KEEL} = \frac{1}{2}\rho C_{D\_K}(\beta) V_B^2 S_K \quad (1.38)$$

$$X_{KEEL} = L_{KEEL} \sin(|\beta|) - D_{KEEL} \cos(\beta) \quad (1.39)$$

$$Y_{KEEL} = -L_{KEEL} \cos(\beta) sign(\beta) - D_{KEEL} \sin(\beta) \quad (1.40)$$

$$N_{KEEL} = Y_{KEEL} \cdot xcef_{KEEL} \quad (1.41)$$

$$K_{KEEL} = -Y_{KEEL} \cdot zcef_{KEEL} \quad (1.42)$$

and similar are the equations for the contribution of the hull.

Roll Damping

A damping coefficient is implemented using an approximation as in Masuyama (2011):

$$K_{\dot{\varphi}} = -2(a_0 + 0.4Fn) m g GM \left(\frac{T}{2\pi}\right)^2 \quad (1.43)$$

where a_0 was estimated from roll tests of similar yachts.

Rudder Forces and Moments

Since the calculations were performed considering a steady state mode, the influence of the trailing vortex sheet of the keel onto the rudder is not included. There is a need so, to introduce a correction factor and treat these effects implicitly. Thus the angle of attack a_R on the rudder is provided by:

$$a_R = \delta - \gamma_R \cdot \beta \quad (1.44)$$

where δ is the rudder deflection angle, while γ_R is the decreasing ratio of inflow angle (Masuyama and Fukasawa, 2011). The rudder contributions on the equations of motions are:

$$L_{RUD} = \frac{1}{2}\rho C_{L\_R}(a_R) V_B^2 S_R \quad (1.45)$$

$$D_{RUD} = \frac{1}{2}\rho C_{D\_R}(a_R) V_B^2 S_R \quad (1.46)$$

$$X_{RUD} = -L_{RUD} \sin(|a_R|) - D_{RUD} \cos(a_R) \quad (1.47)$$

$$Y_{RUD} = -L_{RUD} \cos(a_R) sign(a_R) - D_{RUD} \sin(a_R) \quad (1.48)$$

$$N_{RUDDER} = Y_{RUD} \cdot xcef_{RUD} \quad (1.49)$$

$$K_{RUDDER} = -Y_{RUD} \cdot zcef_{RUD} \quad (1.50)$$

Resistance

Regarding the surge equation of motion only (1.1), the resistance of the yacht can be decomposed to viscous, induced and wave-making parts. Viscous and induced terms are calculated as in Oossanen (1993) with some modifications regarding the contribution of the bulbous part of the keel, where a form factor has been implemented for the keel-bulb, as in Nesteruk & Cartwright (2011).

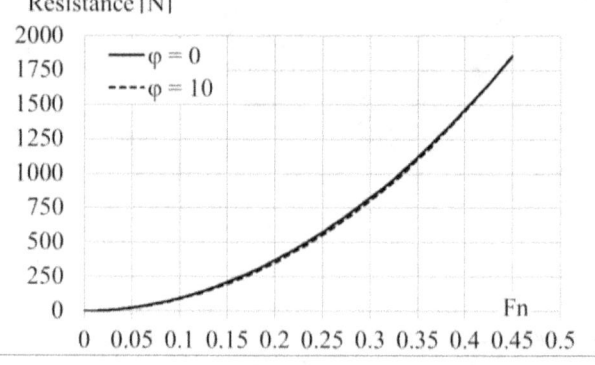

Figure 15 – Resistance curves vs. heel angles.

Wherever included in the above formulation, the wetted surface is calculated from the summation of the areas of the hull panels (Figure 7) that are immersed at that instant, taking into consideration waves, heave, heel, pitch and yaw. Lastly, the wave-making resistance of the yacht is calculated as in Pascual (2007) by an approximation based on the sectional areas the yacht. Total resistance for upright position and a heel angle of 10° is on Figure 15. The difference is insignificant as the wetted area does not change drastically. Larger differences should be expected when the wetted surfaces experiences variation due to wave induced oscillations.

Wave Excitation Forces and Moments

Considering an undisturbed pressure field around the yacht, the wave excitation is limited to Froude-Krylov forces and moments. These are calculated by integrating the unit potential φ_0 (e.g. Belenky & Sevastianov, 2003) on every immersed panel of the hull up to the elevated running waterline, after the panel coordinates have been transformed suitably for the relative position of the hull on the encountered wave.

$$\varphi_0 = \frac{g}{\omega} i \exp\left(k\zeta - i\left(x\cos\mu + y\sin\mu\right)\right) \quad (1.51)$$

$$\vec{F}_W = \mathrm{Re}\left\{-i\omega e^{i\omega t}\zeta \iint_S \varphi_0 \vec{n}_p \, dS\right\} \quad (1.52)$$

CASE STUDIES – EARLY RESULTS

In this section we present the outcome of two simulations; an upwind case where the yacht performs a multiple tacking maneuver in head sea waves and a downwind case where the transitional behavior from asymmetrical surging to surf-riding is examined.

The equation of motions ode's are solved using the 4th order Runge-Kutta method. All calculations involving numerical loops on hull panels and strip theories are perfomed using the OpenMP implementation of the Intel Fortran compiler on Linux Opensuse.

Upwind Simulation

In the first scenario the yacht commences a tacking maneuver with and without the presence of head sea waves of $\lambda=1.0$ L_{WL} and steepness of $h=1/50$. Using a simple PD controller for coursekeeping, the yacht performs tacking maneuvers around earth fixed Y axis and is ordered to keep a heading angle of 20°. When the distance of the trajectory from the Y axis reaches a threshold (set here on 40 meters) the controller orders the yacht to reach a heading of -20° until the bilateral distance criteria is met (Figure 16).

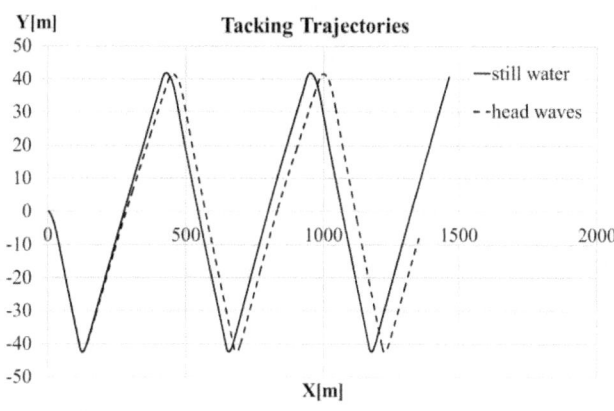

Figure 16 – Tacking Trajectory on calm and wavy seas.

The yacht experiences a surging behavior, however speed losses appear small (Figure 17) and are more easily apprehended by considering the difference in the total track covered by the yacht.

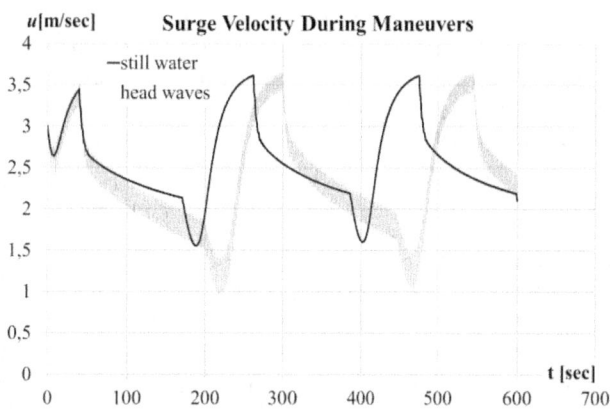

Figure 17 – Surge velocity during tack.

Downwind Simulation

In the second scenario the yacht is sailing under the influence of a purely following true wind (0° off the stern) of constant speed of 10 knots. The yacht is considered to carry a cruising symmetric spinnaker of area $S_{SPIN}=70$ m^2.

Simultaneously it is excited by following harmonic waves of $\lambda=2.0$ L_{WL} and steepness of $h=1/50$. As shown in Figure 18, the yacht experiences surging, which becomes asymmetric if the wave steepness is increased ($h=1/35$). If the incident waves steepness reaches the threshold of $h=1/30$ the yacht adopts surf-riding behavior and oscillates around the wave celerity.

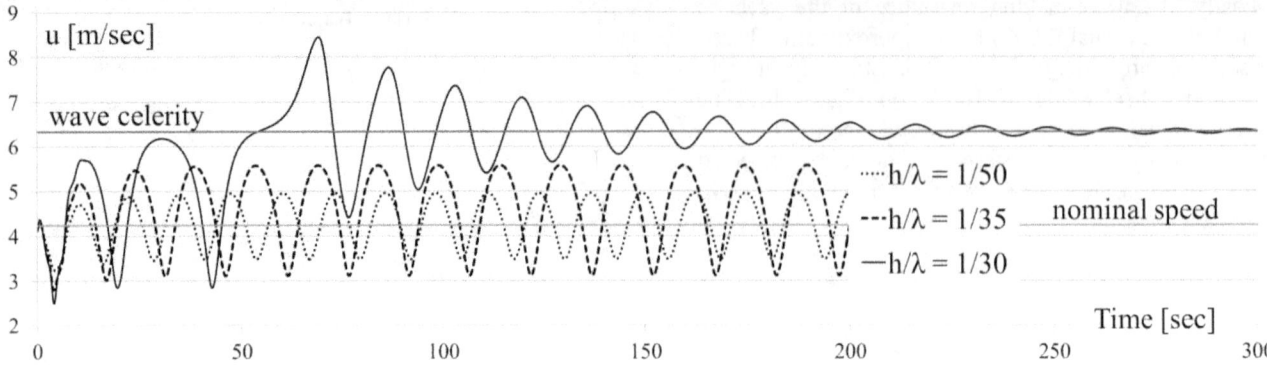

Figure 18 – Asymmetrical Surging and Surf-riding on following waves.

CONCLUSIONS

This study covered the development of a mathematical model for the investigation of directional instabilities of sailing yachts in waves. The character of this tool is modular, providing all parts with a potential for growth in terms of modeling complexity. The model is still under development. Admittedly, some of the sub-models have not had their contribution strongly justified as they are intended to be replaced soon by more sophisticated versions, having been implemented in their current form for the sake of overall model completeness. First trials on instability phenomena, covered in the literature only for motorboats, show qualitatively reasonable predictions. Along with the addition of transient and more detailed formulations regarding the excitations of the sails, as also the implementation of added mass and potential damping terms regarding all modes of motions, validation and verification with experimental and computational data is due to follow. It is the authors' intention to increase the complexity of the implemented sub-models, as also to broaden the model's overall applicability, while retaining the focus on the propensity of sailing yachts for instability phenomena compounding their performance and safety.

ACKNOWLEDGEMENTS

Mr. Angelou acknowledges with gratitude his support by NTUA's PhD Research Fund for Doctoral Candidates (ΕΛΚΕ). The authors also wish to thank the reviewer for his constructive comments.

REFERENCES

Angelou M., Spyrou, K.J., "Simulations of Sails of a Yacht using a Fluid-Structure Interaction Model", 10th HSTAM International Congress on Mechanics, Chania, Greece, 2013.

Angelou M., Spyrou, K.J., "Modeling Sailing Yachts' Course Instabilities Considering Sail Shape Deformations", Proceedings of the 12th International Conference on the Stability of Ships and Ocean Vehicles, Glasgow, UK, 2015.

Batoz J.L., Bathe K.J., Ho L.W., "A Study of Three-Node Triangular Plate Bending Elements". International Journal for Numerical Methods in Engineering, Vol. 15. 1771-1812, 1980.

Hoerner, S.F., "Fluid-Dynamic Drag", Hoerner Fluid Dynamics, 2nd edition, USA, 1965.

Hoerner, S.F., "Fluid-Dynamic Lift", Hoerner Fluid Dynamics, 2nd edition, USA, 1985.

Jones, P., Korpus,R. "International America's Cup Yacht Design Using Viscous Flow CFD", Proceedings of the 15th Chesapeake Sailing Yacht Symposium, Annapolis, MD, January 2001.

Katz J., Plotkin A., "Low Speed Aerodynamics", Cambridge University Press, Second Edition, Cambridge, 2001.

Kagemoto, H., Fujino, M., Kato, Takayoshi, K., Wu, Ch.T., "A Numerical and Experimental Study to Determine the Loads on a Yacht Keel in Regular Wave", Proceedings of the 7th International Conference on the Stability of Ships and Ocean Vehicles, Australia, 2000.

Lasher, W.C., Sonnenmeier, J.R., Forsman, D.R., Zhang, C., White, K., "Experimental Force Coefficients for a Parametric Series of Spinnakers", Proceedings of the 16th Chesapeake Sailing Yacht Symposium, Annapolis, MD, March 2003.

Masuyama, Y. and Fukasawa T., "Tacking Simulation of Sailing Yachts with New Model of Aerodynamic Force Variation During Tacking Maneuver", SNAME Journal of Sailboat Technology, 2011.

Melliere, R.A., "A Finite Element Method for Geometrically Non-linear Large Displacement Problems in Thin, Elastic Plates and Shells", Doctoral Dissertation, Missouri University of Science and Technology, Paper 2102, 1969.

Nesteruk, I. and Cartwright, J.H.E., "Turbulent Skin-Friction Drag on a Slender Body of Revolution and Gray's Paradox", 13th European Turbulence Conference (ETC13), Journal of Physics: Conference Series 318, 2011.

Newman, J.N., "Marine Hydrodynamics", The MIT Press, 9th edition, 1977.

Oltmann, P., Sharma S.D., "Simulation of Combined Engine and Rudder Maneuvers Using an Improved Model of Hull-Propeller-Rudder Interactions", The 15th Symposium of Naval Hydrodynamics, National Academy Press, 1985.

van Oossanen, P., "Predicting the Speed of Sailing Yachts," SNAME Transactions, Vol. 101, 1993.

Pascual, E., "Revised Approached to Khaskind'sMethod to Calculate the WaveMaking Resistance Depending on the Sectional Area Curve of the Ship", Journal of Ship Research, Vol. 51, September 2007, pp. 259-266, 2007.

SNAME, "Nomenclature for Treating the Motion of a Submerged Body Through a Fluid" SNAME Transactions, 1950.

Spyrou, K.J., "Historical Trails of Ship Broaching-To", The Transactions of the Royal Institution of Naval Architects: Part A - International Journal of Maritime Engineering, 152, Part A4, pp. 163-173, 2010.

Tigkas, I., Spyrou, K.J., "Continuation Analysis of Surf-riding and Periodic Responses of a Ship in Steep Quartering Seas", Proceedings of the 11th International Conference on the Stability of Ships and Ocean Vehicles, Athens, Greece, 2012.

Yuanxie, Zhao., "Calculation of Ship Manoeuvring Motions in Shallow Water", Schiffstechnic Bd. 33 1986.

Zienkiewicz O.C., Taylor R.L., "The Finite Element Method, Volume 1 : The Basis", Butterworth – Heinemann, Fifth Edition, Oxford, UK, 2000.

Zienkiewicz O.C., Taylor R.L., "The Finite Element Method, Volume 2 : Solid Mechanics", Butterworth - Heinemann, Fifth Edition, Oxford, UK, 2000.

The SYRF Wide Light Project

Martyn Prince, Wolfson Unit MTIA, University of Southampton, UK
Andrew Claughton, Ben Ainslie Racing, UK (formerly at Wolfson Unit MTIA)

ABSTRACT

Modern racing yacht semi-planing hull forms provide a number of complex challenges for designers and other professionals involved in yacht rating.

The SYRF Wide Light Project was initiated as a means of (1) providing data with which to assess a range of alternative computation methodologies to analyse sailing yacht hydrodynamic forces and moments, (2) making this data available to the entire sailing yacht research community and (3) demonstrating how this type of study can be used to inform the rating process.

This paper presents a comprehensive set of tank test results in both canoe body only and appended configurations to be used as a benchmark for a defined geometry of a modern semi-planing hull.

Five different CFD stakeholders carried out 'blind' CFD analysis on the same test matrix using a range of different computational codes and approaches. The results are presented here along with feedback detailing the software, methods and resources used to generate the results.

This project offers a comprehensive set of public domain data which researchers may use to validate and develop their numerical tools as well as highlighting how successfully commercial CFD codes may be used to confidently predict the variation of the forces on a sailing yacht hull as speed, heel and leeway change.

Finally, discussion will be made on how this first phase of the project may be used to inform handicap rule makers.

NOTATION

CEH	Centre of Effort Height
Fn	Froude Number
LCG	Longitudinal Centre of Gravity
q	Dynamic pressure (N/m^2)
R_I	Induced Resistance
R_U	Upright Resistance
R_{Vapp}	Appendage Viscous Resistance
R_{Vcb}	Canoe Body Viscous Resistance
R_W	Wave Resistance
R_0	Resistance due to heel
SF	Sideforce
SYRF	Sailing Yacht Research Foundation
Te	Effective Draft
V	Velocity
VPP	Velocity Prediction Program
ρ	Density (kg/m^3)
ø	Heel Angle

INTRODUCTION

This paper describes the Wide-Light Project, a project conceived and sponsored by the Sailing Yacht Research Foundation and initiated in 2013.

The "Wide Light" label is meant to describe the design of the modern high performance sailboats. These so called Wide Light boats present many of the hydrodynamic effects that are a challenge to predict: semi-planing hull forms, immersed transom effects, spray creation, keels operating close to the water surface, rudders, dagger-boards and canting keels that generate vertical force, the list goes on. The complexity of these interacting effects is challenging for the designer and a minefield for yacht handicappers who are obligated to handicap all boats equitably, both new and old.

It was therefore the goal of this project to provide data and conclusions of what might be seen as best practice or state-of-the-art modelling methods so as to better inform and equip handicapping systems and box rules to address Wide-Light designs. Projects such as the Delft Systematic Yacht Hull Series (Keuning, 1998) and the nine model series performed at the Canadian National Research Council (Teeters, 2003) have established the current database of hydrodynamics, but they are no longer representative of today's racing fleets. This project is the first step toward expanding the public database to include modern yachts.

The Wide-Light project aligns with the SYRF mission to develop and catalogue the science underlying sailing performance and handicapping, by:

- publishing an assessment of alternative methodologies to specify and analyse sailing yacht hydrodynamic resistance using computational tools,

- making the data available for researchers and students as an accessible experimental data set for a contemporary sailing yacht, and
- demonstrating how this type of study can be used to inform the handicapping process.

It is expected that the project will also be particularly valuable for academics as it will allow students, researchers and lecturers to frame new research projects most effectively.

All the data included in this paper is available from the SYRF website.

METHODOLOGY

This project's methodology draws upon well established procedures in ship hydrodynamics for comparing CFD data with experimental tank data (Larsson, 2010). Unlike in typical commercial and military studies where there are numerous tight controls on the process, for the Wide-Light projects the methods were accelerated to more quickly generate results. All results and information supplied were received within good faith, with no external validation or review process being undertaken.

A "test matrix" was developed to reflect a typical evaluation program for a sailing yacht, including tests of the bare canoe body and the appended hull over a range of speeds, heel and leeway angles. This matrix was distributed to the CFD stakeholders who used their usual procedures to generate data for each point in the matrix. The analysis was performed at model scale to avoid conflating the potential uncertainties of the scaling procedure with the CFD comparison. The stakeholders completed a "CFD Questionnaire" documenting the methodologies used to perform their CFD analysis.

The CFD calculations were delivered to the project leader before the tank testing was performed, therefore ensuring a blind test of the CFD codes. However, due to misinterpretations of the original test matrix, some CFD points were re-worked to correct for mistakes. The towing tank test was performed using the same test matrix as the CFD stakeholders.

PROCESS

The project process was as follows:

- procure a contemporary racing yacht tank test model,
- prepare a test matrix appropriate to this model that could be used in a physical model test and a CFD study,
- create a working group of leading CFD practitioners who were prepared to engage in a blind comparison of data whilst sharing their methodologies,
- CFD practitioners perform analysis and submit results,

- conduct towing tank tests using a state of art facility and methodology,
- prepare comparative data of CFD and model tests,
- prepare a summary report of the work for publication by SYRF, including not only the technical conclusions but also a discussion on how the data generated by the project can best serve the research and educational aims of SYRF, and
- summarise the way that the methodologies developed can be applied to yacht handicapping.

MATRIX

A summary of the test matrix is presented in Table 1 and with additional description in Table 2. The full test matrix can be found in reference SYRF 2015.

Towing tank tests are typically conducted using an assumed sail CEH so that an appropriate bow down trimming moment from the sail thrust can be applied to the model. This is necessary as the model is towed from a vertical location far lower than the CEH and this additional moment induces representative sail trim and heave. To take this approach, an estimate of the hull resistance is required. In this study, each test run had a predetermined LCG to avoid each contributor applying a slightly different sail trim moment. The specified LCG position for each test broadly simulates the effect of the sail trim moment based on an assumed hull resistance curve and appropriate sail plan.

ID	Configuration	TEST	Heel	Fn
CB-1	Canoe Body Only	Upright Resistance	0	0.10-0.80
CB-2		LCG Variation	0	0.35
CB-3		LCG Variation	0	0.5
CB-4		Heel at zero yaw	15	0.25-0.45
CB-5		Heel at zero yaw	25	0.25-0.45
CB-6		Heel with yaw	15	0.35
CB-7		Heel with yaw	25	0.5
HKR-1	Hull Keel & Single Rudder	Upright Resistance	0	0.10-0.80
HKR-2		LCG Variation	0	0.35
HKR-3		LCG Variation	0	0.5
HKR-4		Heel at zero yaw	15	0.25-0.45
HKR-5		Heel at zero yaw	25	0.25-0.45
HKR-6		Heel with yaw	15	0.35
HKR-7		Heel with yaw	25	0.5
HKR-8		Yaw Sweep	15	0.35
HKR-9		Yaw Sweep	15	0.5
HKR-10		Yaw Sweep	25	0.35
HKR-11		Yaw Sweep	25	0.5
HKr-1	Hull Keel & Twin Rudders	Yaw Sweep	15	0.35
HKr-2		Yaw Sweep	25	0.5

Table 1 Summary of Test Matrix

TANK TESTING

A previously tested model canoe body was donated to the project. To protect the intellectual property of the designer and the owner, the hull was modified to lines supplied by SYRF, re-faired, painted and marked up. The model was re-commissioned for testing and fitted with a previously used keel fin, bulb and movable single rudder.

The principal dimensions for this model, designated as model number M1108 by the Wolfson Unit, are presented in Table 3 and the hull lines in Figure 1.

ID	Description
CB-1	Basic resistance test on the unappended hull.
CB-2	Effect of shifting the LCG, i.e. changing fore and aft
CB-3	trim.
CB-4	Resistance test with the hull heeled.
CB-5	
CB-6	Change of leeway with rudder on centerline, positive
CB-7	and negative leeway.
HKR-1	Basic resistance test on the appended hull.
HKR-2	Effect of shifting the LCG, i.e. changing fore and aft
HKR-3	trim.
HKR-4	Resistance test with the hull heeled.
HKR-5	
HKR-6	Change of leeway with rudder on centerline, positive
HKR-7	and negative leeway.
HKR-8	
HKR-9	Leeway Sweeps and Rudder Variations at fixed speed
HKR-10	and heel angle.
HKR-11	
HKr-1	Leeway Sweeps and Rudder Variations at fixed speed
HKr-2	and heel angle.

Table 2 Test Matrix Description

M1108	Metric
Overall length	4.88 m
Design waterline length	4.60 m
Displacement (appended)	215 kg
Displacement (canoe body)	197 kg
Maximum beam	1.28 m
Draft to datum	1.15 m

Table 3 Model Principal Dimensions

The tank testing was carried out by the Wolfson Unit in the QinetiQ #2 towing tank at the Haslar Technology Park, Gosport, UK. The tank is 12m wide x 5.5m deep x 270m long. The model was towed using a purpose designed 3-post dynamometer which allowed the model freedom to heave and pitch, but provided restraint in yaw, sway and roll.

The model was ballasted to the specified displacement and trimmed to float parallel to DWL. For each test, a trim moment was applied that equated to the change in LCG outlined in the SYRF test matrix.

Measurements were made of the resistance, trim, heave, sideforce, roll moment and yaw moment.

To eliminate the effects of asymmetry, the majority of the heeled (and yawed) tests were made on both port and starboard tacks; these aspects and those relating to repeatability are detailed in Brown, 2002.

Measurements were also made of the wave height within the wake at a position 970mm aft of the transom on the tank centreline (equivalent to vessel centreline with zero leeway) using a sonic wave measuring device.

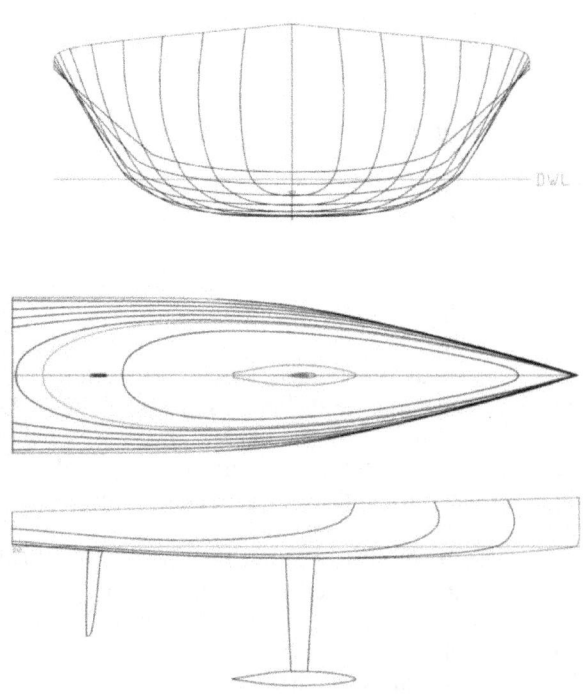

Figure 1 Hull Lines

To ensure consistent model boundary layer conditions, the canoe body was fitted with turbulence inducing studs (3.2mm diameter and 2.5mm high) around a section girth 300mm aft of datum. The appendages were in their previously tested condition. The bulb used carborundum grain strip at 20% of the chord length while the keel fin and rudder were fitted with smaller studs (1mm diameter and 0.7mm high) at 25% of the chord length. The model and appendage resistance data were corrected for the resistance of these studs and for the viscous effects to a standard water temperature of 15°C, mirroring that used for the CFD analysis.

Figure 2 Photos of the model under test in a heeled, appended condition

137

COMPUTATION FLUID DYNAMICS

An invitation was extended to practitioners who were active in the racing yacht field. Those who felt able to support the project were allowed to participate. In total, five individuals were designated as CFD stakeholders. Fortunately, this list represents a large segment of the CFD community, from commercial and open source RANS codes to computationally less intensive panel codes, which are detailed in Table 4.

Software	Contributor	Affiliation	Type
FINE™/ Marine	Benoit Mallol, Jason Ker	Numeca, Ker Yacht Design	RANS
FlowLogic	David Egan	Ennova Technologies	Panel Code
OpenFOAM	Sandy Wright	Wolfson Unit MTIA	Open Source RANS
SHIPFLOW	Lars Larsson, Michal Orych	Chalmers University of Technology, FLOWTECH Int. AB	Combination Panel Code & RANS
Star-CCM+	Rodrigo Azcueta, Matteo Lledri	Cape Horn Engineering	RANS

Table 4 Details of CFD Approaches

RESULTS

Practicalities

The test matrix for the project was developed with the CFD work in mind. In comparison to the computational process, the experimental (tank test) process is more complex and requires additional processing steps to allow for the comparison between the experimental and computational data. The rationale behind the processing of the experimental data follows.

A typical upright resistance curve, expressed as a drag area: $A_D = drag/q$ and is shown in Figure 3. The first few runs of the upright resistance curve are used to align the model to produce a minimum amount of sideforce and yaw moment. It is impossible to set the model up so accurately that it runs with no sideforce or yaw moment, but the data is corrected for the presence of these forces and moments.

For the heeled and yawed tests, a series of runs over a range of yaw angles are conducted at each speed and heel angle combination. These tests are implemented on both port and starboard tacks.

The Wolfson Unit's best practice uses an experimental set up that allows testing on both tacks. This provides the ability to determine the forces from a mean value from both tacks and delivers a more reliable test data set.

The yaw angles should be chosen so that the sailing sideforce for the heel angle is spanned. Using a

predetermined leeway angle sequence that remains the same for all speeds and heel angles will produce a substantial number of test points that are far removed from the 'sailing' condition. Also, for the heeled tests it is convenient to determine the down thrust of the sails (SF tanø) based on estimated sailing sideforce for the yacht. This may be applied regardless of the leeway angle set as it is not only easier from a practical view point but also avoids fluctuations in displacement affecting the determination and interpretation of induced drag.

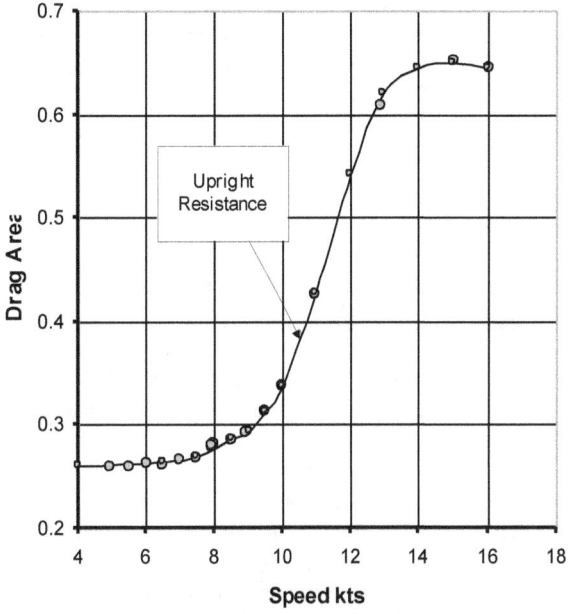

Figure 3 Typical Drag Area Curve for an Upright Resistance test

Typical test data is shown in Figure 4. There is usually some difference in the sideforce for a given leeway on the two tacks. At higher speeds and heel angles a difference of resistance tack to tack at the same sideforce, *not* the same leeway, of 1–2% is acceptable. Greater differences than this indicate a misalignment of the centre planes between the hull, keel and rudder.

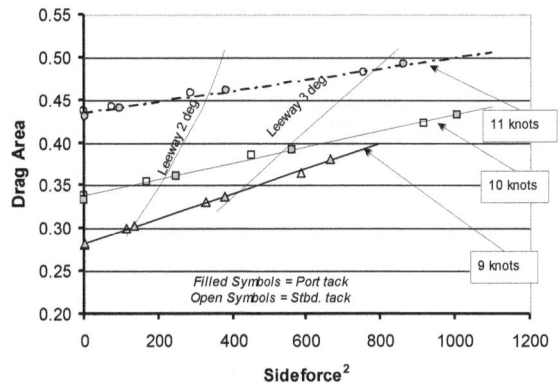

Figure 4 Typical heeled and yawed resistance test results

Typically 3-4 tests per tack are done at each speed. It should be noted that although it is tempting, from an analysis perspective, to make one of these a zero yaw point, there is often some non-linearity close to zero sideforce that makes it more useful to begin the tests at some low leeway angle. This process is repeated for all the speed/heel angle combinations in the test matrix.

As part of the tank analysis, the data from both tacks is analysed to determine the mean line through the data from both tacks and to average out the sideforce differences at the same nominal leeway on each tack.

Sailing Yacht Resistance Breakdown

The total hydrodynamic resistance of the yacht is assumed to be the sum of the following components:

$$R_{TOT} = R_U + R_0 + R_I \qquad \text{Eqn 1}$$

with:
$$R_U = R_W + R_{Vapp} + R_{Vcb} \qquad \text{Eqn 2}$$

R_0 is the increase in resistance above R_U at zero sideforce from the fitted line to the test data.

Because this study is not extrapolating the data to full scale it is not necessary to consider the de-construction of the resistance into viscous and gravitational components.

The combination of the aforementioned drag components is shown graphically in Figure 5 and is detailed in other sources (Claughton, 1998, for example). At a speed V and heel angle ø the resistance value can be determined by the intersection of the resistance against sideforce squared line with the equilibrium sailing sideforce line (shown as a dashed vertical line in Figure 5). The requirement of a typical Velocity Prediction Program (VPP) hydrodynamic force model is to determine these three resistance components — this approach must be adopted to assist with the comparison of the experimental and computed data.

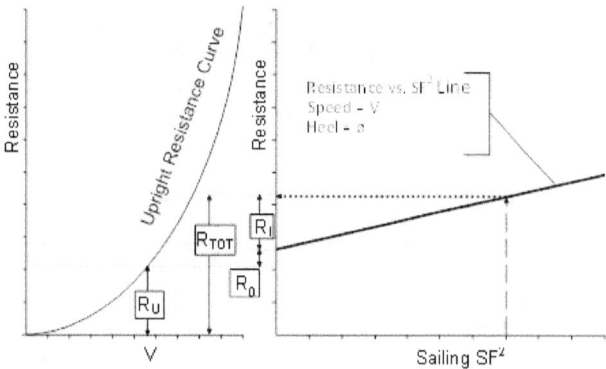

Figure 5 Resistance Breakdown

The first step is to fit a cubic spline curve to the upright resistance results using a least squares fit. This allows the upright resistance to be determined at any speed. Looking at the heeled and yawed results, unless stall is occurring, a

straight line can usually be fitted to the drag versus SF² data points at each speed and heel angle. The slope of the line is determined by the induced drag characteristics of the keel and rudder combined with the wavemaking effects. The slope of the line may be expressed as an effective draft (*Te*) derived from the formula:

$$Te = \sqrt{\frac{1}{\left(\frac{dR_I}{dSF^2}\right)\rho\pi V^2 \cos^2\phi}} \qquad \text{Eqn 3}$$

dR_I/dSF^2 = slope of resistance versus sideforce² line

The intercept of the straight line with the zero sideforce axis (R_0) determines the resistance due to heel, and may be expressed as a ratio to the R_U. Thus, for each tested speed and heel angle, the hydrodynamic behaviour can be expressed as an effective draft (*Te*) and a heel drag ratio (R_H/R_U), where:

$$R_H = R_U + R_0 \qquad \text{Eqn 4}$$

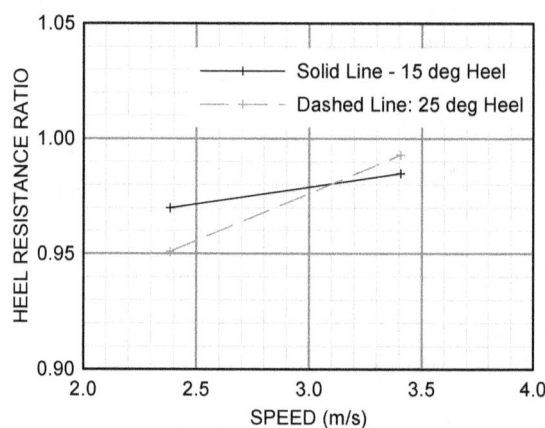

Figure 6 Variation of R_0/R_U versus speed for model derived from tank tests

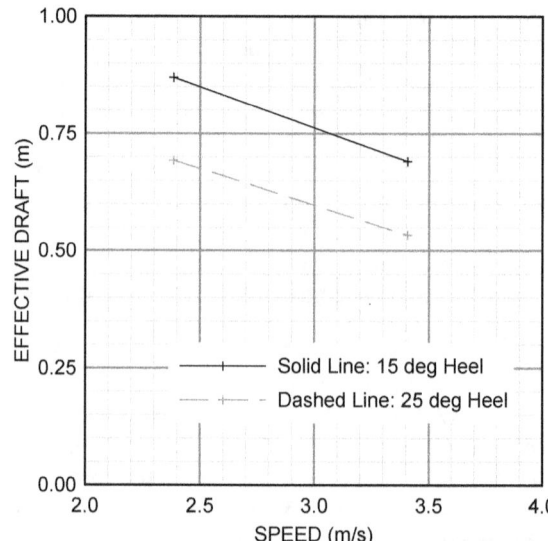

Figure 7 Variation of Te versus speed for the model derived from tank tests

139

Figure 6 illustrates this approach and shows the results from tests HKR 8-11 expressed as heel resistance ratio and Figure 7 the effective draft plotted against model speed in m/s.

The previous plots show typical behaviour — heel resistance increases both with heel angle and speed while effective draft reduces with increasing speed and heel angle as the keel root comes closer to the water surface.

Presenting the data in this form gives a much clearer insight into the quality of the results than simply comparing on a point by point basis. It allows us to see if the computational results actually match the trends of the physical test results, even if the "absolute" values are somewhat different.

CFD Results

Each participant submitted computational results for their specific CFD program. In addition to submitting results via spreadsheet, each participant provided a brief summary of his methodology and results. The CFD results are comprehensive and have not been presented in detail in this paper. The submitted files from each participant are available via reference SYRF, 2015.

CFD Questionnaire
CFD package?
Mesh Generation Software?
Solver?
Post process and visualization software?
Towing point?
Free surface tracking/capturing method?
Turbulence model?
Numerical Ventilation/Streaking
Ventilation observed?
Correction/Type?
Mesh type?
Computational Domain Size?
Local refinement?
Speed treatment?
Cell resolutions (in mm)?
Mesh sizes (in million cells)?
Time to prepare meshing?
Time to mesh each case?
CPU requirements?
Computer?
MeshSize/meshing/Solving/cores?
Cost effectiveness?

Table 5 Sample of CFD participant questionnaire

For the sake of brevity, this paper contains a summary of the results in the section below. All the results and associated files are available on the SYRF Technical Resources Library.

ANALYSIS

The previous sections describe an approach to viewing the hydrodynamic behaviour of a yacht hull that can be applied to any sailing vessel. Applied to Wide-Light designs, three fundamental features are apparent: the upright resistance increases with speed, the appendages represent a significant proportion of the total resistance, and the effective draft of the hull and keel reduce as speed and heel angle increase. The model tank test results can be compared with the CFD results with regards to the two following questions:

a) Do the computational results capture the general behaviour?

b) Do the absolute values of the data points agree?

Although CFD results are acknowledged to differ point by point with the experimental data, their relative differences can still be used as a reliable comparator. This approach can still be adopted for this study even though there is only a data set for one hull.

It must be borne in mind that the aim of this paper is to present the work carried out on the Wide light Project and as such as a summary analysis of the results from all the data sources (complete results in SYRF, 2015) has been undertaken. In time, more in depth analysis will be undertaken and it is hoped that other parties will also participate to this and add to it.

Results Comparison
Upright: Trim and Heave

The trim and heave curves for the tank test and CFD calculations are summarised in Figures 8 and 10 for the bare canoe body and Figure 9 and 11 for the appended model.

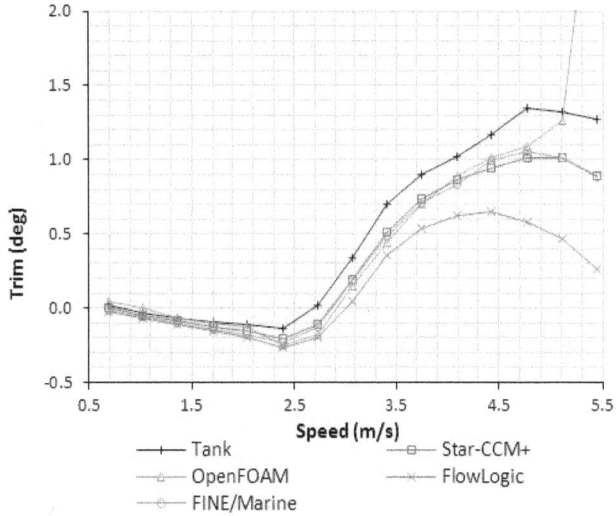

Figure 8 Canoe Body Only: Upright Trim Comparison

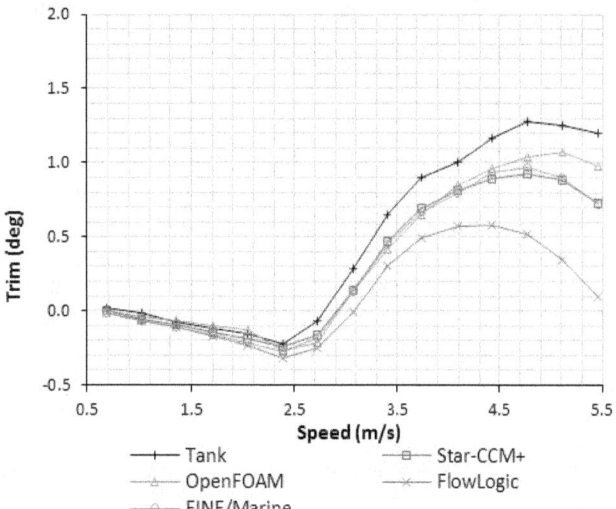

Figure 9 Appended: Upright Trim Comparison

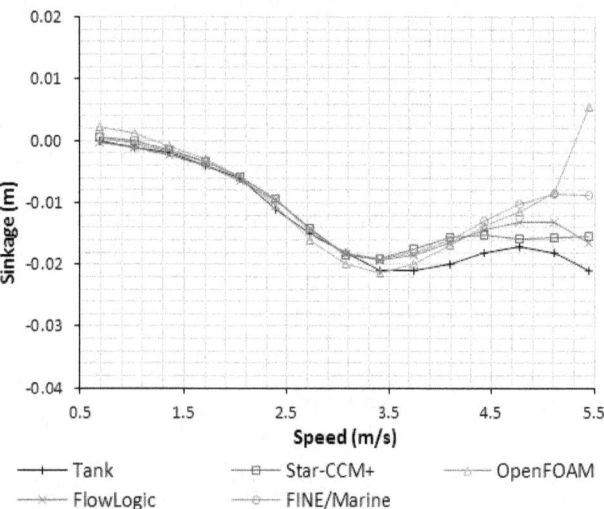

Figure 10 Canoe Body Only, Upright Sinkage Comparison

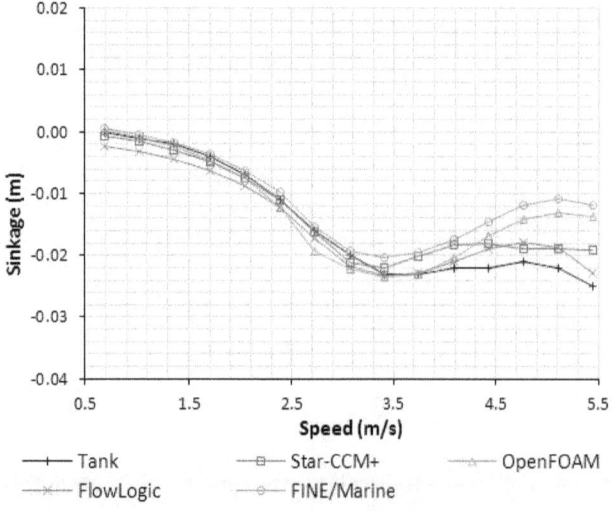

Figure 11 Appended: Upright Sinkage Comparison

Upright: Resistance

The upright resistance curves from the tank results and the computed results are shown in Figure 12. CB1 defines the canoe body only results; HKR1 defines the results including the hull, keel and rudder.

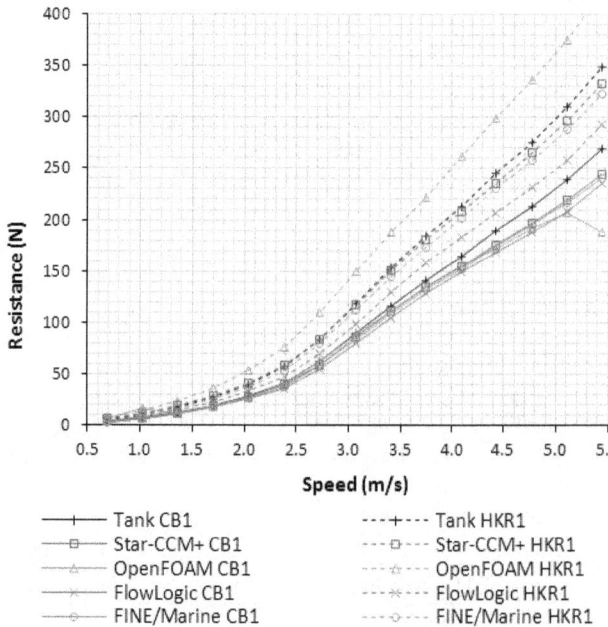

Figure 12 Canoe Body and Appended Upright Resistance Comparison

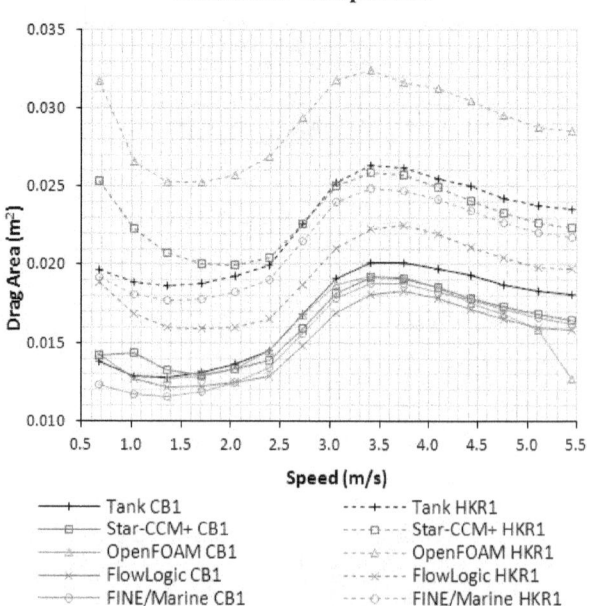

Figure 13 Canoe Body and Appended Upright Resistance Area Comparison

Figure 13 presents the Drag Area, where:

$$Drag\ Area = \frac{Resistance}{{}^1/_2 \rho V^2} \qquad \text{Eqn 5}$$

In Figure 13 it is important to note the suppressed zero.

141

The resistance at low speed is due to surface friction resistance, and the increase in resistance above that is due to residuary (wave making) resistance. For the canoe body, only the friction and residuary resistance are of similar magnitude. For the appended model, the friction resistance is nearly always greater than the residuary resistance. The friction resistance can be calculated from published data using the wetted area and waterline length, while the residuary and pressure form resistances require model tests or CFD to predict.

A "ratio plot" approach is used to more clearly evaluate the magnitude of the differences between the results. The results are expressed as the ratio of:

$$\frac{Resistance\ Comparator}{Resistance\ Baseline}$$

for each test point. Where tank test data is available, the tank drag is used as the baseline and therefore the tank test resistance ratio is constant at 1.00. The results for the upright resistance tests are shown in Figures 14 and 15.

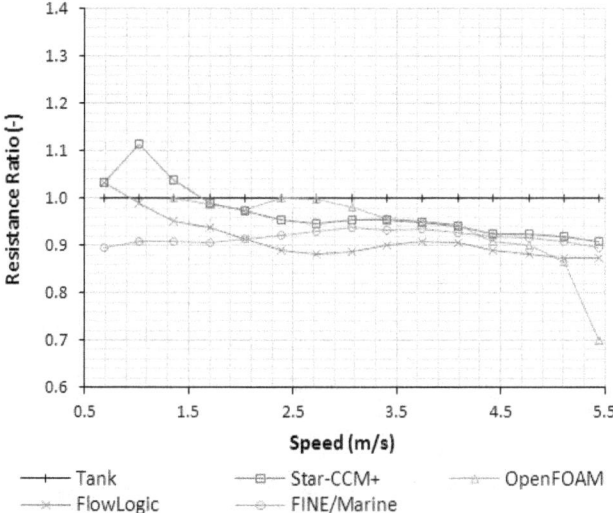

Figure 14 Canoe Body Only Upright Resistance, Ratio of CFD to Tank Test Results

Upright Run Summary

In summary, above a speed of 2.5 m/s, the CFD trim values depart from experimental data, at the higher speeds by approximately 0.4° for the RANS based codes and 0.8° for the Panel code. The OpenFOAM result shows a discontinuity at the highest speed which is reflected in the sinkage value.

The sinkage characteristics are similar for all methods up to 3.5 m/s, above this the various methods diverge, by up to 0.01m.

The body motion comparative trends presented in Figures 8 -11 are similar between the CFD methods and tank test results, for both the canoe body only and the appended results therefore in broad terms showing reasonable agreement.

Figure 14 for the CB-1 (canoe body only) case shows that most CFD points, for resistance prediction, lie within

7-8 % of the tank value. However, Figure 15 reveals that the agreement for OpenFOAM and FlowLogic is less in the HKR-1 (appended) configuration. The lack of agreement for OpenFOAM is largely because the computational scheme inherent in OpenFOAM meant that the computational time needed to accurately capture the drag of the keel and rudder was far beyond that which could be devoted to this project. As such, a simplified mesh was employed that reduced the accuracy of the appendage resistance predictions, but which did not affect the prediction of heeled resistance ratio and effective draft.

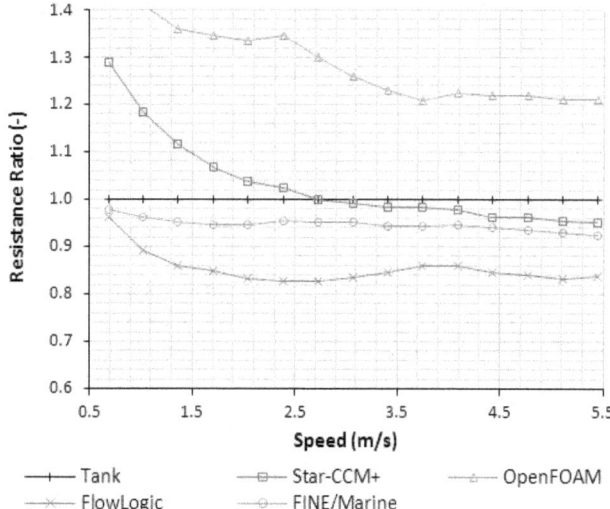

Figure 15 Appended Upright Resistance, Ratio of CFD to Tank Test Results

LCG Variation

The ratio plots for the LCG shift tests on the hull alone are shown in Figure 16 for a speed of 2.385 m/s (Fn = 0.35) and Figure 17 for a speed of 3.407 m/s (Fn = 0.5).

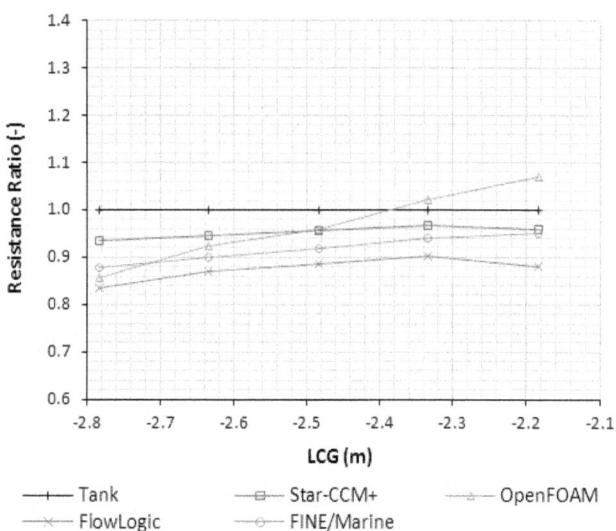

Figure 16 Canoe Body Only Upright LCG Shift at 2.385 m/s, Ratio to Tank Test Results

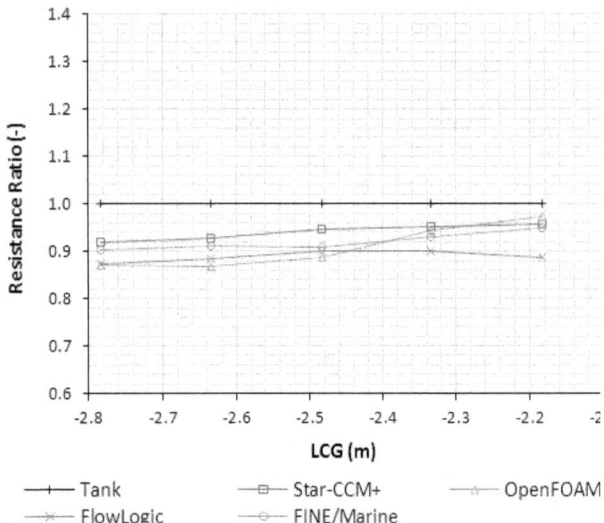

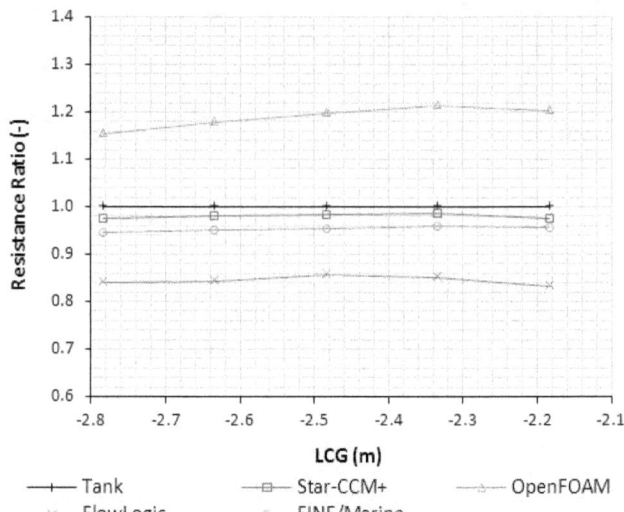

Figure 17 Canoe Body Only Upright LCG Shift at 3.407 m/s, Ratio to Tank Test Results

The ratio plots for the LCG shift tests on the appended hull are shown in Figure 18 for 2.385 m/s and Figure 19 for 3.407 m/s.

Figure 20 from the tank test shows typical behaviour for a change of LCG – there is a discernible resistance minimum for each speed at a given LCG position.

The CFD results mirror the tank behaviour, although the rate of change of resistance moving away from the minimum varies slightly as shown by the ratio plots, Figure 18 and 19.

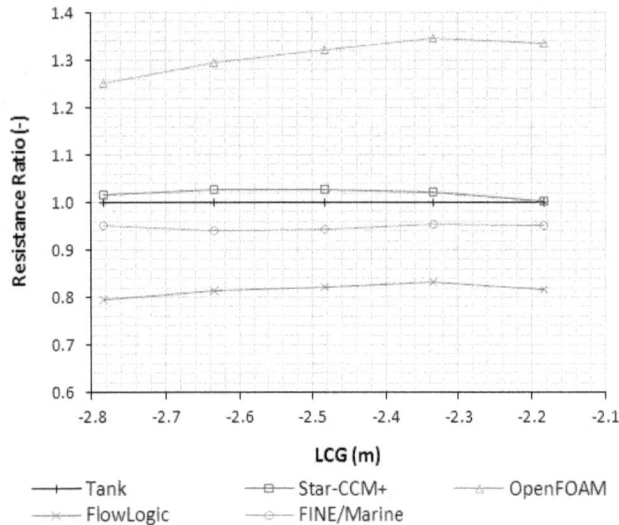

Figure 18 Appended Upright LCG Shift at 2.385 m/s, Ratio to Tank Test Results

Figure 19 Appended Upright LCG Shift at 3.407 m/s, Ratio to Tank Test Results

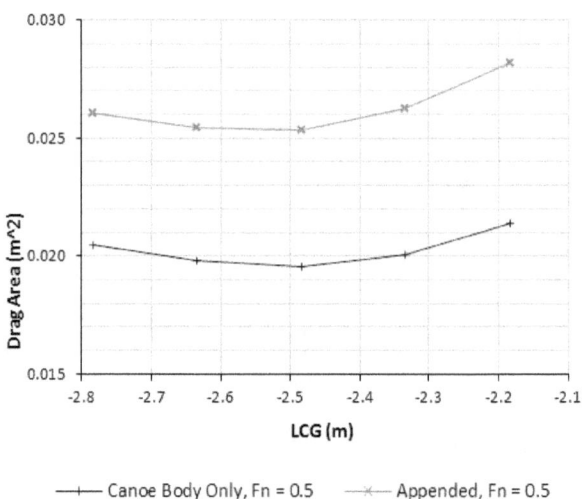

Figure 20 Drag Area Change with LCG from Tank Tests

Heeled: Resistance

The results of the heeled resistance tests are best expressed as a ratio of $\frac{Resistance\ at\ Heel}{Upright\ Resistance}$ at each test speed. This "Heel resistance ratio" for the tests are summarised in Figures 21 - 24. For example, for the canoe body tests at 15 degrees heel the Heel resistance ratio for the Star-CCM+ result at 1.7 m/s is 0.947; in other words, when heeled, the drag of the hull is 5.3% less than with the hull upright.

Although it is not a condition that the yacht can ever sail in, the resistance of the yacht when heeled at zero yaw angle provides a reliable indicator of the hull's characteristics. With this understanding, Figures 21 and 22, which only illustrates CFD results for the canoe body, shows consistent hull behaviour, with the heeled hull consistently having less resistance than the upright hull because the wetted area reduces as heel angle increases. A similar behaviour is shown for the appended hull, Figures 23 and 24, but the resistance reduction is somewhat reduced because the keel and rudder wetted surface do not change with heel.

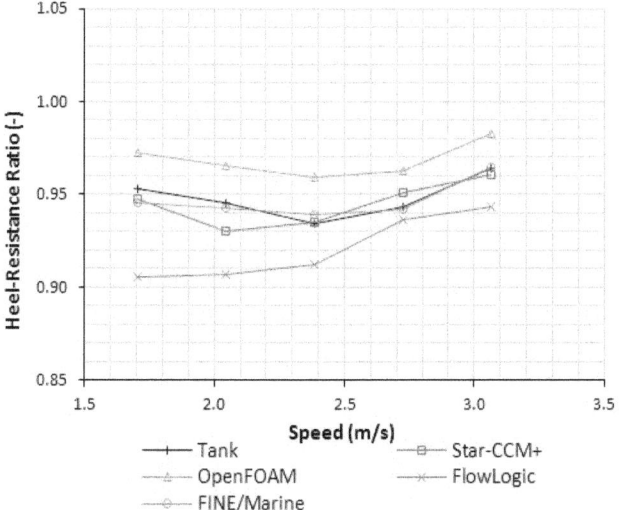

Figure 21 Canoe Body Only, Heel Resistance Data at 15° Heel

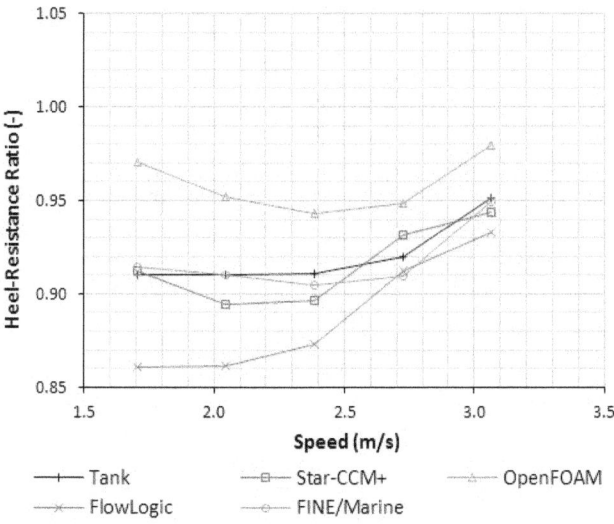

Figure 22 Canoe Body Only, Heel Resistance Data at 25° Heel

The towing tank results also show a trend of heeled resistance increasing as speed increases, and this is generally captured by the CFD results. Again, the ratio plots show that the trends are captured, but the absolute values differ to the same degree as the upright resistance results, as can be seen in Figures 25 - 28.

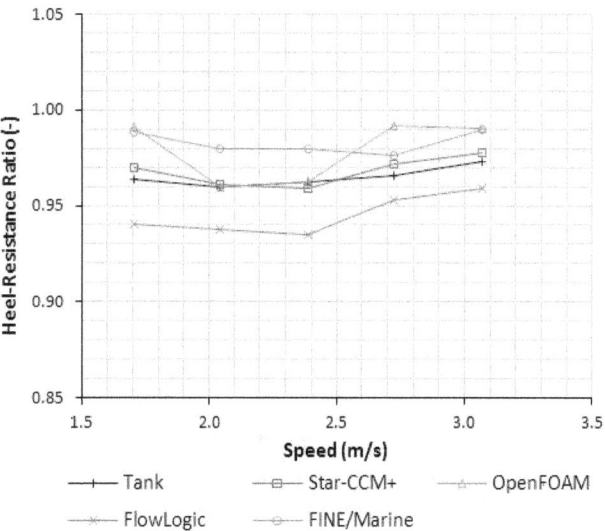

Figure 23 Appended, Heel Resistance Data at 15° Heel

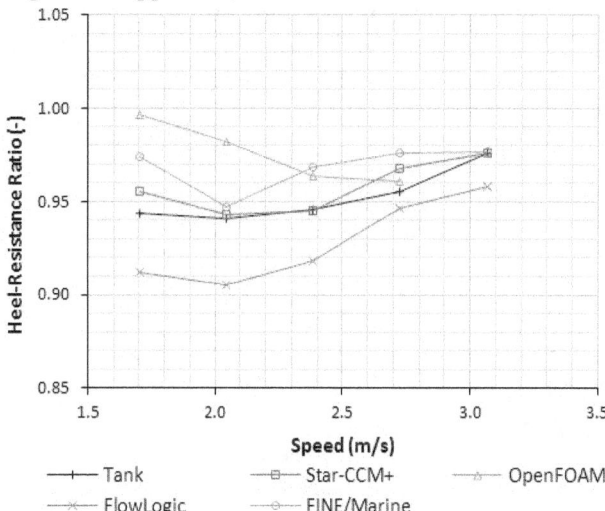

Figure 24 Appended, Heel Resistance Data at 25° Heel

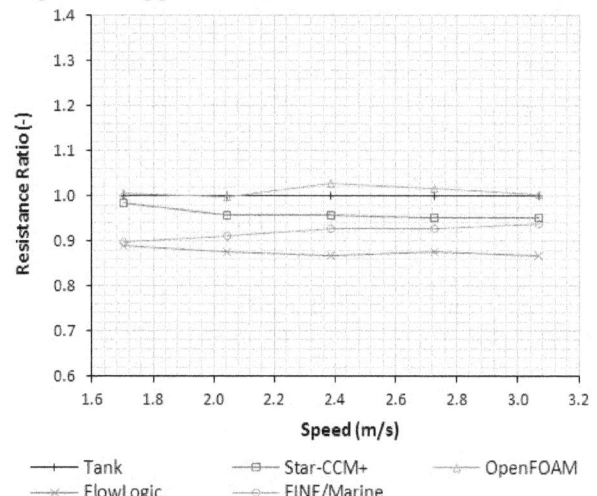

Figure 25 Canoe Body Only, 15° Heel, Ratio of CFD to Tank Test Results

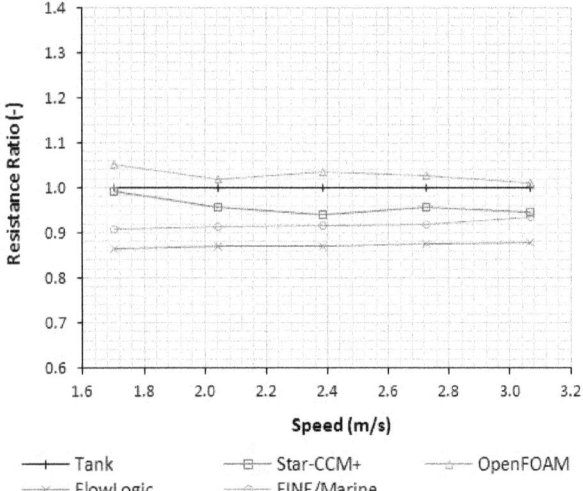

Figure 26 Canoe Body Only, 25° Heel, Ratio of CFD to Tank Test Results

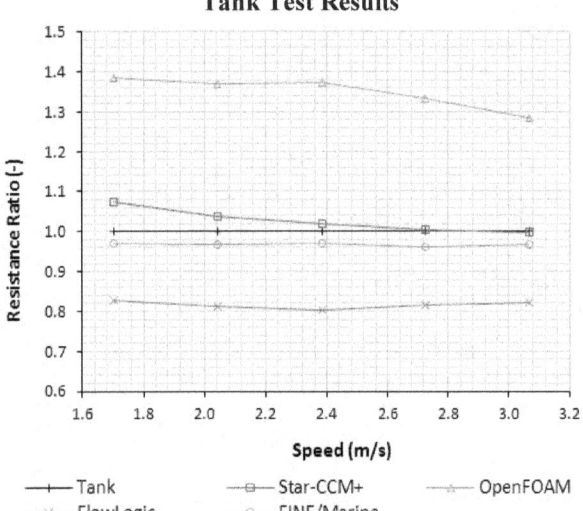

Figure 27 Appended, 15° Heel, Ratio of CFD to Tank Test Results

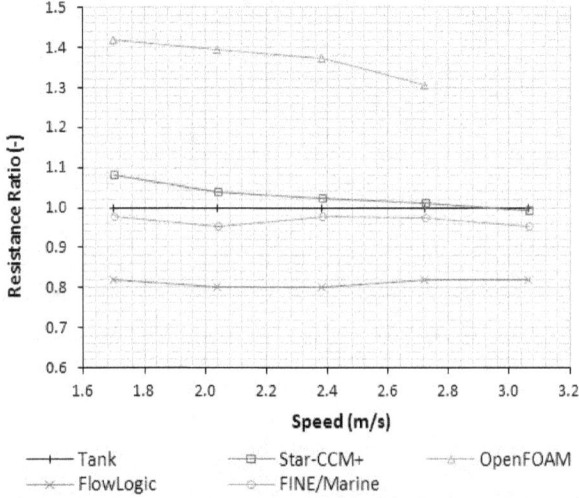

Figure 28 Appended, 25° Heel, Ratio of CFD to Tank Test Results

Heel with Yaw: Canoe Body Only

The results of the yaw sweeps on the canoe body are presented in Figure 29 as plots of resistance vs. yaw angle and in Figure 30, Sideforce vs. yaw angle. Because the model is heeled these do not show a discernible minimum at zero yaw. In fact the drag minimum is at a negative yaw angle because this aligns the centreline of the heeled waterplane more nearly with the hulls direction of travel through the water.

The yaw sweep results are presented as ratio plots to the tank data in Figure 31 and 32.

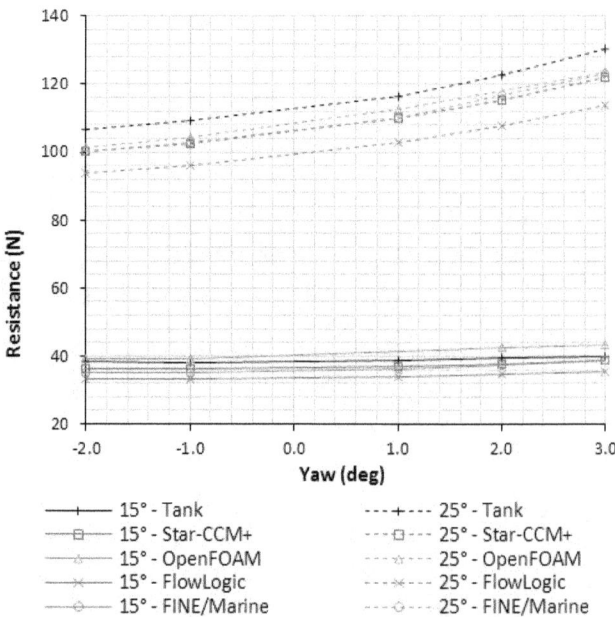

Figure 29 Canoe Body Only: Yaw Sweep Resistance Data

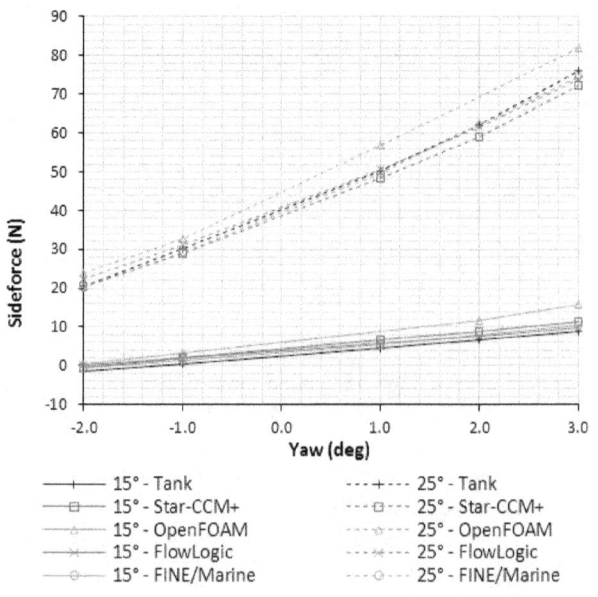

Figure 30 Canoe Body Only: Yaw Sweep Sideforce Data

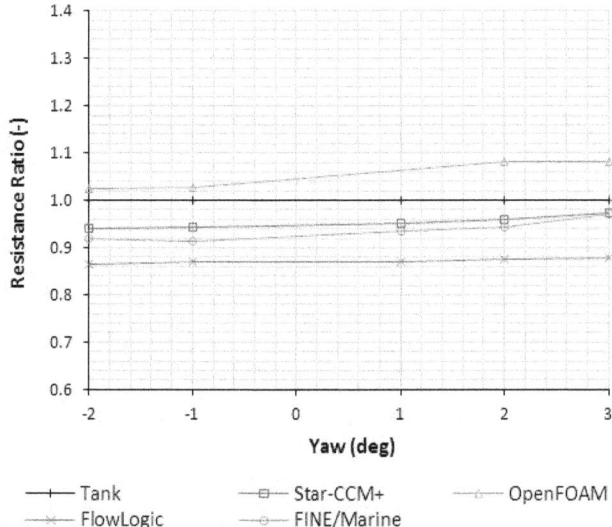

Figure 31 Canoe Body Only: 15° Heel, Yaw Sweep Resistance Ratio to Tank Results

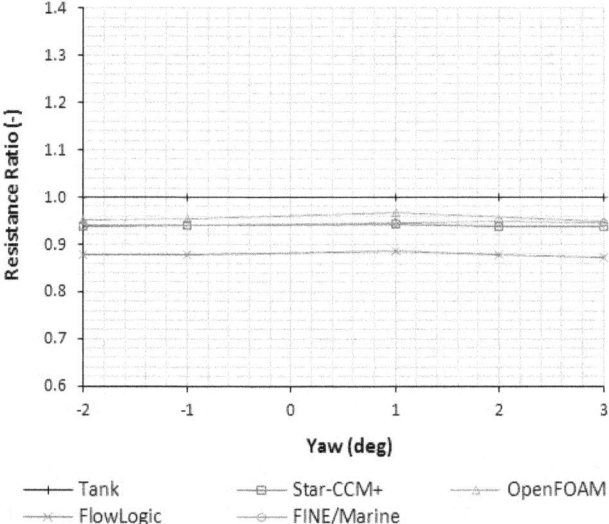

Figure 32 Canoe Body Only: 25° Heel, Yaw Sweep Resistance Ratio to Tank Results

The results of the yaw sweeps with the keel and rudder (at zero rudder angle) fitted are presented in Figure 33 as plots of resistance vs. yaw angle and in Figure 34 with sideforce vs. yaw angle. Here the resistance minimum is closer to zero yaw because this corresponds to the minimum Sideforce from the appendages. The yaw sweep results are presented as ratio plots to the tank data in Figure 35 and 36. Only a partial set of OpenFOAM results were supplied at 25° heel due to convergence issues experienced at the higher yaw angles.

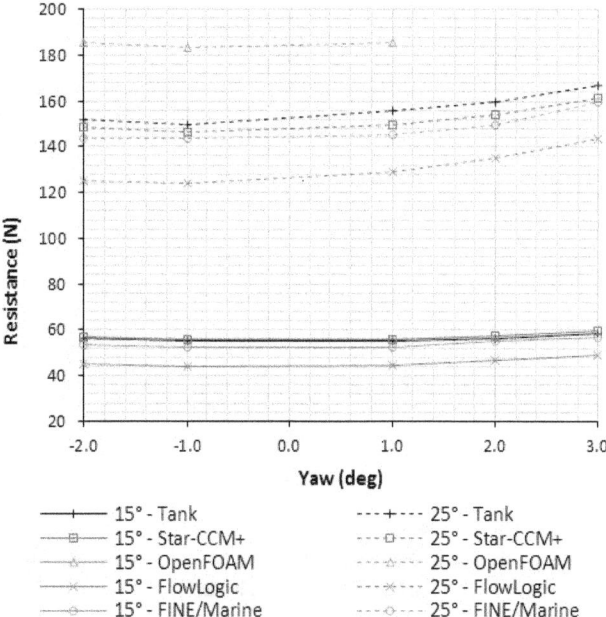

Figure 33 Appended: Yaw Sweep Resistance Data

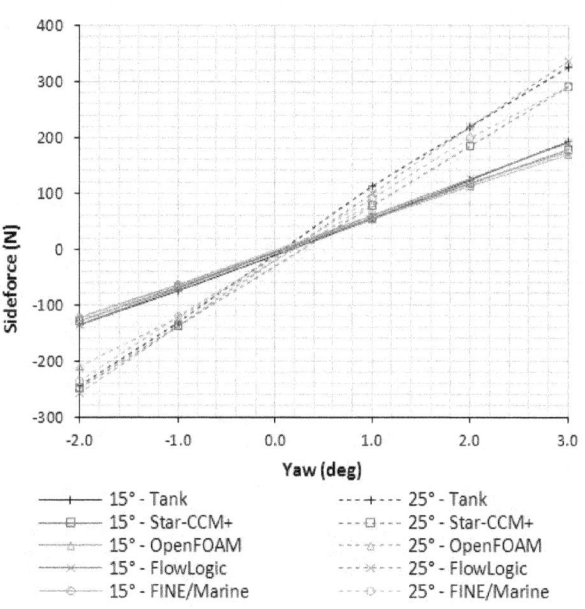

Figure 34 Appended: Yaw Sweep Sideforce Data

Just as with the previous figures, the ratio plots in Figures 35 and 36 show that the CFD results capture the trends shown by the tank test results, with similar divergence of absolute results that are seen consistently throughout the tests.

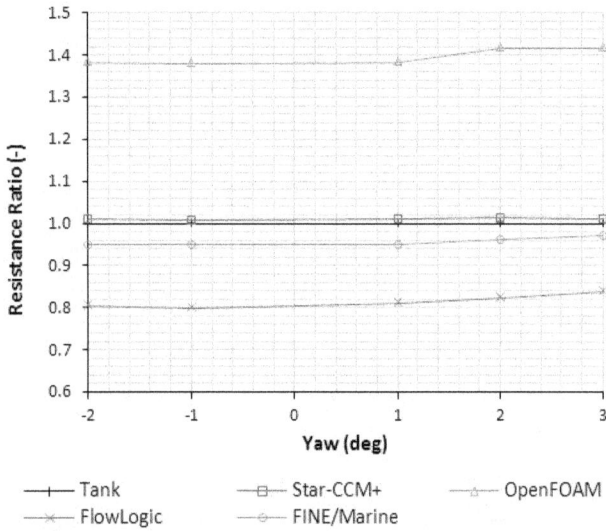

Figure 35 Appended: 15° Heel, Yaw Sweep Resistance Ratio to Tank Results

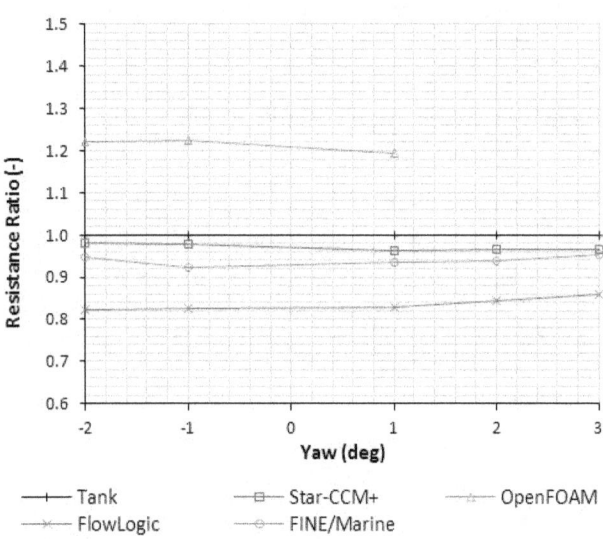

Figure 36 Appended: 25° Heel, Yaw Sweep Resistance Ratio to Tank Results

Heel with Yaw: Appended

As shown in Figure 4, the results of the yaw sweep tests can be captured in a single straight line on a resistance vs Sideforce^2 plot. The yaw sweep data are presented in Figures 37 and 38.

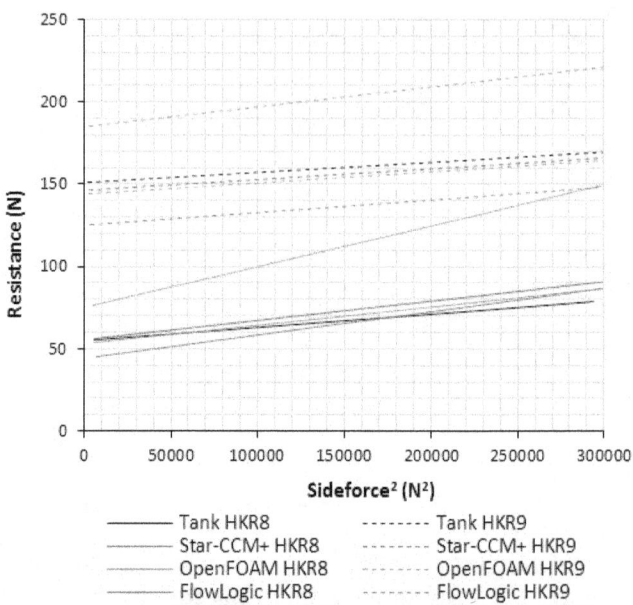

Figure 37 Appended, Summary of Fitted Lines to Yaw Sweep Data, 2.385 m/s & 3.407 m/s at 15° Heel

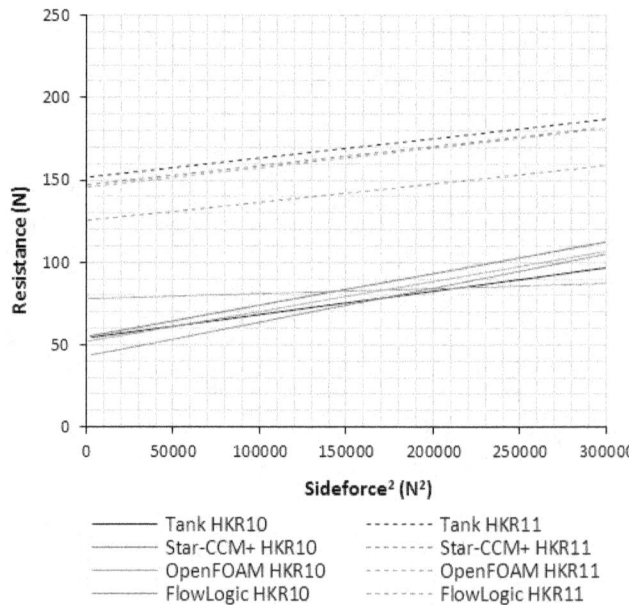

Figure 38 Appended, Summary of Fitted Lines to Yaw Sweep Data, 2.385 m/s & 3.407 m/s at 25° Heel

As described in a previous section, the character of the Resistance vs Sideforce^2 plots can be expressed as a heeled resistance ratio R_H/R_U and an effective draft, expressed in metres. Figures 39 - 42 shows the data expressed in this way for the heel angles of 15° and 25°.

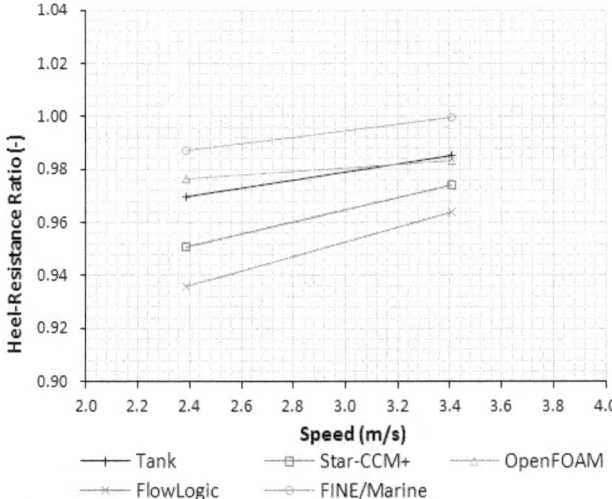

Figure 39 Appended, Heel Resistance Ratio Comparison at 15° Heel

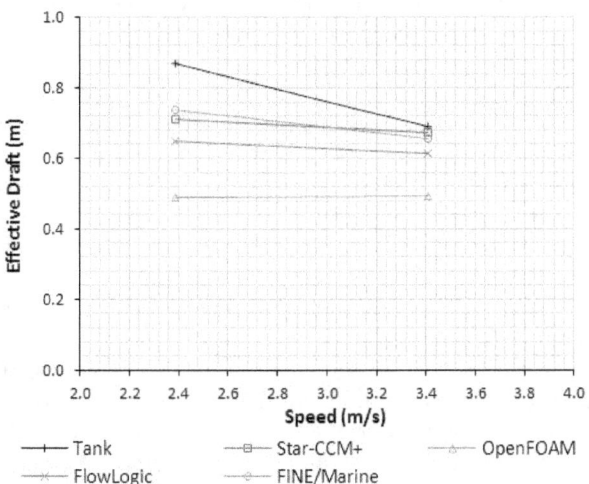

Figure 40 Appended, Effective Draft Comparison at 15° Heel

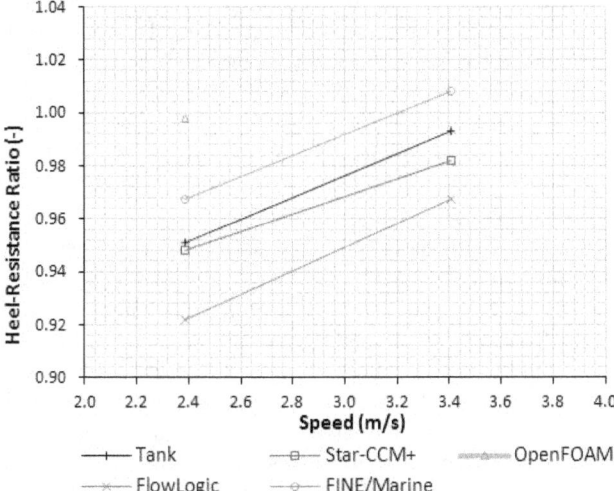

Figure 41 Appended, Heel Resistance Ratio Comparison at 25° Heel

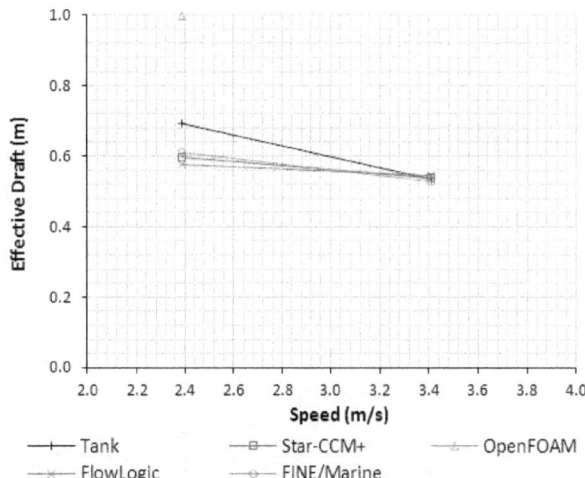

Figure 42 Appended, Effective Draft Comparison at 25° Heel

In broad terms, the Star-CCM+ and FINE/Marine codes are closest to the towing tank results in both absolute terms and capturing the slope of the Drag vs. Sideforce^2 line. The FlowLogic results capture the slope, but the OpenFOAM results are at some variance to the experimental result for the reasons discussed previously.

Heel with Yaw: Appended; Twin Rudder

Figure 43 shows the Resistance Sideforce^2 plots for the twin rudder results derived from the CFD calculations. No tank tests were made with the model in this configuration.

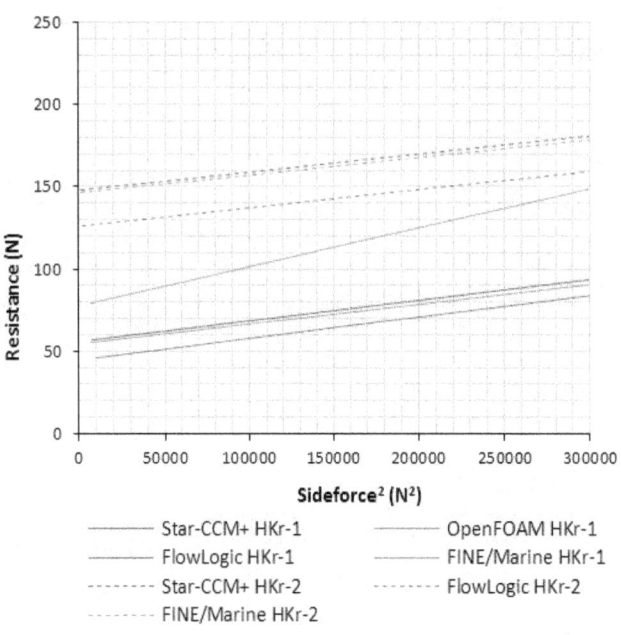

Figure 43 Comparison of CFD Results for Twin Rudder Configuration

Lift Curve Slope

This paper has focused on the comparison of resistance and side force with speed and heel angle, as these are the

148

main drivers of performance. However the data was also analysed to determine a simple lift slope from the appended heel and yawed tests. These results are presented in the Table 6.

The value was determined from the slope of the line SF/V^2 versus yaw angle where SF is in N and Speed in m/s.

These results reflect a satisfactory degree of consistency – the trend of lift slope reducing as heel angle increases is captured by CFD. However, these results reveal that the reduction of lift slope as speed increases is not captured.

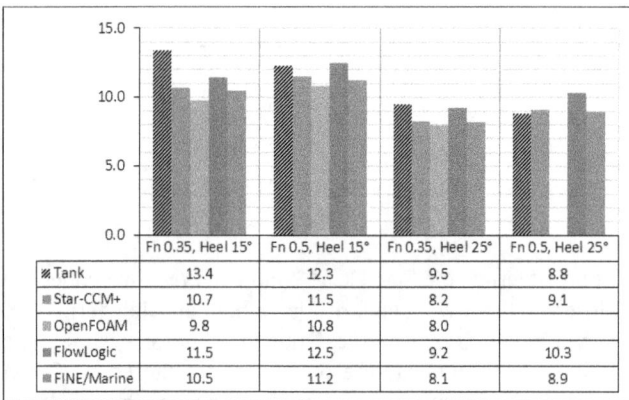

	Fn 0.35, Heel 15°	Fn 0.5, Heel 15°	Fn 0.35, Heel 25°	Fn 0.5, Heel 25°
Tank	13.4	12.3	9.5	8.8
Star-CCM+	10.7	11.5	8.2	9.1
OpenFOAM	9.8	10.8	8.0	
FlowLogic	11.5	12.5	9.2	10.3
FINE/Marine	10.5	11.2	8.1	8.9

Table 6 Lift Slope Comparison

DISCUSSION

The fundamental aim of this project was to explore how computational methods might be employed to improve force models used in VPP based handicap rules. The fundamental components of a sailing yacht's resistance are idealised in equation 1. Detailed below is a summary relating to those individual components.

Upright Resistance

Figures 14 and 15 show that the computational methods are tolerably capable of predicting the upright resistance for both the bare canoe body and the appended hull. Generally there is an offset of less than 10% when averaged across the entire speed range, with several methods being significantly closer at a number of the speeds tested, and the CFD generally under predicts the resistance. The computational methods are also sensitive to the effects of shifting the centre of gravity fore and aft.

Heeled Resistance

Figures 27 and 28 show that the advanced RANS codes (Star-CCM+ & FINE/Marine) are able to accurately predict the effects of heel on resistance.

Resistance due to Yaw

Figures 31 and 32 show that the computational methods are all able to predict the effect on sideforce of changing yaw angle.

Heel Drag Ratio

Figures 39 and 41 show that the computational methods

are all able to predict the increase of drag at zero sideforce as speed increases, although of course the "offset" found in the upright resistance curves is carried through into the results.

Induced Resistance

Figures 40 and 42 show that the tank test derived effective draft reduces as speed increases, this trend is not well captured by the computational codes. For handicapping work it is crucial that the force models capture this effect because it is where designers can seek to find a performance advantage by optimising the hull and keel volume distributions.

Applicability to Handicappers

This paper highlights the benefits of using the formulations, as per equation 1, as a means of comparing CFD results for sailing yachts, in general. It offers a route or common language for all contributors of sailing yacht force data to adopt for ease of data comparison and evaluation, in particular yacht handicappers. For example, if one particular approach, either under or over predicts the absolute resistance that does not render the data redundant. Provided this trend is maintained throughout the data set then the relative changes in resistance with speed, heel and sideforce will be unaffected.

These results, especially with the tank to CFD comparisons show that the CFD approaches, in general, are capturing many of the properties of the sailing yacht force breakdown. They do highlight areas where there are notable differences.

For the development of handicap rules, the use of CFD has already been used in the development of improved residuary resistance models (ORC, 2015) and these results support this as a suitable method to guide regression models.

The test matrix adopted in this study was extensive and necessitated significant computational resource, each individual RANS run took anywhere from 3 hours to 30 hours on multiple processors to complete. Therefore, if the process of data breakdown described previously, using heel drag ratio and effective draft were applied then a reduced set of runs would be adequate to define the general resistance characteristics of a particular hull form. With a reduction in the number of runs, then a number of hull geometries could be investigated on an economically viable basis. This is important, especially as yacht design is tending in the direction of "wide light" designs. These vessels bring particular handicapping challenges resulting from the range of displacement and LCG conditions that they can be raced in. This can include the use of moveable, changeable ballast options and the influence of dynamic lift devices. Therefore having the ability to investigate these properties with a viable method allowing for better estimation of the hydrodynamic force characteristics will lead to more refined rules.

The results of this paper will inform handicap rule makers as they support the use of particular CFD methods as appropriate tools for the estimation of various hydrodynamic resistance components.

It must be borne in mind that computational codes and hardware are constantly being upgraded, both in terms of numerical refinement, accuracy and efficiency of use.

CONCLUSIONS

This first phase of the Wide Light project brought together several respected CFD technicians with experience in evaluating the hydrodynamics of sailing yachts. They were able to collaborate on setting up the tank test program for an existing model of a boat fitting the parameters of Wide and Light design. Using the geometry of the boat and the design of the tank test program they performed pre-test evaluations of that model. CFD results were compared with the tank test results.

The following broad conclusions may be drawn:

- A body of physical test data relating to a defined geometry has been published and is available for validation of other data.
- Commercial CFD codes may be used to confidently predict the variation of the forces on a sailing yacht hull as speed heel and leeway change. These studies do not need prohibitively large mesh density to achieve valid results.
- Less computationally heavy codes, e.g. FlowLogic can produce data to capture the typical global behaviour of a sailing yacht hull.

FUTURE WORK

A second phase of the Wide Light project is planned to build upon the lessons from Phase 1. Specifically, Phase 2 is intended to take one or two of the promising CFD programs of Phase 1, provide their contributors with the geometry of a small fleet of designs and let them evaluate those designs and the combinations of speed/heel/leeway that are of interest to handicap rule-makers. The fleet of boats included in Phase 2 will represent realistic variations from a baseline design in the critical parameters that drive performance: displacement and beam for a fixed length are an obvious choice of variations.

ACKNOWLEDGEMENTS

This project was funded by SYRF and the authors would like to thank the SYRF Board, Project Management & Advisory Committee team; Judel/Vrolijk & Co. who kindly made one of their towing tank models available for the project without which the project would not have been possible and the CFD stakeholders who supported the project.

REFERENCES

Larsson, L., Stern, F., Visonneau, M., "Numerical Ship Hydrodynamics, an assessment of the Gothenburg 2010 Workshop" ISBN 978-94-007-7188-8.

Teeters, J., Pallard, R., Muselet, C., "Analysis of Hull Shape Effects on Hydrodynamic Drag in Offshore Handicap Racing Rules" 16th Chesapeake Sailing Yacht Symposium, March 2003.

ORC Technical Committee., "ORCi Documentation" http://www.orc.org/rules/ORC%20VPP%20Documentation%202015.pdf.

Keuning, J.A., Sonnenberg, U.B., "Developments in the Velocity Prediction based on the Delft Systematic Yacht Hull Series" The Modern Yacht Conference March 1998.

http://sailyachtresearch.org/19-library/47-wide-light SYRF Technical Resource Library, 2015.

Claughton, A, Wellicome, J., Shenoi, R., "Sailing Yacht Design: Theory" 1998, ISBN 0-582-36856-1

Brown, M., Campbell, I., Robinson, J., "The Accuracy and Repeatability of Tank Testing, From Experience of ACC Yacht Development" High Performance Yacht Design Conference, December 2002.

Numerical Simulations of a Surface Piercing A-Class Catamaran Hydrofoil and Comparison against Model Tests

Thilo Keller, TU Berlin, Berlin, Germany

Juryk Henrichs, DNV GL SE, Potsdam, Germany

Karsten Hochkirch, DNV GL SE, Potsdam, Germany

Andrés Cura Hochbaum, TU Berlin, Berlin, Germany

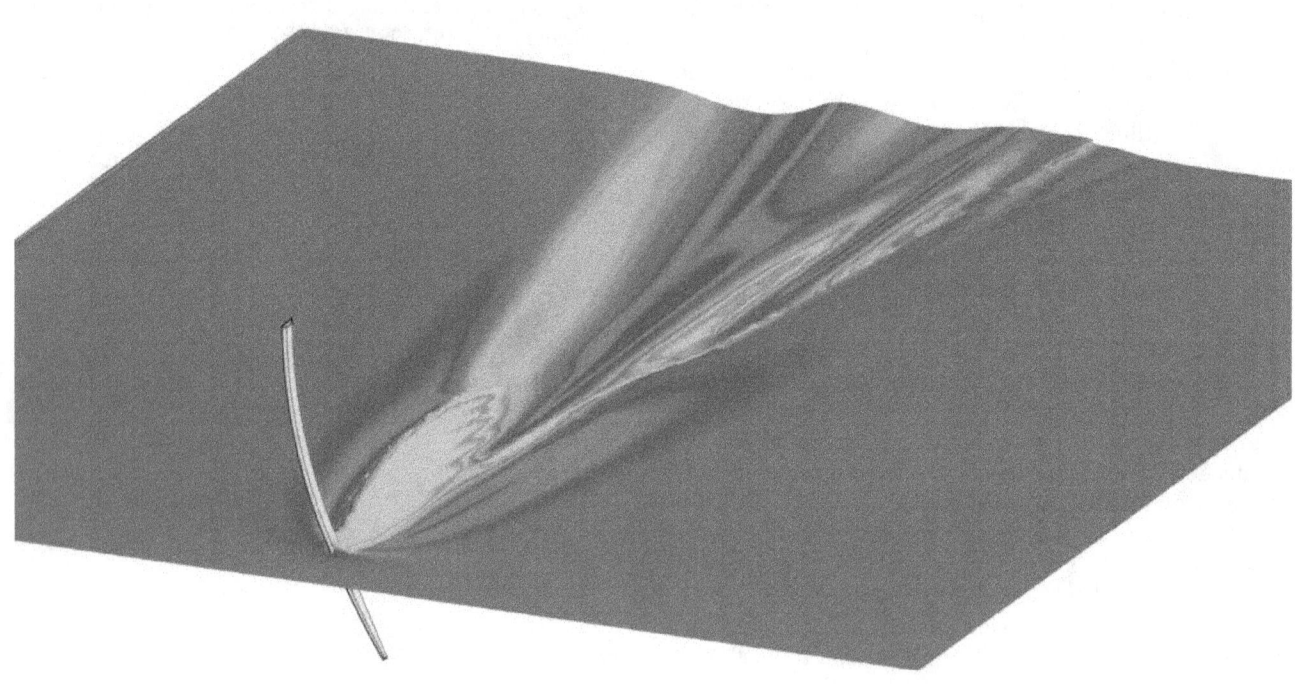

ABSTRACT

Hydrofoil supported sailing vessels gained more and more importance within the last years. Due to new processes of manufacturing it is possible to build slender section foils with low drag coefficients and heave stable hydrofoil geometries are becoming possible to construct. These surface piercing foils often tend to ventilate and undergo cavitation at high speeds. The aim of this work is to define a setup to calculate the hydrodynamic forces on such foils with a RANS based CFD method and to investigate whether the onset of ventilation and cavitation can be predicted with sufficient accuracy.

Therefore, a surface piercing hydrofoil of an A-Class catamaran is simulated by using the RANS code FineMarine with its volume of fluid method for predicting two phase flows. The C-shaped hydrofoil is analysed for one speed at Froude Number 7.9 and various angles of attack (*AoA*) by varying rake and leeway angle in ranges actually used while sailing. In addition model tests were carried out in the K27 cavitation tunnel of the Technical University of Berlin, for the given hydrofoil and in the same conditions as simulated with CFD to provide data for validation.

Based on the CFD calculation this paper shows how the rake and leeway angles influence the foil's lift to drag ratio.

The simulations have been verified by extensive analyses, including domain size verification for unrestricted water, mesh refinement and $y+$ verification. The influence of the dimensions of the cavitation tunnel on the flow around the hydrofoil and the wave system is also considered, as the test section of the K27 influences the flow around the foil, the forces and the wave elevation.

Finally the CFD results are compared with experiments conducted in the K27 in order to validate the used method.

NOTATION

CFD   Computational Fluid Dynamics
EFD   Experimental Fluid Dynamics
ITTC   International Towing Tank Conference

α	Angle of attack (AoA)	[°]
β	Leeway angle (in figures: L)	[°]
φ	Rake angle	[°]
ω	Cant angle	[°]
ρ	Density of water	[kg/m³]
Λ	Aspect ratio	[-]
c	Chord length	[m]
m	Tank blockage coefficient	[-]
s	Span	[m]
t	Acceleration time	[s]
v	Speed	[m/s]
A	Projected foil area	[m²]
A_x	Area of tank cross section	[m²]
AoA	Angle of Attack for 2D	[°]
Fr	Froude number	[-]
L_{ref}	Reference length	[m]
Re	Reynolds number	[-]
T	Draft	[m]

INTRODUCTION

Within the last years we have seen more and more sailing vessels using the technology of hydrofoils, either to support the hull or to use them as a fully foiling system. Especially the 34th America's Cup has shown how much potential this technology offers. It shall be noted that there are several different systems developed to get boats fully foiling. The widely used Moth's technology with a T-foil setup and an adjustable flap is probably the most often built foiling system. But on catamarans like the A-Class the rule book forbids foils like the T-foils on the Moth. The A-Class rules [1] only allow foils to be inserted from top and with a minimum distance between the foils of 1.5m under the waterline. That is why only surface piercing lifting foils are seen nowadays in this class. This type of foil shows some additional challenges in the design process. First of all they show a significant amount of wave making which is causing extra drag compared to deeply submerged foils. Secondly, ventilation and cavitation are likely to occur at increasing speeds and larger AoA's. Ventilation means that air is sucked down from the free water surface towards the tip of the foil. This leads to a sudden loss of lift, which

typically cannot be compensated fast enough to continue a controlled flight.

Within this work RANS calculations have been carried out for a representative segment of an A-Class catamaran lifting foil, to investigate if the hydrodynamic forces on such foils can be correctly predicted with a CFD code in a computationally cost effective setup and if the onset of cavitation and ventilation can be predicted with acceptable accuracy. In order to validate the calculations experimental investigations have been carried out for the same geometry and conditions in the K27 cavitation tank of the TU Berlin.

All computations have been done blind, i.e. they have been carried out prior to the model tests and were preceded by intensive studies on the numerical effects of e.g. computational domain and grid size as well as time step size.

To allow for the most straight forward comparison between experiment and CFD and to avoid any uncertainties due to scale effects, a segment of the A-Cat foil small enough to fit in the test section of the K27 was used.

For the numerical simulations the well-known software package FineMarine was chosen. This software package includes the meshing tool Hexpress to generate unstructured 3D grids needed for the RANS simulation.

The aim of this paper is to present the methods used as well as the results derived from experiment and CFD in order to find a feasible approach for this kind of calculations.

THE TEST OBJECT

As test object, an A-Class catamaran hydrofoil has been chosen. Typically these boats sail with both foils immersed at the same time. Due to the large distance between the two foils no direct interaction is expected and therefore only the starboard foil has been investigated.

The foil itself has a span wise bend with a constant radius which makes it a so called C-foil. The chord and section is constant within that bended span region. Towards the tip, the foil is straight over the last 0.3m and shows an elliptical taper on the leading edge with a straight trailing edge (cf. Figure 1). The straight part of the foil is angled 30° (cant angle) to the vertical. The chord length of the foil is $c=0.16$m tapering to $c_{Tip}=0.06$m at the tip.

In order to fit the foil in the measuring section of the K27 cavitation tank only a relatively small immersion of $T=0.3$m could be tested, which represents a rather limiting operational case.

The resulting aspect ratio of only $\Lambda=2.645$ is rather small and significantly changes the overall performance and behavior of the foil as well as its stall characteristics.

With an immersion of $T=0.3$m the span equals $s=0.34$m and projected foil area is $A=0.0437$m².

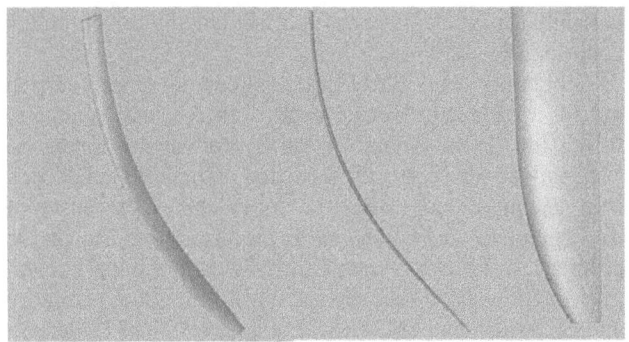

Figure 1 - Hydrofoil views, left: perspective, middle: front, right: side

The section of the foil is a Selig/Donovan SD7032-099-88 and is kept the same over the whole span. It is designed for low Reynolds numbers. The main characteristics of the section are: Thickness 10% of c at 26.6% of c and a camber of 3.4% of c at 45.1% of c (cf. Figure 2). The angle of attack α for zero lift is $\alpha_{CL=0}$=-4.08°

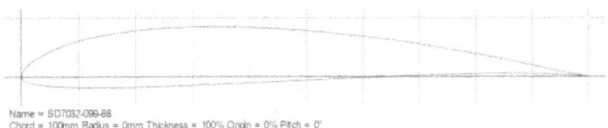

Name = SD7032-099-88
Chord = 100mm Radius = 0mm Thickness = 100% Origin = 0% Pitch = 0°

Figure 2 – Foil section [2]

THE TEST CASES

For all CFD simulations the draft T=0.3m, the cant angle ω=30° and the speed v=10m/s were kept constant to simplify the test matrix. The rake φ was varied between +2°< φ <-8° in steps of 2° resembling the rake adjustment in the center board case of the boat. The leeway angle β was varied from -2°< β <+20° in steps of 4°.

The results, however, will be presented with respect to the resulting 2-dimensional *AoA* the straight part of the foil will experience and a Cross-Flow-Angle indicating the amount of flow along the span.

THE TEST FACILITY

The K27 cavitation tank of Technical University of Berlin offers ideal conditions for the experimental investigations of the hydrofoil. It provides high enough flow speeds (with a maximum speed of 12 m/s) while offering a free surface at the same time and a comparatively large test section, big enough to fit the chosen segment of the foil in full scale.

Figure 3 shows the principle layout of the K27 facility. The test section is 3.5m long with a rectangular cross section of 0.6m depth (to the free surface) and 0.6m width. The foil is placed in the center plane of the test section 0.3m from the inlet.

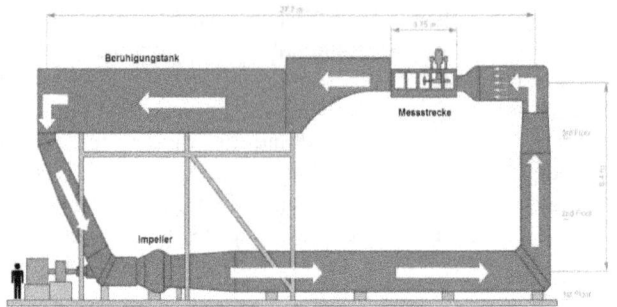

Figure 3 - K27 side view [3]

The forces on the foil were measured using a 6 component force balance. The balance itself can be rotated around the *y*-axis to change the rake of the foil. To adjust the leeway angle an additional adapter was applied which allowed a rotation around the pitched *z*-axis. The foil was fixed to that adapter with a cant angle of 30° to the normal of the adapter plate.

It has to be noted that the rotation sequence is different to that typically used in ship dynamics. To sum up the performed rotation: 1st the rake angle was applied to the foil, 2nd the leeway angle and 3rd the cant angle.

Figure 4 - Test setup in the K27

Figure 4 shows the test setup, including the leeway adjuster plate and the 6 component force balance. The black marker in the background indicates the height of the free surface during the test runs. The waterlines on the foil are spaced every 50mm from the tip and the sections every 20mm from the trailing edge.

Force reference point

The intersection of the trailing edge with the free surface was chosen as reference point for the force vector and as origin for the CFD computations. The x-axis is facing forward against the free stream, the y-axis is facing to port and the z-axis is facing upwards (see Figure 5)

NUMERICAL COMPUTATION

The numerical simulation was performed using the commercial software package FineMarine Version 3.1-3 which comprises the RANS CFD solver ISIS, the grid generator Hexpress and the post-processing tool CFView.

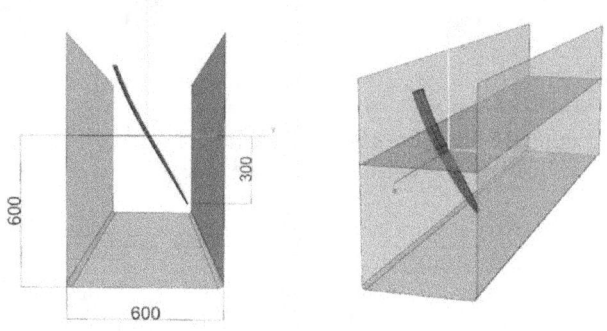

Figure 5 – Foil in the test section of the K27 and origin

Domain

CFD calculations for the foil have been carried out for two situations: placed in the measuring section of the towing tank and, to determine the effect of blockage, in unrestricted waters.

In case of the latter a domain size study has been carried out varying domain width and length to make sure the domain size does not affect the flow and resulting forces. Independency from the domain size was found with the outlet placed 25 chord length behind and the inlet placed 4 chord length forward of the foil trailing edge. The domain sides needed to be at a width of 12 chord length to each side of the foil.

Figures 5 and 6 show the domain for the computations in the cavitation tunnel (Figure 5) and for the open water case (Figure 6).

Grid Generation

Hexpress generates non-conformal strictly hexahedral unstructured meshes. Mesh generation is based on an initial background grid which is gradually refined towards surface patches, curves, or control volumes. The level of refinement determines the number of subdivisions of the initial grid.

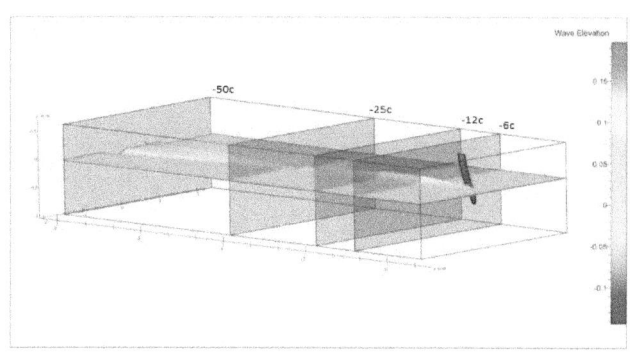

Figure 6 – Domain length variation

Table 1 shows the grid refinements on the various surface patches of the foil (see also Figure 7) as well as the resulting cells size on the surface and the number cells for different mesh sizes investigated.

Table 1 – Refinement levels and cell count for different mesh sizes

Mesh	Ref. side	Ref. LE/TE	Ref. LE-curve	Cell size on surface	Nb. of cells in mesh
	[-]	[-]	[-]	[m]	[-]
Very coarse	5	6	0	0.02	3285245
Coarse	6	7	0	0.01	3679304
Medium	7	8	10	0.005	3775107
Fine	8	9	12	0.0025	4790642
Very fine	9	10	14	0.00125	9216295

To capture the high pressure and velocity gradients near the leading and trailing edge as well as on the tip more refinements have been applied here compared to the rest of the foil. Figure 7 shows some mesh details for the "fine" mesh.

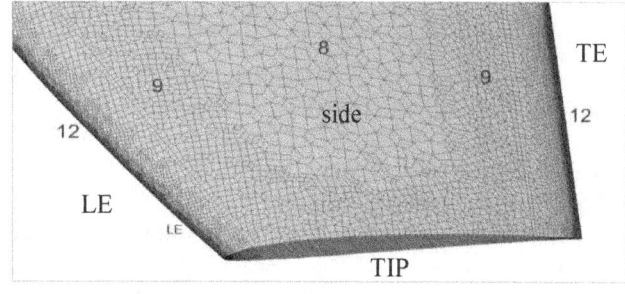

Figure 7 – Mesh refinements on the foil

Figure 8 shows the change in side force of the foil for different mesh refinements. Grid independency is achieved for cell sizes smaller $0.02c=0.0025$m (fine and very fine grid). The settings of the fine mesh were therefore chosen for the actual simulations.

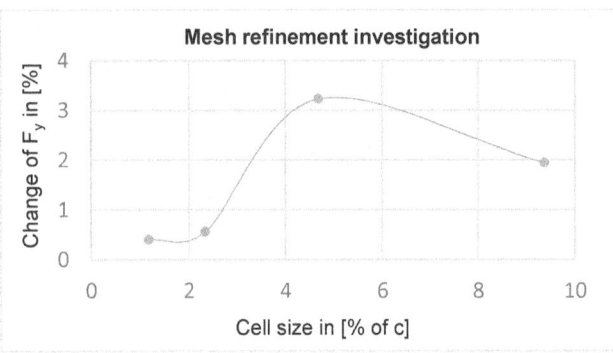

Figure 8 – Mesh refinement vs. side force

For a sufficient resolution of the free surface two different refinements were applied to the static water plane, first the z-direction was refined to have a sharp borderline between the two fluids and secondly a patch was refined in x- and y-direction to allow for a better capturing of the wake behind the foil. Some details of the mesh used for the simulations in the K27 are shown in Figure 9.

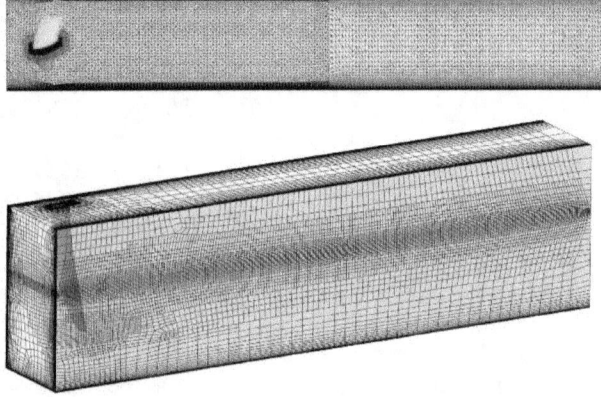

Figure 9 – Mesh for K27-simulations

To capture the boundary layer on the foil as well as on the walls of the cavitation tunnel, a high Reynolds number turbulence model with wall functions was chosen. This requires the non-dimensional distance of the first cell layer normal to the wall (y^+) to be roughly about 100 [4].

Figure 10 shows the achieved y^+ distribution for various target values of y^+ (following the Blasius boundary layer velocity profile) on the pressure side of the foil (top row) and of the suction side of the foil (bottom row). For a target value of y^+=50, the achieved y^+ near the leading edge falls below a value of 30. To ensure the correct application of the wall functions a target value of y^+=150 was chosen. For the fine grid this resulted in the insertion of 3 viscous layers on the foil sides and 4 layers at the leading and trailing edge.

For the walls of the test section of the K27 the target value of y^+ was set to 300, without intending to resolve the boundary layer, to reduce mesh size.

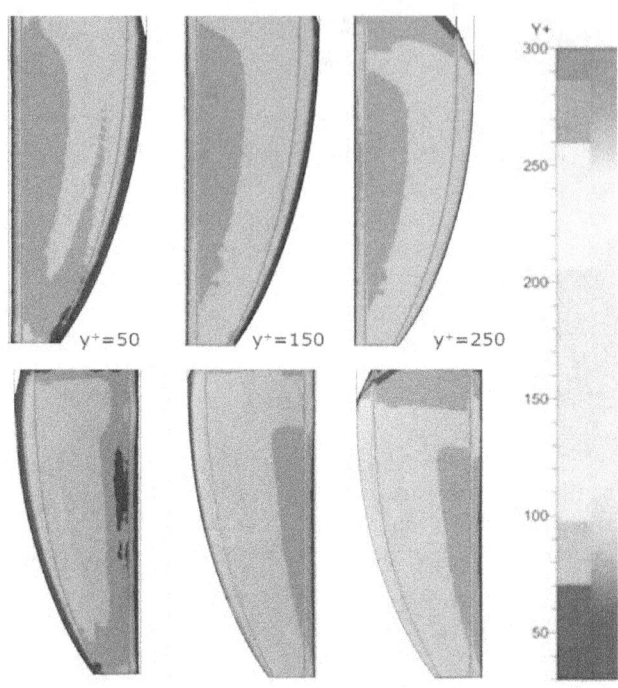

Figure 10 – y^+ allocation on the foil, top: suction side, bottom: pressure side

Turbulence Model

The Menter-k-ω-SST-Model was chosen for turbulence closure showing generally a good accuracy while keeping computation times small. It is a two equation turbulence model using a blending between near wall flows (using k-ω-model) and far away from the wall flows (using k-ε-model). The turbulent kinetic energy k and the turbulence frequency ω can be preset in FineMarine. Since no data was available for the turbulence intensity in the K27, the recommended values have been used [4].

Boundary Conditions

For the simulations in the K27 the walls and the bottom of the K27, as well as the foil itself, have been set as "Solid Wall", which means that the velocities are zero there.

The inlet and outlet have been set to "Far Field", which is a Neumann-Boundary-Condition, stating that the gradient of the flow variables is equal to zero.

The top was set as "Updated Hydrostatic Pressure" which is the only pressure boundary condition used. This is a Dirichlet-Boundary-Condition constantly reinitializing the pressure.

Initial Condition and Time Step

To accelerate the convergence of the forces, the calculations are started at rest and the flow is accelerated to target speed following a quarter-sinus function. The acceleration time is calculated from equation (2), which is a rule of thumb. L_{ref} is the reference length of the simulated body, in this case the chord length, while v represents the target speed of the free flow.

$$t = 20 \frac{L_{ref}}{v} \qquad (2)$$

The simulation is performed as pseudo unsteady simulation using a fixed time step. In order to reduce the simulation time a variation of different time steps has been investigated. The largest possible time step allowing convergence of the simulation is 0.01s (cf. Figure 11). Some of the simulations performed with a bigger time step of 0.05s diverged.

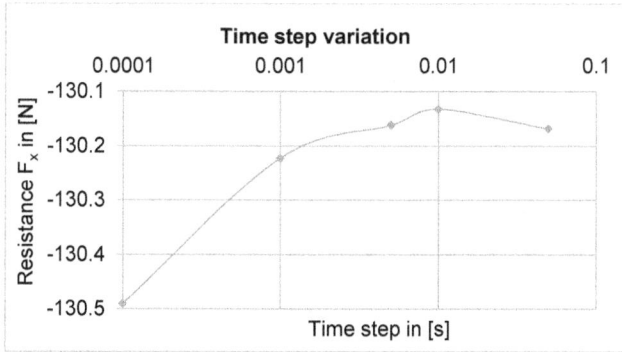

Figure 11 - Time step variation

CFD RESULTS

All results will be presented in terms of the dimensionless lift and drag coefficients (Equations 3 and 4) and are based on the 2-dimensional *AoA*.

$$c_L = \frac{L}{\rho/_2 v^2 A} \qquad (3)$$

$$c_D = \frac{F_X}{\rho/_2 v^2 A} \qquad (4)$$

Comparison K27 versus open water

To quantify a possible blockage effect CFD calculations have been carried out for unrestricted open water and considering the actual geometry of the K27 measurement section.

Table 2 shows the blockage *m* for a series of different leeway angles at a constant rake of $\varphi = -8°$. At a leeway angle of $\beta = 20°$, the blockage has a value of 0.077. This means that the projected area of the foil is 7.7% of the cross sectional area of the measurement section.

Figure 12 shows the lift coefficient C_L and the lift/drag ratio C_L/C_D for this series - for open water (ow) and in the K27. The deviation in C_L between the runs for open water and in the K27 is increasing with increasing *AoA*. This is most probably a consequence of the increasing blockage effect for increasing *AoA*.

Table 2 – Blockage of the K27 for different leeway angles

Rake	Leeway (β)	2D AoA	Cross flow angle	m=Ax/A
[°]	[°]	[°]	[°]	[-]
0	0	0	0	0.026
0	8	6.9	-4.0	0.041
0	20	17.5	-10.3	0.077

At small *AoA*'s only C_L increases as an effect of the blockage while C_D is nearly unaffected. This results in a noticeable increase of C_L/C_D for K27 compared to open water.

With larger *AoA*'s and hence a larger blockage the effect on C_D increases as well. Therefore C_L/C_D at larger *AoA*'s is similar for K27 and open water.

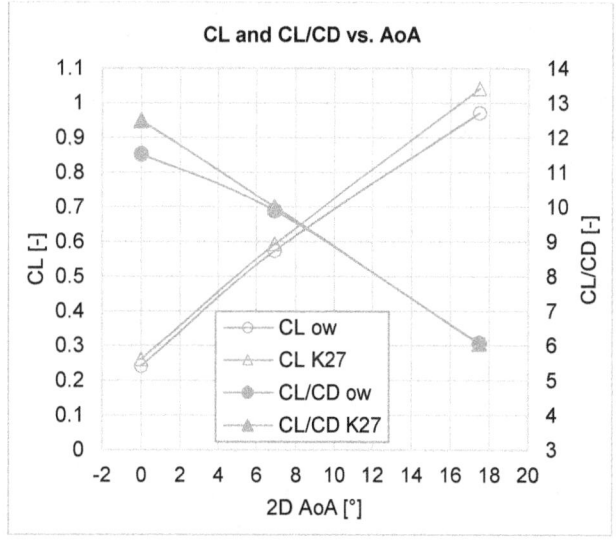

Figure 12 – Effect of blockage on C_L and C_D

Figure 13 shows the wave pattern in unrestricted open water (left) and in the K27 (right) for β=0° (top row) and β=8° (bottom row). It can be seen the blockage effect on wave making is significantly larger at higher *AoA*'s.

This indicates that the effect of blockage on C_D is mainly contributed to an effect on wave making resistance and induced resistance is not affected.

As this was not the focus here, it will be left to future work to investigate whether or not typical corrections for blockage in towing tank or wind tunnel experiments can be applied for this kind of investigation.

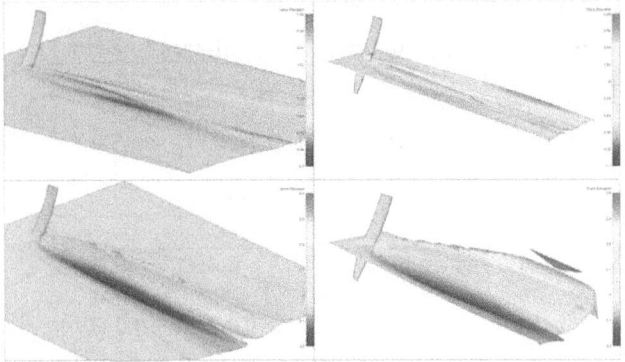

Figure 13 - Wave elevation: open water (left), K27 (right) for β=0° (top row) and β=8° (bottom row)

Polars (K27 domain)

Figure 14 shows C_L over C_D polars plotted for a number of constant leeway angles in the range between -2° and 20°. For each leeway angle the rake was varied from +2° to -8° with the leftmost point on each polar representing φ = +2°.

Each value of C_L can be derived by a number of different combinations of leeway and rake. For a constant C_L it can be seen that the polar with more leeway reduces well drag.

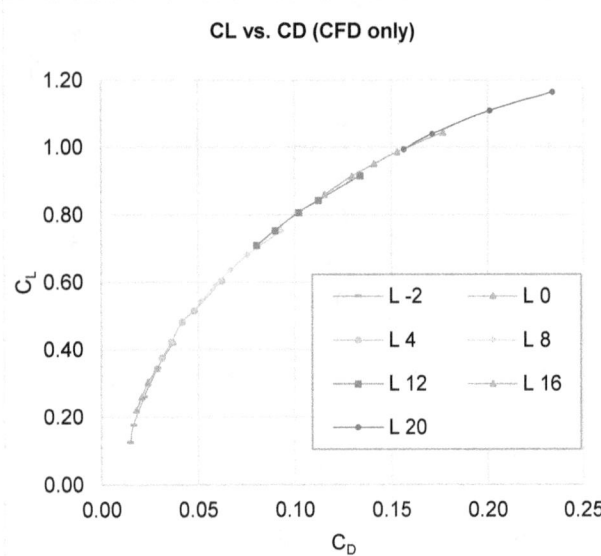

Figure 14 - Lift vs. drag polars for varying leeway and rake

To translate this to more practical terms for sailors: Increasing leeway is more efficient than increasing rake in terms of the pure hydrodynamics; where less rake means, that the foil is moved forward in the top. The maximum saving on drag for a given lift, observed from the CFD simulations, can be larger than 5%!

However, towards the smaller lift coefficients the polars

are closing in to each other and will cross each other if extrapolated to smaller values of rake and therefore C_L. This indicates that there is a limit to increasing leeway if this is traded off by too large a decrease in rake.

Figure 15 shows the lift coefficient over the *AoA* compared to the 2D lift characteristic of the used section. As expected for a low aspect ratio foil (Λ=2.645), the gradient of the 3D curve is significantly smaller and the stall angle is significantly higher than those of the 2D curve [5]. Furthermore the gradient is not linear for small *AoA* in the 3D case. These effects are the result of the induced drag due to the tip vortex, which does not exist in the 2D case and whose influence increases with decreasing aspect ratio.

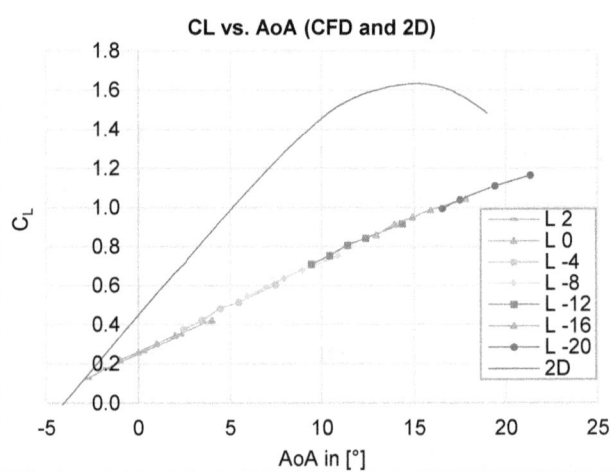

Figure 15 - Lift polar, CFD and 2D

Ventilation and cavitation inception

For almost all *AoA*'s from 10° onwards (all rake variations for leeway angles β >= 12°) regions were found near the leading edge and the tip where the absolute pressure value drops below the vapor pressure of water. This is a strong indication that for those cases cavitation is likely to occur during the experimental investigations.

However, no clear indication for the onset of ventilation was found.

Ventilation inception requires three relevant prerequisites to exist: (1) a region of sub-atmospheric pressure, (2) a path of low impedance to air flow from the free surface into this region (e.g. separated flow) and (3) a disturbance in the free surface [6][7].

Regions of sub-atmospheric pressure are present due to flow acceleration near the position of maximum thickness of the profile and along the span. But only at leeway angles of *β* > 20°, significant amounts of flow separation were observed, fulfilling the second prerequisite. The location of these regions was close to the tip rather than to the free surface, hence not providing for the third prerequisite.

Additionally mesh size and resolution were chosen to balance accuracy and computational effort. Although a comparatively large mesh was applied, the resolution was certainly not fine enough to capture effects as e.g. the formation of free surface filament vortices which could initiate ventilation [6].

K27 RESULTS

The experiments have been conducted following the setup described above.

Ventilation inception

While it was the intention to investigate the same rake and leeway combinations at the same speed as in CFD, it was found, that ventilation at v=10 m/s flow speed occurred already at AoA's of approximately 3° (or at leeway angles of 4°). Only very few measurement points are therefore available at this speed. In order to avoid ventilation and to determine the lift and drag characteristic of the foil for varying rake and leeway angles smaller flow speeds of 5m/s and 7.5m/s have been chosen for additional measurements.

Figure 16 shows the ventilation on the foil and the size of the ventilated cavity. The ventilation was very stable in position and size, but for a given AoA it was not reproducible at which flow speed ventilation occurred, when accelerating to the target flow speed. However, a clear dependency on the amount of acceleration was found and for very careful acceleration ventilation did sometimes not occur at all.

Figure 16 - Ventilated foil in K27

The early inception of ventilation in the K27 was not predicted by CFD but seems to be a result of the quality of the free surface in the K27. For higher speeds the free surface became very blurred showing a layer of white water with a thickness of up to 60mm (cf. Figure 16). This provided a significant disturbance to the free surface (3rd prerequisite for ventilation inception) as well as local separation close to the free surface (2nd prerequisite).

Three different regimes can be distinguished with respect to ventilation: fully attached or fully wetted (FW), instable or partially ventilated (PV) and fully ventilated (FV) [7],[8].

As indicated by [7] the FV regime is marked by the stall angle, which is reached at about 20° AoA due to the small aspect ratio. Hence the PV or instable regime applies for most of the measured conditions, which explains why ventilation inception was experienced rather random. The dependency to flow acceleration must be contributed to

dynamic interactions of the flow and the tank geometry leading to larger perturbations.

Ventilation can occur in a process of self-stabilization in which the required flow separation does not need to exist initially over the entire surface but is induced by the ventilated cavity itself during ventilation inception [7]. This may be the primary inception mechanism as observed in the K27. Figure 17 shows a typical transition process from FW to FV. Clearly ventilation starts at the free surface and stabilizes in a cavity which covers the entire foil.

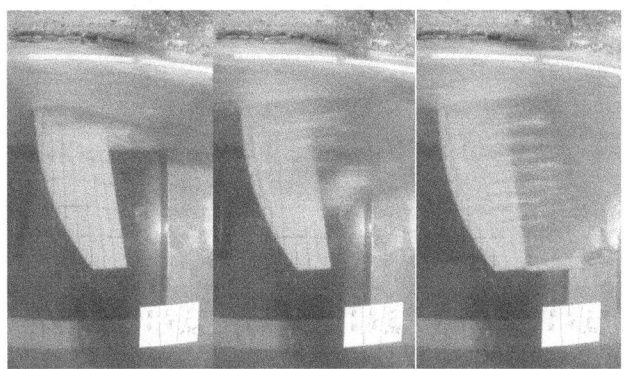

Figure 17 – Ventilation inception (sequence in transition from FW to FV regime)

Polars

Eventually model tests have been performed for speeds of 5m/s, 7.5m/s and 10m/s. Figure 18 shows the lift coefficient over the AoA for all measured points. For a leeway of β = -2° (L-2) and β = 4° (L+4) the whole range of rakes could be measured for the speeds of 5m/s and 7.5m/s. The red circles mark those rake and leeway combinations which could be measured at all three speeds including 10m/s.

It is found that the lift coefficient decreases with increasing speed. This effect is larger for small AoA and decreases for larger AoA's. Additionally the zero lift angle becomes larger with increasing speed.

The change in wave elevation due to the increased speed might explain these effects. The higher the flow speed is, the lower is the pressure on the suction side and the deeper is also the resulting wave trough. This in return decreases the wetted surface and reduces the available area for lift generation.

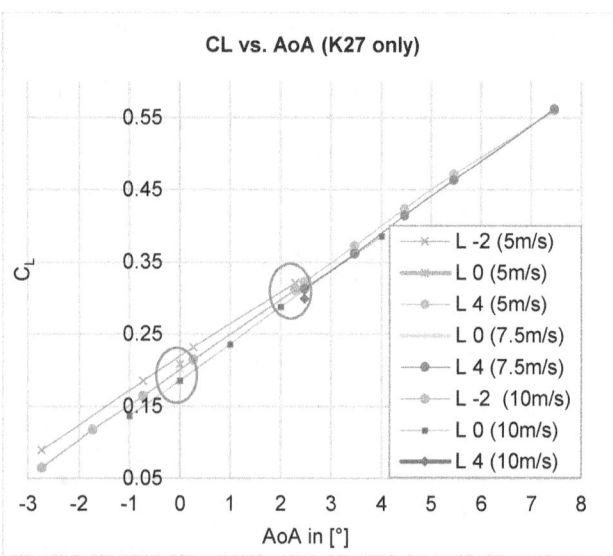

Figure 18 - CL over 2D AoA for K27 results

COMPARISON OF CFD AND EFD

Corrections of the experimental data

Two effects have been observed during the tests which affect the results of the measurements and need to be corrected for a sensible comparison with the CFD results. These are:

- A reduction in wetted surface due to a blurred interface between air and water
- A change in *AoA* due to flexibility of the foil and the mounting frame.

As Figure 16 shows there is a noticeable layer of white water at the free surface. In order to correct the area in the calculation of C_L and C_D (Equations 3 and 4), additional measurements have been carried out at zero rake and leeway and at 10m/s: one with increased draft of 350mm and one with a decreased draft of 250mm. The result of the 250mm draft run compared well with the corresponding CFD results (at 300mm draft). Therefore all C_L and C_D values determined from the K27 measurements are calculated with the reduced projected area corresponding to the 250mm draft.

During the CFD simulations the structure of the foil was assumed to be infinitely stiff. However, the real foil showed some significant amount of flexibility. For the highest lift conditions (maximum *AoA* without ventilation) a sideward deflection of the foil tip of up to 60mm was estimated. In order to correct for this flexibility an offset depending on C_L was applied to the cant angle which was used to calculate the 2D-*AoA* of a section and the cross-flow-angle.

Polars

The full range of rake angles could be measured for only two leeway angles (β = -2° and β = 0°) in the K27 cavitation tank at a speed of 10m/s. These two are compared to the CFD results after application of the above corrections.

Figure 19 shows the lift coefficient over the 2D-*AoA* comparing both, the EFD and CFD results. The overall values match reasonably well for higher *AoA*'s with a deviation of 5% or less. For *AoA*s smaller than 1° (and respective small values of C_L) however the difference increases up to 44%. Additionally the EFD polars also show a slightly steeper slope than those of the CFD.

Unfortunately these effects could not be evaluated for higher angles of attack, but it is likely this trend will persist.

While the correction of the wetted area results in a vertical shift of the EFD polars (increasing C_L) and a decrease of the slope, the correction of the cant angle affects the slope of the polars only. Since the deflection of the foil was not accurately measured there is some uncertainty in the amount of correction necessary. However, values which would allow for a better agreement between the EFD and CFD results are unreasonably high. Hence other effects may exist which are not covered by the above corrections.

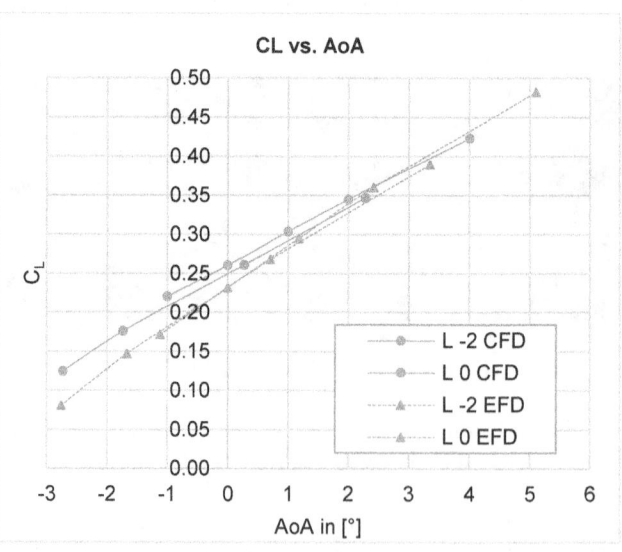

Figure 19 - Comparison CFD vs. EFD Lift

Figure 20 shows the comparison of the drag coefficient plotted over the *AoA* for the EFD and CFD results. A large difference can be observed between the EFD and CFD results, especially for *AoA*'s larger than -1° (for instance 25% at 2°). The data sets for the two CFD cases and for the two EFD cases are close together. Therefore another systematic error is suspected to be present either in the experiment or in the CFD simulation.

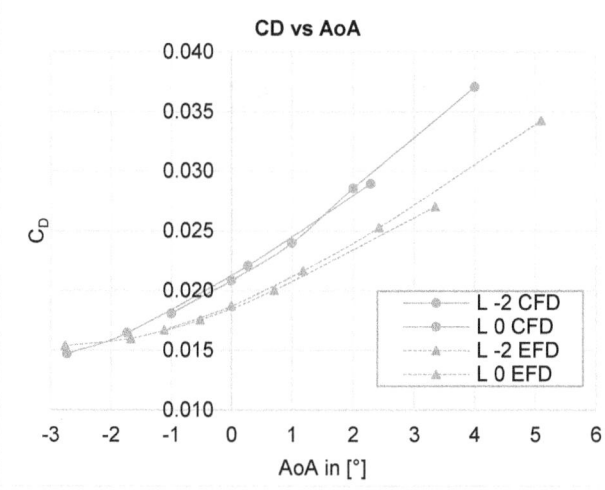

Figure 20 - Comparison CFD vs. EFD Drag

CONCLUSION

The comparison of the results derived from experimental investigations and RANS CFD computations for an A-Class catamaran hydrofoil show a good agreement for C_L for relevant *AoA*'s but significant deviations in C_D. The latter may be a result of the assumption of fully turbulent flow in the CFD calculations or some difference in the respective setups of calculation and experiment. The section in use is a so called laminar profile intended to decrease frictional drag by shifting the laminar-turbulent transition of the boundary layer aft and thus providing a significant amount of laminar flow along the section. A laminar-turbulent transition model was not available in Fine/Marine 3.1-3. However, future investigations should investigate this effect.

The experiments revealed a significant amount of flexibility in the section of the hydrofoil. Future experiments should either be conducted on a sufficiently stiff test object or provide measurements of the foil deflection. For numerical investigations, Fluid-Structure-Interaction provides means to capture flexibility of the foils.

No indication was found from the CFD simulations for the early ventilation inception observed in the K27 cavitation tunnel.
To capture the flow in the necessary detail to allow for ventilation, substantially finer and hence larger meshes are needed [7][9]. This is computationally expensive and unattractive if looking at series of runs for different operational conditions. As an alternative, adaptive grid refinement [9] may provide means to balance effort for future investigations.

The free surface in the K27 was highly disturbed and is likely to have caused the early ventilation inception. This disturbance is not given in CFD. It would be worthwhile to investigate if a disturbance with a similar effect could be applied in CFD to induce ventilation inception – an adequate grid resolution provided.

Alternatively, it would be interesting to repeat the experiments in a towing tank with an undisturbed free surface. This would probably be closer to real life condition where a much smoother free surface can be expected than in the circulating tank and it is also likely to delay ventilation inception to larger *AoA*'s.

However, this might be difficult since towing tank carriages capable of reaching 10m/s speed are similarly rare as cavitation tanks with the possibility to investigate a free surface.

REFERENCES

[1] IACA, "International A-Class Catamaran Class Rules", http://www.a-cat.org/sites/default/files/ISAF %20A%20CAT%20rules%202010_0_0.pdf , 2013

[2] Airfoiltools: http://airfoiltools.com , 29.03.2015

[3] Fritz, J.; FG DMS – TU Berlin: „Vorlesungsfolien Propeller und Kavitation", https://isis.tuberlin.de/ pluginfile.php/225358/mod_resource/content/1/ PropellerKavitation_7.pdf. TU Berlin: 14.07.2015

[4] Numeca International: „User Manual – FineMarine v3.1". Brüssel, Belgien: 2013

[5] Hoerner, S.F., "Fluid-Dynamic Lift", Second Edition, Bakersfield, California, USA, 1985.

[6] Barden, T.A., "On the Road to Establishing Ventilation Probability for Moth Sailing Dinghies", 18th Australasian Fluid Mechanics Conference, Launceston, Australia, 2012.

[7] Harwood, C.M., "Experimental and Numerical Investigation of Ventilation Inception and Washout Mechanisms of a Surface-Piercing Hydrofoil", 30th Symposium on Naval Hydrodynamics, Hobart, Tasmania, Australia, 2014.

[8] Fridsma, G., "Ventilation Inception on a Surface-Piercing Dihedral Hydrofoil with Plane-Face Wedge Section", Report 952, Davidson Laboratory, Stevens Institute of Technology,Hoboken New Jersey, USA, 1963.

[9] Visonneau, M., "Anisotropic Grid Adaptation for RANS Simulaton of Ship Flow", 11th International Conference on Fast Sea Transportation, Honululu, Hawai, USA, 2011.

THE 22nd CHESAPEAKE SAILING YACHT SYMPOSIUM
ANNAPOLIS, MARYLAND, MARCH 2016

Advanced CFD-Simulations of free-surface flows around modern sailing yachts using a newly developed OpenFOAM solver

Janek Meyer, Yacht Research Unit Kiel, Kiel, Germany

Hannes Renzsch, FluidEngineeringSolutions GmbH & Co. KG, Schleswig, Germany

Kai Graf, Yacht Research Unit Kiel, Kiel, Germany

Thomas Slawig, Kiel University, Kiel, Germany

ABSTRACT

While plain vanilla OpenFOAM (OF) has strong capabilities with regards to quite a few typical CFD-tasks, some problems actually require additional solvers and numerical methods for efficient computation of high-quality results. One of the fields requiring these additions is the computation of large-scale free-surface flows as found e.g. in naval architecture. This holds especially for the flow around typical modern yacht hulls, often planing, sometimes with surface-piercing appendages. Particular challenges include, but are not limited to, breaking waves, sharpness of interface, numerical ventilation (aka streaking) and a wide range of flow phenomenon scales. A new OF-based application including newly implemented discretisation schemes, gradient computation and rigid body motion computation is described. The new code is validated against published experimental data; the effect on accuracy, computational time and solver stability is shown by comparison to standard OF-solvers (interFoam / interDyMFoam) and Star-CCM+. The code's capabilities to simulate complex "real-world" flows are shown on a well-known racing yacht design.

NOTATION

Latin letters

BICS	Blended Interface Capturing Scheme
BRICS	Blended Reconstructed Interface Capturing Scheme
CBC	Convective Boundedness Criterion
CDS	Central Differencing Scheme
Co	Courant-Friedrichs-Lewy Number
CoE	Center of Effort
DDS	Downwind Differencing Scheme
GDS	Gamma Differencing Scheme
HRIC	High-Resolution Interface Capturing
IGDS	Inter-Gamma Differencing Scheme
LCG	Longitudinal Center of Gravity
MULES	multi-dimensional limiter for explicit solution
NV	Numerical Ventilation
NVA	Normalized Variable Approach
NVD	Normalized Variable Diagram
OF	OpenFOAM
PISO	Pressure Implicit with Splitting of Operator
SHM	snappyHexMesh
SIMPLE	Semi-Implicit Method for Pressure Linked Equations
UDS	Upwind Differencing Scheme
VOF	Volume-of-Fluid
a	Matrix coefficient
$\mathbf{e}$	Explicit distance vector
$\mathbf{g}$	Gravity vector
g	Z-component of gravity vector
$\mathbf{n}$	Surface normal vector
N	Numerical Ventilation
p	Total pressure
q	Source term
$\mathbf{s}$	Area vector
s	Surface
t	Time
$\mathbf{u}$	Velocity vector
$\hat{\mathbf{u}}$	Velocity vector of old timestep
V	Volume

Sub- and Superscripts

c	Main diagonal or central cell
C	Central cell
D	Downwind cell
f	Cell face
i	Ith cell or row of the matrix
L	Face owner cell
n	Neighbor element in the matrix
R	Face neighbor cell
U	Upwind cell

$^{-}$ Face owner subdomain

$^{+}$ Face neighbor subdomain

Greek symbols

α Volume fraction

Γ Free surface

θ Angle between face normal and free surface normal

λ_f Pressure switch

μ Dynamic viscosity

μ_e Effective dynamic viscosity

ρ Density (kg/m^3)

$\hat{\rho}$ Reverse interpolated face value of density

ϕ Generic quantity

$\tilde{\phi}$ Normalized generic quantity

Φ Face flux

Operators

$\mathscr{D}()$ Divergence operator

$\mathscr{I}()$ Interpolation operator

$\mathscr{R}()$ Reconstruct volume field operator

∇ Nabla Operator

INTRODUCTION

OpenFOAM (OF) used in various projects in academia and industry was found to be a powerful CFD-package. Unfortunately, using standard OF VOF-solvers ((LTS)inter(DyM)FOAM) on typical naval architectural problems, two major issues arose, namely lack of stability and exceedingly long turn-around times. Based on previous experience using commercial CFD programs like CFX or Star-CCM+ and developing and using proprietary codes like FreSCo^{+}, it was decided to implement new solvers and numerical methods based on general industry practice. Based on an extensive review of literature and existing programs various methods were implemented and tested and the most successful are described in the following. Within the scope of this paper we call the new solver *OurSolver*.

METHODS

Governing Equations and Solution Algorithm

For the calculation of the flow the incompressible unsteady Reynolds-averaged Navier-Stokes equations are solved. The momentum conservation equation is defined as (employ the eddy-viscosity hypothesis for closure)

$$\frac{\delta \rho \mathbf{u}}{\delta t} + \nabla \cdot (\rho \mathbf{u} \mathbf{u}) - \nabla \cdot \mu_e \left(\nabla \mathbf{u} + (\nabla \mathbf{u})^T \right) \\ = -\nabla p + \rho \mathbf{g} \tag{1}$$

The mass conservation equation is

$$\nabla \cdot \mathbf{u} = 0 \tag{2}$$

Integrating equation (2) over the volume and applying the Gauss Theorem leads to

$$\int_s \mathbf{u}_s \cdot \mathbf{n} ds = 0 \tag{3}$$

or in a non-integral form using Divergence operator $\mathscr{D}()$

$$\mathscr{D}(\mathbf{u}) = 0 \tag{4}$$

For the calculation of the free surface the Volume-of-Fluid (VOF) method introduced in Hirt et al. (1981) is used. Here, an additional transport variable for the volume fraction is introduced. Its conservation equation is given as

$$\frac{\delta \alpha_i}{\delta t} + \nabla \cdot (\alpha_i \mathbf{u}) = 0 \tag{5}$$

with the volume fraction α_i for the ith fluid. The flow properties are then calculated by

$$\rho = \sum_i \rho_i \alpha_i \quad \mu = \sum_i \mu_i \alpha_i \quad 1 = \sum_i \alpha_i \tag{6}$$

The free surface is defined by the volume fraction $\alpha = 0.5$. Using implicit euler discretisation for time the semi-discretized, linearized momentum equation yields

$$\frac{\rho}{\Delta t} \mathbf{u}_c + \frac{1}{V} a_c \mathbf{u}_c + \frac{1}{V} \sum_n a_n \mathbf{u}_n \\ = \mathbf{q} + \rho \mathbf{g} - \mathscr{R}((\nabla p)_f \cdot \mathbf{s}_f) + \frac{\rho}{\Delta t} \hat{\mathbf{u}}_c \tag{7}$$

For cell c a_c is the matrix coefficient of the main diagonal and a_n are the respective off diagonal elements resulting from the neighbor cells of the implicit part of the convective and diffusive terms. The subscript f indicates a value at the face. The time discretisation needs the velocity of the old time step $\hat{\mathbf{u}}_c$. The source term $\mathbf{q}$ contains all other explicit parts. For the *In the spirit of Rhie-Chow* interpolation generally used in OpenFOAM, the pressure gradient is calculated at the cell-faces. The reconstruct volume field operator $\mathscr{R}()$ reconstructs a volume field from the face flux field, where $\mathbf{s}_f$ is the face area vector.

$$\mathscr{R}(\phi_f) = \left(\sum_f \frac{\mathbf{s}_f}{|\mathbf{s}_f|} \mathbf{s}_f \right)^{-1} \cdot \left(\sum_f \frac{\mathbf{s}_f}{|\mathbf{s}_f|} \phi_f \right) \tag{8}$$

Rearranging (7) to $\mathbf{u}_c$ yields the velocity equation:

$$\mathbf{u}_c = \left(\frac{\rho}{\Delta t} + \frac{1}{V} a_c \right)^{-1} \\ \left(\mathbf{q} + \rho \mathbf{g} - \mathscr{R}((\nabla p)_f \cdot \mathbf{s}_f) + \frac{\rho}{\Delta t} \hat{\mathbf{u}}_c - \frac{1}{V} \sum_n a_n \mathbf{u}_n \right) \tag{9}$$

Substituting and rearranging (9) into (2) yields a poisson equation for the pressure:

$$\mathscr{D} \left(\alpha_f^{-1} \nabla p \right) = \mathscr{D} \left(\alpha_f^{-1} (h_{tf} + h_{sf}) \right) \tag{10}$$

with

$$\alpha_f = \left(\frac{\mathscr{I}(\rho)}{\Delta t} + \mathscr{I}\left(\frac{1}{V} a_c \right) \right) \quad (11)$$

$$h_{tf} = -\frac{\mathscr{I}(\rho)}{\Delta t} \mathscr{I}(\hat{\mathbf{u}}_c) \quad (12)$$

$$h_{sf} = -\mathscr{I}(\rho)\mathbf{g} + \mathscr{I}\left(\frac{1}{V} \sum_n a_n \mathbf{u}_n - \mathbf{q} \right) \quad (13)$$

and the interpolation operator $\mathscr{I}()$ which interpolates a face field from a volume field using a central differencing interpolation scheme. To avoid a steady state solution depending on the timestep Δt the term for the time is excluded from the interpolation and built on the face directly. Therefore each term has to be interpolated on its own. This is not done in the original interFoam solver, where all terms are interpolated mutually. After solving the pressure equation the fluxes can be updated with

$$\Phi = \frac{\mathbf{s}_f}{\alpha_f} \cdot (h_{tf} + h_{sf} - (\nabla p)_f) \quad (14)$$

The equations are solved in a segregated algorithm shown in figure 1.

High Resolution Schemes

Keeping a sharp interface using the unmodified transport-equation for the volume-fraction (5) requires a special discretisation scheme for the convective term. The scheme should transform every gradient into a step function to guarantee sharpness. The solution for α has to be bounded between zero and unity. Additionally, the scheme should allow local Courant-Friedrichs-Lewy Numbers (Cos) larger than unity to decrease computational costs without losing stability. Furthermore the scheme should prevent numerical oscillation. These requirements result in a huge family of discretisation schemes. We implemented three schemes, the High-Resolution Interface Capturing (HRIC) scheme presented in Muzaferija et al. (1998), the Blended Interface Capturing Scheme (BICS) described in Queutey et al. (2007) and the Blended Reconstructed Interface Capturing Scheme (BRICS) presented in Wackers et al. (2010). Tests showed that the latter has small advantages.

This chapter first describes the principle of the Normalized Variable Diagram (NVD). Afterwards the high resolution scheme BRICS is being described.

Normalized Variable Diagram

The NVD is a common way to define High-Resolution schemes. Considering three neighboring cells the scheme for a generic quantity ϕ can be formulated as

$$\phi_f = f(\phi_U, \phi_C, \phi_D). \quad (15)$$

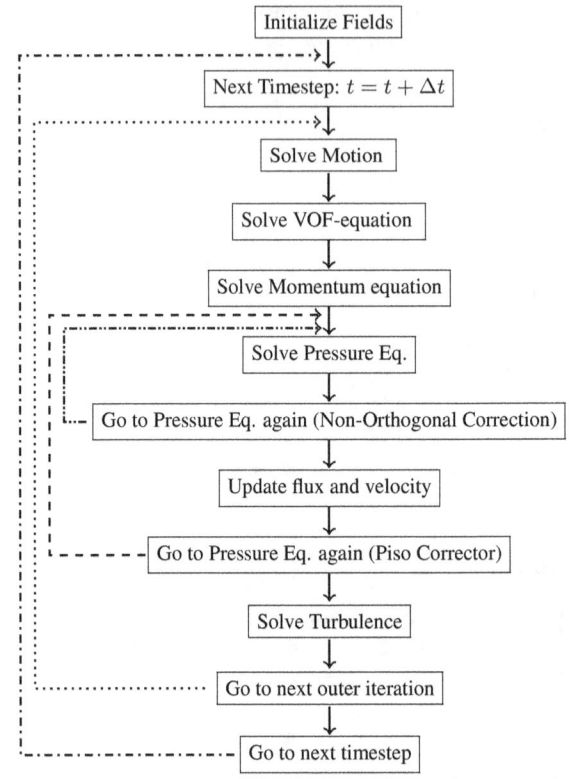

Figure 1: Solution Algorithm

Here the subscripts U, C and D stands for the upwind, central and downwind cell and depend on the flow direction, see figure 2. The subscript f stands for the face between the central and downwind cell. On structured grids the Normalized Variable Approach (NVA) defines the normalized variable $\tilde{\phi}$ as:

$$\tilde{\phi} = \frac{\phi - \phi_U}{\phi_D - \phi_U} \quad (16)$$

Based on this formula one can calculate the normalized values for the three cells:

$$\tilde{\phi}_U = 0 \quad (17)$$

$$\tilde{\phi}_C = \frac{\phi_C - \phi_U}{\phi_D - \phi_U} \quad (18)$$

$$\tilde{\phi}_D = 1. \quad (19)$$

The basic idea of the NVD is to define a scheme depending on $\tilde{\phi}_C$. Now the definition is simplified to

$$\tilde{\phi}_f = f(\tilde{\phi}_C) \quad (20)$$

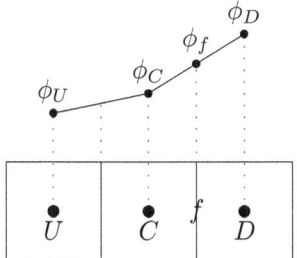

Figure 2: NVD cell notations

but still considers the three neighboring cells. The final face value is obtained by denormalizing $\tilde{\phi}_f$

$$\phi_f = \tilde{\phi}_f(\phi_D - \phi_U) + \phi_U. \qquad (21)$$

This principle can be implemented easily for structured grids. For arbitrarily unstructured meshes it is not obvious how to determine the far upwind cell and the concept becomes quite complicated. The calculation of the upwind cell is explained at the end of this section.

Blended Interface Capturing Scheme

The BICS scheme starts from the Inter-Gamma Differencing Scheme (IGDS).

$$\tilde{\phi}_{IGDS} = \begin{cases} \tilde{\phi}_C & \text{if } \tilde{\phi}_C < 0 \\ h_{IGDS} & \text{if } 0 \leq \tilde{\phi}_C < \beta_{IGDS} \\ 1 & \text{if } \beta_{IGDS} \leq \tilde{\phi}_C \leq 1 \\ \tilde{\phi}_C & \text{if } 1 < \tilde{\phi}_C \end{cases} \qquad (22)$$

with

$$h_{IGDS} = -\frac{\tilde{\phi}_C^2}{\beta_{IGDS}} + (1 + \frac{1}{\beta_{IGDS}})\tilde{\phi}_C \qquad (23)$$

The IGDS is very similar to the base function of the HRIC scheme presented in Muzaferija et al. (1998). For a sharp interface it introduces downwind differences. To fulfill the Convective Boundedness Criterion (CBC) it blends to upwind differences depending on the normalized variable. The scheme parameter β_{IGDS} is usually set to 0.5 to guarantee stability. The IGDS is limited to a local Courant Number of 0.3. Therefore the BICS blends to the Gamma Differencing Scheme (GDS) (Jasak et al., 1999), see equation (24), depending on the local Courant number.

$$\tilde{\phi}_{GDS} = \begin{cases} \tilde{\phi}_C & \text{if } \tilde{\phi}_C < 0 \\ h_{GDS} & \text{if } 0 \leq \tilde{\phi}_C < \beta_{GDS} \\ 0.5\tilde{\phi}_C + 0.5 & \text{if } \beta_{GDS} \leq \tilde{\phi}_C \leq 1 \\ \tilde{\phi}_C & \text{if } 1 < \tilde{\phi}_C \end{cases} \qquad (24)$$

with

$$h_{GDS} = -\frac{\tilde{\phi}_C^2}{2\beta_{GDS}} + (1 + \frac{1}{2\beta_{GDS}})\tilde{\phi}_C \qquad (25)$$

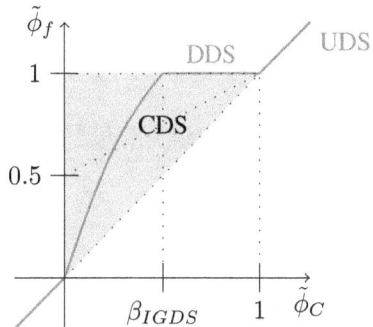

Figure 3: Inter-Gamma Differencing Scheme

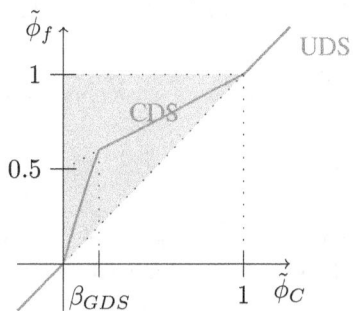

Figure 4: Gamma Differencing Scheme

As one can see in Figure 4 the GDS is second order accurate like the Central Differencing Scheme (CDS) but blends to upwind differences to fulfill the CBC. It has no Courant number limitations. Usually the scheme parameter β_{GDS} is set to 0.1. Lower values may lack stability and higher values will reduce the accuracy. The Courant number dependency of the BICS is given in following equation.

$$\tilde{\phi}_{BICS^*} = \begin{cases} \tilde{\phi}_C & \text{if } \tilde{\phi}_C < 0 \\ h_{BICS^*} & \text{if } 0 \leq \tilde{\phi}_C < \beta_{BICS} \\ p\tilde{\phi}_C + (1-p) & \text{if } \beta_{BICS} \leq \tilde{\phi}_C \leq 1 \\ \tilde{\phi}_C & \text{if } 1 < \tilde{\phi}_C \end{cases} \qquad (26)$$

with

$$h_{BICS^*} = -\frac{1-p}{\beta_{BICS^2}}\tilde{\phi}_C^2 + (p + \frac{2(1-p)}{\beta_{BICS}})\tilde{\phi}_C \qquad (27)$$

and with the slope p depending on the local Courant number Co, see equation (28), and the scheme parameter β_{BICS} depending on the slope p, see equation (31). The scheme is using a quadratic variation for the part below β_{BICS} and a linear variation for the region above β_{BICS}.

The slope p is calculated by

$$p(Co) = \alpha_p(Co)p_{IGDS} + (1 - \alpha_p(Co))p_{GDS} \qquad (28)$$

with

$$\alpha_p(Co) = \begin{cases} 1 & \text{if } Co \leq 0.3 \\ \frac{Co-0.3}{e^{(Co-0.3)}-1} & \text{if } 0.3 < Co \end{cases} \qquad (29)$$

164

and

$$p_{GDS} = 0.5; \quad p_{IGDS} = 0.$$ (30)

The scheme parameter β_{BICS} is calculated by

$$\beta_{BICS}(p) = a_0 + a_1 p(Co)$$ (31)

with

$$a_1 = \frac{\beta_{GDS} - \beta_{IGDS}}{p_{GDS} - p_{IGDS}}; \quad a_0 = \beta_{IGDS} - a_1 p_{IGDS}$$ (32)

with

$$\beta_{GDS} = 0.1; \quad \beta_{IGDS} = 0.5$$ (33)

as the standard parameters.

Finally the classical angle correction from the HRIC scheme is applied.

$$\tilde{\phi}_{BICS} = \tilde{\phi}_{BICS*}(\cos\theta)^{C_0} + \tilde{\phi}_{GDS}[1 - (\cos\theta)^{C_0}]$$ (34)

where θ is the angle between the normal of the face and the normal of the free-surface. Here, the difference between the BICS and HRIC scheme is that the BICS reduces to the GDS instead of the upwind scheme for the angle correction.

Calculation of the Upwind Cell for the Normalized Variable Approach

For arbitrarily unstructured grids the calculation of the upwind value U is not straight forward. The upwind cell is not known explicitly. One might have multiple cells or no existing cell for the upwind region. This chapter will describe three ways for the calculation of the upwind value.

The classical approach is to extrapolate to an imaginary upstream node U by the use of the gradient projection method.

$$\phi_U = \phi_C - \vec{CU} \cdot \vec{\nabla}\phi_C, \quad with \quad \vec{CU} = -\vec{CD}.$$ (35)

Another approach given in Jasak et al. (1999) renounces the upwind value completely and uses a modification of the NVD criterion. Instead of using the upwind and downwind cell, one uses the upwind and downwind face, see figure 5. The new definition of $\tilde{\phi}_C$ is

$$\tilde{\phi}_C = \frac{\phi_C - \phi_f^-}{\phi_f^+ - \phi_f^-} = 1 - \frac{\phi_f^+ - \phi_C}{\phi_f^+ - \phi_f^-}.$$ (36)

After some transformation which is explained in detail in Jasak et al. (1999) one obtains the final equation

$$\tilde{\phi}_C = 1 - \frac{\phi_D - \phi_C}{2(\nabla\phi)_C \cdot \vec{CD}}.$$ (37)

For both methods shown the extrapolation may lead to values outside the interval [0, 1]. Therefore the values will be bounded to the given interval once they have been

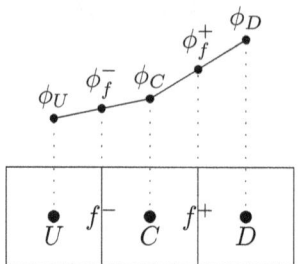

Figure 5: Modified approach for the NVD criterion

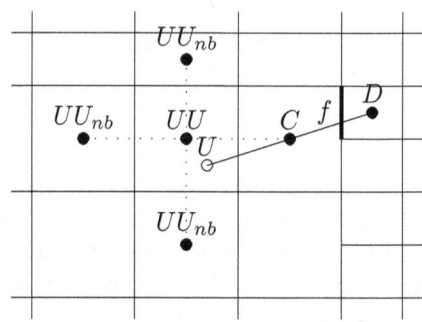

Figure 6: Reconstruction of the upstream node U

calculated. The methods are satisfactory for continuous quantities. For the discontinuous volume fraction these approaches may lead to big errors and stability problems.

Another way of obtaining the upstream node U is based on a reconstruction. It is introduced as an Extension of the BICS in Wackers et al. (2010). The resulting scheme is called BRICS. The difference to the original BICS lies in the reconstruction of the upstream node U as explained below.

The BRICS first searches the cell UU containing the imaginary point U, see figure 6. The search consist of a loop over the neighboring cells of C. The cells UU can be precomputed. If the mesh is not changing the search algorithm has to be called only once. After the cell UU is identified the value for the point U can be calculated with a weighted interpolation based on the cell UU and its neighboring cells UU_{nb}

$$\phi_U = \sum_{N=UU,UU_{nb}} \frac{\phi_N}{||\vec{NU}||} / \sum_{N=UU,UU_{nb}} \frac{1}{||\vec{NU}||}.$$ (38)

Implementation of the High-Resolution Schemes

The High-Resolution schemes allow maximal Cos larger than unity without losing a sharp interface. Indeed the solver lacks stability with higher Cos. To stabilize the solution the scheme is implemented as a deferred correction scheme.

Pressure Gradient Reconstruction

The total pressure p of equation (1) consist of the static pressure $\rho g h$ and dynamic pressure p_d.

$$p = \rho g h + p_d \tag{39}$$

with the z-component g of the gravity vector $\mathbf{g}$ and the z-coordinate h. Substituting the total pressure of (1) with (39) leads to a momentum equation depending on the dynamic pressure

$$\frac{\delta \rho \mathbf{u}}{\delta t} + \nabla \cdot (\rho \mathbf{u}\mathbf{u}) - \nabla \cdot \mu_e \left(\nabla \mathbf{u} + (\nabla \mathbf{u})^T \right)$$
$$= -\nabla(p_d) - gh\nabla(\rho) \tag{40}$$

Both momentum equations presented here can be used for calculation. The interFoam solver uses equation (40) and OurSolver uses equation (1). The reasons for our decision are explained in this section.

Considering the behavior of pressure and its gradient at the free surface it is not possible to use the dynamic pressure for the momentum equation. Doing this will lead to a term containing the gradient of the dynamic pressure and a gradient of the density, see equation (40). The pressure gradient has a jump at the free surface and the density gradient is undefined for a sharp interface. Still the dynamic pressure is used for the original interFoam solver.

Numerical smearing at the interface avoids the jump behavior. For cells containing the lighter fluid the error is especially large. For density ratios typical for water and air the values could be around 500 times larger. These wrong values lead to unphysical high velocities in the free surface cells containing the lighter fluid. Having a satisfying sharp interface, due to a High Resolution Scheme and/or due to grid refinement, will enforce this behavior. The velocity overshoots decrease the stability of the solver. Often, simulations with grids fulfilling the requirements for a sufficient sharp interface diverge using the standard interFoam solver. Sometimes a solution is possible with adjusted solver settings (smaller time step, more iterations) but then the solver is slow compared to other well known commercial solvers. One solution is to limit the velocities to a user specified value every time it is calculated. This leads to a more stable solver but still does not address the root cause. Therefore we have implemented a method for the reconstruction of the pressure and its gradient at the face for arbitrary unstructured grids given in Queutey et al. (2007).

Four equations, the momentum equation (1), the pressure equation (10), the velocity equation (9) and the equation for the flux (14), contain the surface normal gradient of the pressure which requires the reconstruction. This gradient is on the explicit right hand side three times and one time (for the pressure equation) it is in the implicit part.

The reconstruction is explained in detail in Queutey et al. (2007) and only the important parts of our implementation are described here. The method assumes that the free surface Γ is exactly at the face f. It requires a known function, which is continuous on each subdomain Ω^+ and Ω^- and discontinuous across the interface. Here the density fulfills these requirements, if a High-Resolution scheme is used for the discretisation of the volume fraction. The reconstruction is only possible using the total pressure. Otherwise in case of dynamic pressure we will get the undefined density gradient at the free surface. This is the reason why our solver is using the total pressure and not the dynamic pressure like interFoam.

Reconstructing $\left(\frac{\nabla p \cdot \mathbf{n}}{\rho} \right)_f$ instead of $(\mathbf{n} \cdot \nabla p)_f$ is the only solution. The equation given for the surface normal gradient is

$$\left(\frac{\nabla p \cdot \mathbf{n}}{\rho} \right)_f = \frac{1}{\hat{\rho}} \frac{p_R - p_L}{h} \boxed{ + \frac{\nabla p_L \cdot \mathbf{e}^- + \nabla p_R \cdot \mathbf{e}^+}{\hat{\rho} h} } \tag{41}$$

with the explicit distance vectors $\mathbf{e}^+$ and $\mathbf{e}^-$ as shown in figure 7 for the correction of grid non-orthogonality. The boxed term is treated explicitly. The equation given for the face value is

$$p_f = \frac{h^+ \rho^+ p_L + h^- \rho^- p_R}{h^+ \rho^+ + h^- \rho^-}$$
$$\boxed{ + \frac{\rho^+ \rho^-}{\hat{\rho}} \left(\frac{h^- \mathbf{e}^+ - h^+ \mathbf{e}^-}{h} \right) \cdot \left(\frac{h^+}{h} \left(\frac{\nabla p}{\rho} \right)_L + \frac{h^-}{h} \left(\frac{\nabla p}{\rho} \right)_R \right) }$$
$$\tag{42}$$

with the face value $\hat{\rho}$ calculated by a reversed linear interpolation

$$\hat{\rho} = \frac{h^-}{h} \rho^- + \frac{h^+}{h} \rho^+ \tag{43}$$

Again, the boxed term is treated explicitly. It is not obvious how to calculate the explicit, boxed part of equation (41). The pressure gradients can be calculated using the Gauss Theorem $\int_v \nabla \phi dv = \int_s \phi d\mathbf{s}$. The Gauss Theorem requires the face values. Using the linear interpolated face values is possible, but numerical tests showed a small negative influence on to the result. Therefore we use the reconstructed face values calculated with (42). But, Equation (42) itself requires the cell gradients (calculated with the Gauss Theorem). Thus, it is a coupled system of equations and the implementation is not straightforward. Here it is implemented with the following iterative approach:

1. Calculate p_f^n using ∇p^{n-1}

2. Calcuate ∇p^n using p_f^n

3. Limit ∇p^n

4. Update iterator $n = n + 1$. Go to step 1

The limitation at Step 3 is essential for convergency. Unfortunately this limitation has a significant influence on the

166

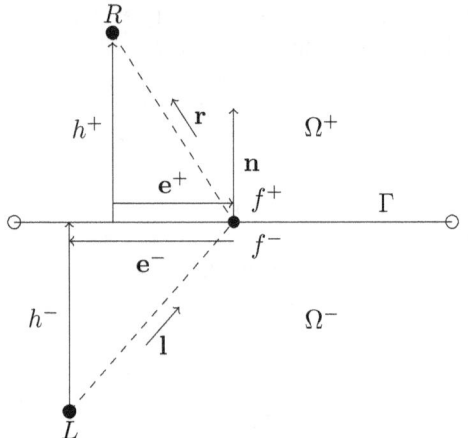

Figure 7: Notations for discontinuity reconstruction

computational time. Numerical tests with different numbers of iterations were done and showed that one iteration only is enough for a steady state result.

The final implementation consist of only one iteration and calculates the reconstructed face value with the old pressure gradient. Since there are outer iterations of the SIMPLE-algorithm there is still an iterative process updating the face value and cell gradient. It should be mentioned that a method without this gradient reconstruction also requires a limitation of the pressure gradient for sufficient stability.

For the reconstruction the flow equations have to be modified as follows (modifications are highlighted in red):

Momentum equation:

$$
\frac{\rho}{\Delta t}\mathbf{u}_c + \frac{1}{V}a_c\mathbf{u}_c + \frac{1}{V}\sum_n a_n\mathbf{u}_n \\
= \mathbf{q} + \rho\mathbf{g} - \rho\mathscr{R}((\nabla p)_f \cdot \mathbf{s}_f/\hat{\rho}) + \frac{\rho}{\Delta t}\hat{\mathbf{u}}_c
$$
(44)

Pressure equation:

$$
\mathscr{D}\left(\frac{\mathscr{I}(\rho)}{\hat{\rho}}\alpha_f^{-1}\nabla p\right) = \mathscr{D}\left(\alpha_f^{-1}\left(h_{tf} + h_{sf}\right)\right)
$$
(45)

Velocity equation:

$$
\mathbf{u}_c = \left(\frac{\rho}{\Delta t} + \frac{1}{V}a_c\right)^{-1} \\
\left(\mathbf{q} + \rho\mathbf{g} - \rho\mathscr{R}((\nabla p)_f \cdot \mathbf{s}_f/\hat{\rho}) + \frac{\rho}{\Delta t}\hat{\mathbf{u}}_c - \frac{1}{V}\sum_n a_n\mathbf{u}_n\right)
$$
(46)

Flux:

$$
\Phi = \frac{\mathbf{s}_f}{\alpha_f} \cdot \left(h_{tf} + h_{sf} - \frac{\mathscr{I}(\rho)}{\hat{\rho}}(\nabla p)_f\right)
$$
(47)

Motion

An important part of the drag prediction for yachts is the correct determination of flotation. While flow around a yacht is usually simulated in an Eulerian coordinate system, the yacht moves with respect to this coordinate frame. The motion itself depends on fluid forces as well as external forces and moments either fixed or resulting from fluid forces. To correctly capture the behavior of highly dynamic vessels (e.g. if foil-supported), time accuracy is of utmost importance. As the *sixDoFRBM*-Method present in OpenFOAM has been shown to be unsuitable (see Devolder et al. (2015)), a robust method has been implemented.

Time Integration

The computation of linear acceleration is based on a fully dynamic model with acceleration computed from instantaneous force:

$$
\mathbf{f} = \mathbf{M} \cdot \mathbf{a}
$$
(48)

Time integration to compute velocities and position is achieved by the trapezoidal rule (see Hadžić et al. (2005)):

$$
\mathbf{v}_{n+1}^{i+1} = \mathbf{v}_n + \frac{\mathbf{f}_n + \mathbf{f}_{n+1}^{i+1}}{2m}\Delta t
$$
(49)

$$
\mathbf{r}_{n+1}^{i+1} = \mathbf{r}_n + \frac{\mathbf{v}_n + \mathbf{v}_{n+1}^{i+1}}{2}\Delta t
$$
(50)

Similar equations hold for rotational motion.

To achieve (semi-) implicit coupling between flow and motions, the solution of above equations is integrated into the SIMPLE- or outer loop of the CFD solver (see Figure 1).

Stability

While implicit trapezoidal integration of the equations of motion could be considered to be generally stable, some decoupling between flow and motion takes place due to the quasi-explicit integration in the outer loop. This decoupling gives rise to a virtual added mass effect (see Förster et al. (2007)), potentially destabilizing the solution if virtual added masses are equal to or larger than the vessel's mass. Hadžić et al. (2005) have shown that a per iteration under-relaxation of velocities is comparable to the application of an added mass to the force defect between iterations. When the outer loop converges, the effect of this under-relaxation on results vanishes.

While this approach is simple and robust, if the under-relaxation factors are well chosen, even it does not guarantee optimal convergence behavior. More efficient approaches (e.g. Soeding (2001)) are currently under investigation.

To improve convergence to steady state if transient behavior is not of interest, velocity-proportional damping can be applied to the motion.

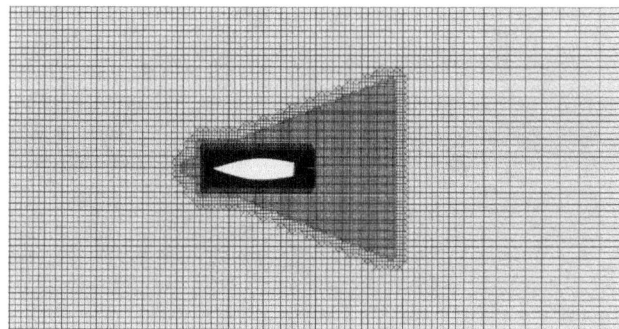

Figure 8: Sysser60 grid top view

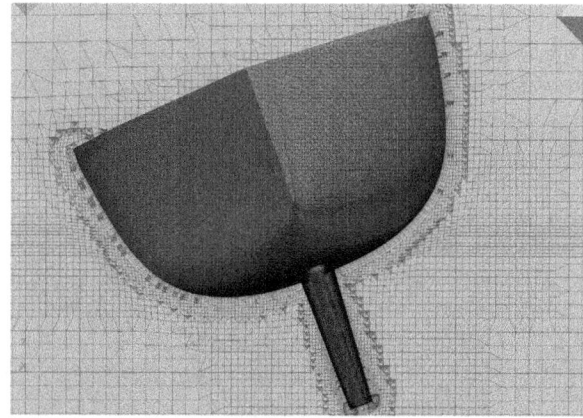

Figure 9: Sysser60 grid front view

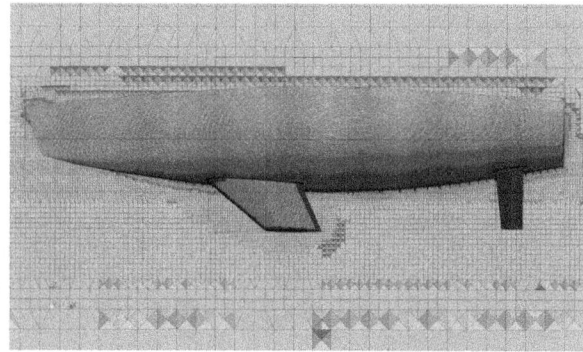

Figure 10: Sysser60 grid side view

External Forces

To properly include the effects of sail or propulsive forces and moments these can be applied at any given point (e.g. CoE), either as fixed values or depending on flow forces and moments. The point of attack and, if required, the direction are defined in boat-fixed frame of reference.

VALIDATION

Sysser60

Description

In order to compare our new solver with interFoam (of OpenFOAM 2.4.x) and the commercial solver Star-CCM+ v9.06 the Sysser 60 test case of the Delft Systematic Yacht Hull Series of Delft University of Technology (2013) has been chosen. The simulations are carried out without motion to better distinguish between phenomena induced by flow solver and motion solver. Particular focus during this test was on the phenomenon of numerical ventilation. To evaluate the solvers capabilities in this regard a flotation with the bow knuckle above water was chosen. The simulation is done in model scale. The hull has a waterline length of $L = 2.16m$. The flow speed is 1.806 m/s corresponding to a Froude no. of 0.39. The hull is heeled by $20°$ and yawed by $3°$.

Figure 8 to 10 show the grid. It was generated with snappyHexMesh (SHM) and has 2.7 million cells. The waterline was refined anisotropically in the z direction. The cells at the outlet were stretched to avoid reflections of the waves. Figure 8 only shows a part of these stretched cells. The domain has an length of $3L$ in front of the hull and $3L$ behind the hull before stretching the cells at the outlet. The height is $1.5L$ below waterline and $1.0L$ above waterline. In the area of the Kelvin wave pattern additional refinement is applied up to a length of $1.4L$ behind the hull The domain is $4.3L$ wide. The grid was exported to Star-CCM+ so that all simulations are done on exactly the same grid.

Solver Setups

The interFoam solver uses the same algorithm as shown in figure 1. For the VOF-method an additional compressive term is used to keep a sharp interface. The equation is solved using the multi-dimensional limiter for explicit solution (MULES). It uses an implicit predictor based on upwind differencing for the convective term. Afterwards the solution is corrected with an explicit corrector using the vanLeer discretization scheme. Despite a lot of time consuming tests it was not possible to get a stable setup for the given grid with the original interFoam solver. Therefore a function for the limitation of the velocity has been added. Each time the velocity changes, the cell velocities are limited to a user specified value. In this case the velocity is limited to $15\frac{m}{s}$. In fact this modification harms the conversation of mass. But tests have shown that this has no influence onto the results. For clarity this solver is called interFoamMod in this paper. Similar to OurSolver, interFoam allows the use of the α-subcycling technique. This method allows subdivision of the timestep in smaller subtimesteps for the solution of the VOF-transport equation which improves more stability and accuracy.

Star-CCM+ also uses an algorithm similar to figure 1. It is not known if any additional iterations for the correction of mesh non-orthogonality are applied. Also, it is not possible to prescribe any kind of α-subcycling. The HRIC scheme is used for the convective term of the VOF-equation. To suppress numerical ventilation the blend to upwind differencing has been switched off as recommended in Boehm et al. (2014) for steady state simulations.

All three solvers are adjusted to the same settings as close as possible, see Table 1. The goal was to use a time step of $\Delta t = 0.02s$, 5 outer iterations, 1 Piso iteration, 1 additional iteration for the correction of non-orthogonality and no alpha-subcycling. The averaged Co is 0.123. Although the modified solver interFoamMod has been used, additional Piso iterations and α-subcycles were necessary to get sufficient solver stability here. For the sake of clarity, this is only done to get solver stability. Nevertheless additional iterations and subcycles generally have a positive influence onto the quality of the result.

In all simulations the k-ω-SST was chosen. A total time of $30s$ is simulated.

Star-CCM+ is using some unknown method to suppress an unphysical splash wave at the beginning of the simulation. For OurSolver the velocity is ramped over the first $0.2s$ to avoid the same unphysical behavior. The simulations were also done without ramping, showing that the ramping is not required for solver stability.

Wave pattern

The transversal wavecuts of figure 11 are obtained by cutting the free-surface with an increment of $\frac{x}{L} = 0.1L$. The longitudinal wavecuts of figure 12 are obtained by cutting with an increment of $\frac{y}{L} = 0.05L$. The wavecuts show good agreement in the region close to the hull. At the bow only the Star-CCM+ simulation produces a different result with three breaking waves which is discussed later in this section. Both figures show that the waves of the interFoam-Mod simulation are increasingly damped the further from the hull. Other simulations (not shown here) with a much longer Kelvin refinement of about $6L$ behind the hull prove this behavior. Independent of the refinement, the interFoam-Mod waves are much flatter far from the hull. This emphasizes that the discretization of the convective term of equation (5) requires a High Resolution scheme for a correct convection of the waves. Considering future simulations including ocean waves this is an important point to avoid damping of these waves. Figure 13 also illustrates the damping characteristics of the interFoamMod simulation. It is an important point that the interFoamMod simulation is using five alpha-subcycling due to stability reasons whereas both other solvers do not use any additional subcycles. Subcycling supports the accuracy of the simulation for both wave sharpness and wave convection. Despite this, interFoam-Mod is less accurate.

The waves at the bow of the Star-CCM+ simulation are significantly different than the results of OurSolver and inter-

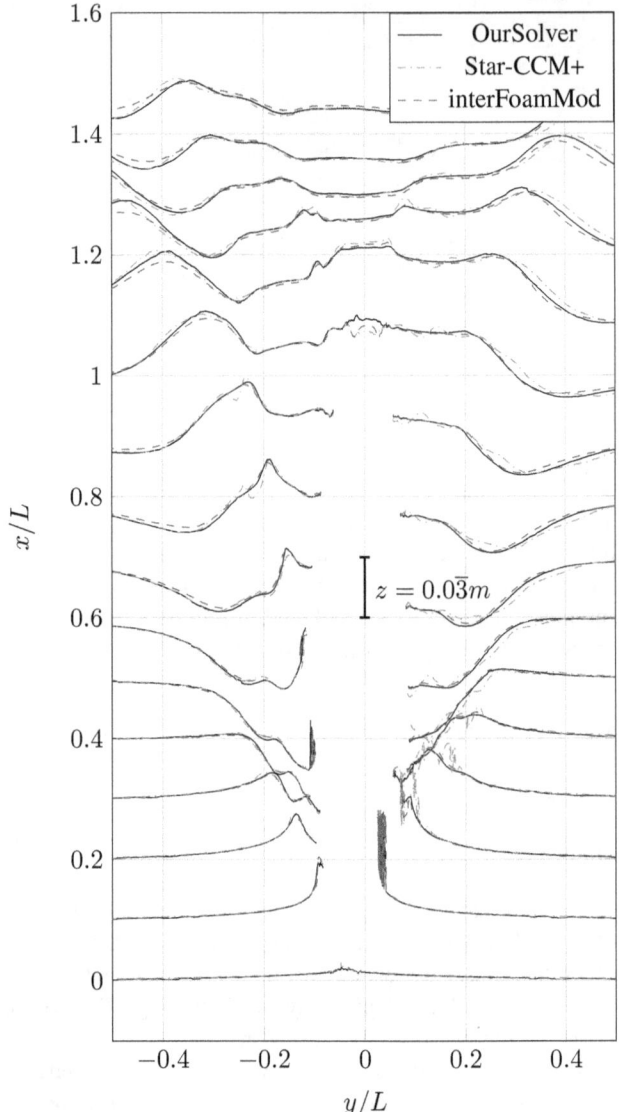

Figure 11: Sysser60 transversal wavecuts

FoamMod. Figure 14 shows that Star-CCM+ generates a wave breaking three times. The other two solvers have one to one and a half breaking waves depending on the chosen timestep. Increasing the cell size to 11 million cells has no effect on this behavior and the wave count is still the same for OurSolver and interFoamMod. Using the finer grid for Star-CCM+ boosts this behavior to four clearly breaking waves. To the authors knowledge the four waves of Star-CCM+ have not been observed for real flow. As Star-CCM+ is a closed code, we do not have any knowledge about the implemented models beyond the description in the theory guide. Finally this phenomenon might need additional investigations.

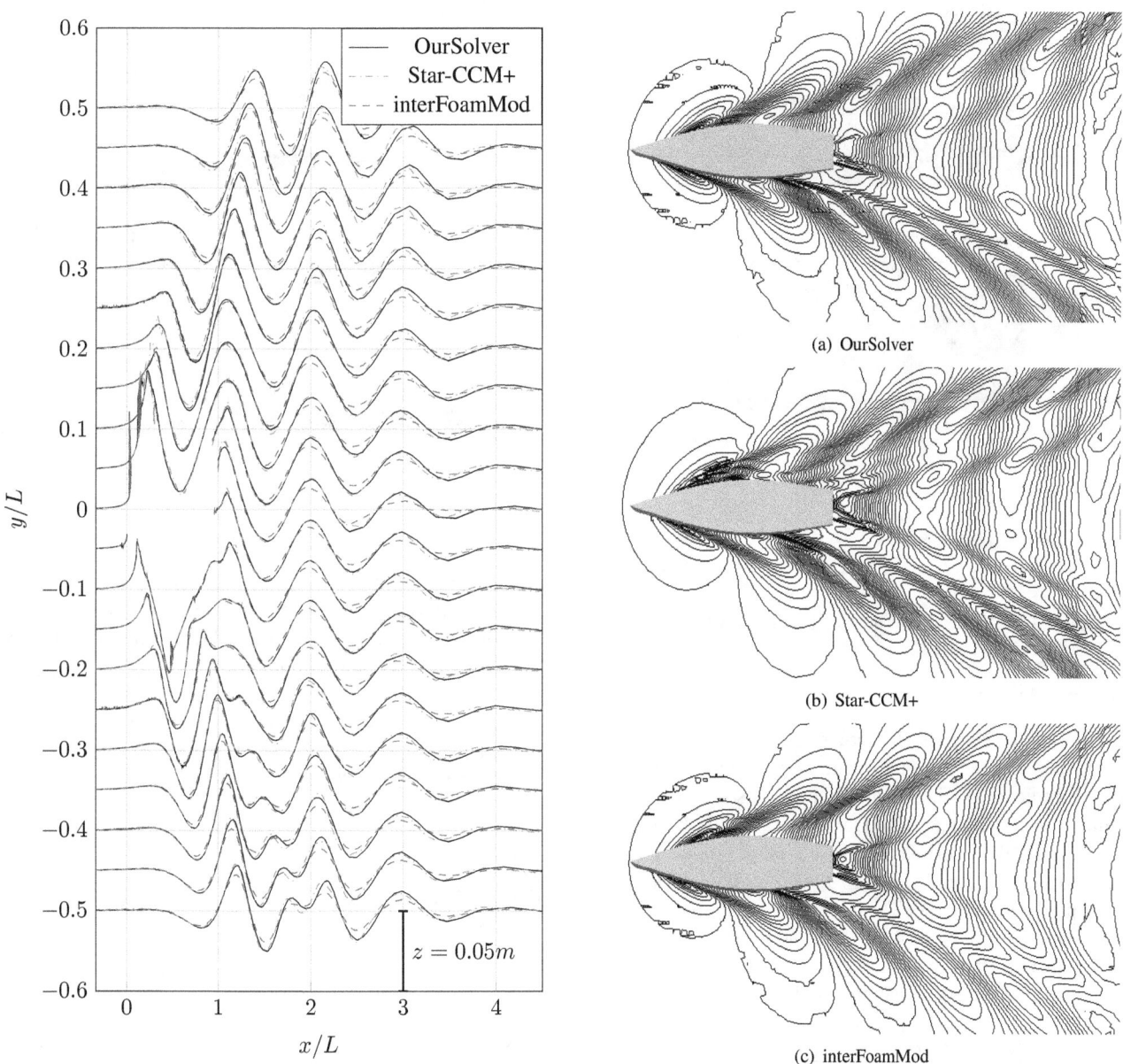

Figure 12: Sysser60 longitudinal wavecuts

(a) OurSolver

(b) Star-CCM+

(c) interFoamMod

Figure 13: Wavepattern of the Sysser60 test case

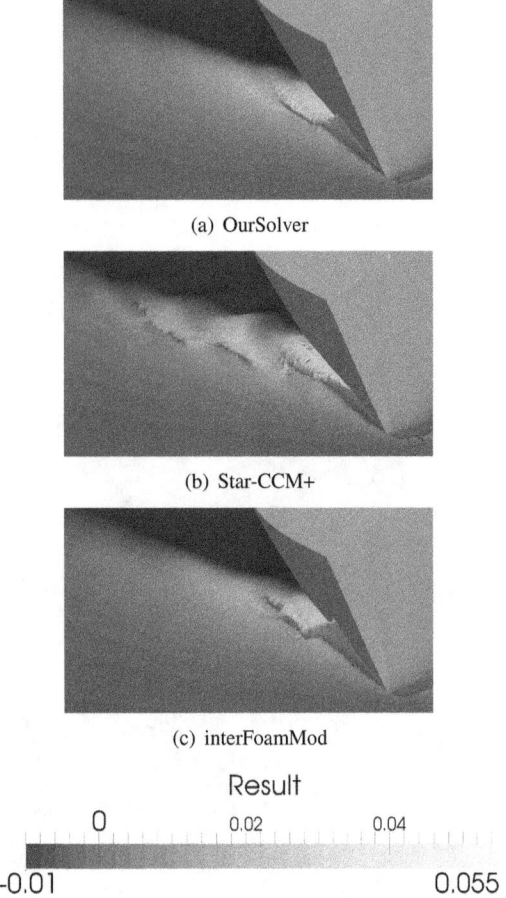

(a) OurSolver

(b) Star-CCM+

(c) interFoamMod

Result

0   0.02   0.04

-0.01   0.055

Figure 14: Waves at the bow of the Sysser60 test case

Numerical ventilation

The amount of Numerical Ventilation (NV) is shown in figure 15 plotting the volume fraction on the hull. For a distinct differentiation the volume fraction is scaled from 0.85 to unity. While the results of OurSolver and Star-CCM+ are quite similar the interFoamMod simulation produced much more air under the hull. Figure 15 is not sufficient for an reliable predication of NV. The NV changes with the time due to a periodic oscillation of the flow and the chosen timestep might have a significant influence on the result.

To estimate the NV equation (51) is introduced.

$$N = \frac{\sum_{f=1}^{n} \lambda_f |\mathbf{s}_f| |(1 - \alpha_w f)|}{\sum_{f=1}^{n} \lambda_f |\mathbf{s}_f|} \tag{51}$$

where $\mathbf{s}$ is the face area vector of the face f, n is the number of faces defining the hull patch, α_w is the volume fraction of the water and λ is a switch depending on the pressure

$$\lambda_f = \begin{cases} 1 & \text{if } p_f > p_l \\ 0 & \text{if } p_f \leq p_l \end{cases} \tag{52}$$

Where the limiting pressure p_l is set to $150\frac{N}{m^2}$. The pressure switch excludes faces above and at the free surface,

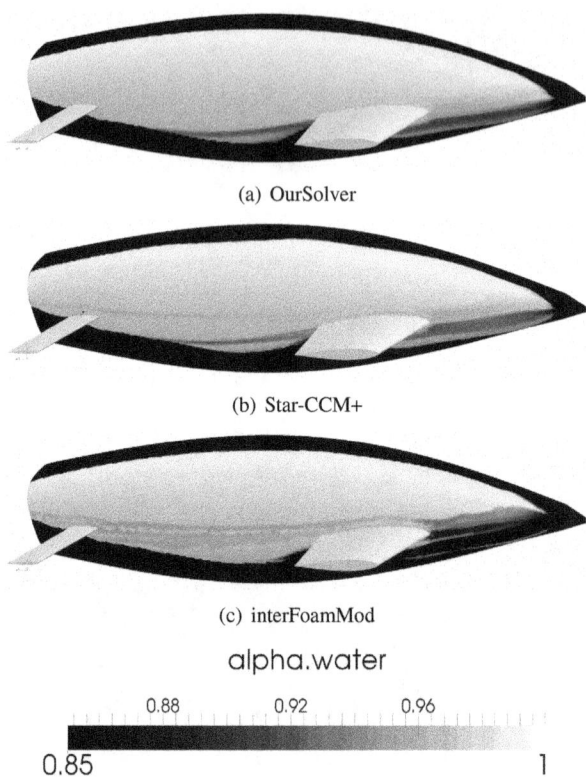

(a) OurSolver

(b) Star-CCM+

(c) interFoamMod

alpha.water

0.88   0.92   0.96

0.85   1

Figure 15: Numerical ventilation of the Sysser60 test case

respectively. Therefore physically correct spray does not have a negative effect on the assessment of the NV. This equation does not produce an absolutely correct evaluation but allows a rough approximation.

Figure 16 shows the NV plotted over time. It shows that the result of interFoamMod has twice the NV compared to OurSolver and Star-CCM+.

Velocity overshoots

Figure 17 shows the maximum velocities for each timestep. The maximum velocity of the interFoamMod simulation is mostly at the prescribed limit of 15m/s. The results of OurSolver and Star-CCM+ are physically more correct. OurSolver only reaches the limiting velocity for the first timesteps. Star-CCM+ does not have such limitation and only exceeds this value for the first timesteps and at $t = 16s$. Overall the maximal velocity of Star-CCM+ is smaller than OurSolver. Concluding it can be said that the velocities of both solvers lie in a reasonable range whereas the interFoamMod velocities have unacceptably high values and never drop down to plausible values. For the sake of clarity, the velocity limitation is not included in the original interFoam solver. Without this limitation the velocities would reach much higher values and the solution will diverge until the solver crashes in the first timesteps.

The simulation with OurSolver has been done with and

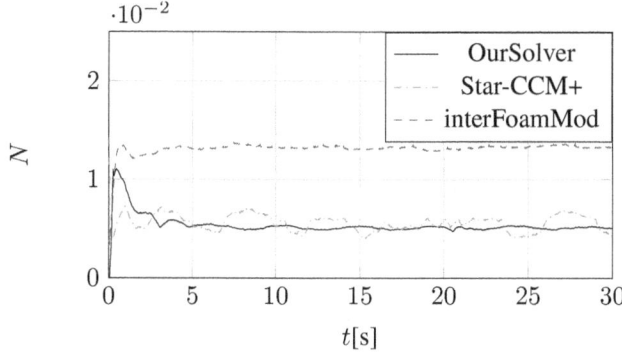

Figure 16: Numerical ventilation

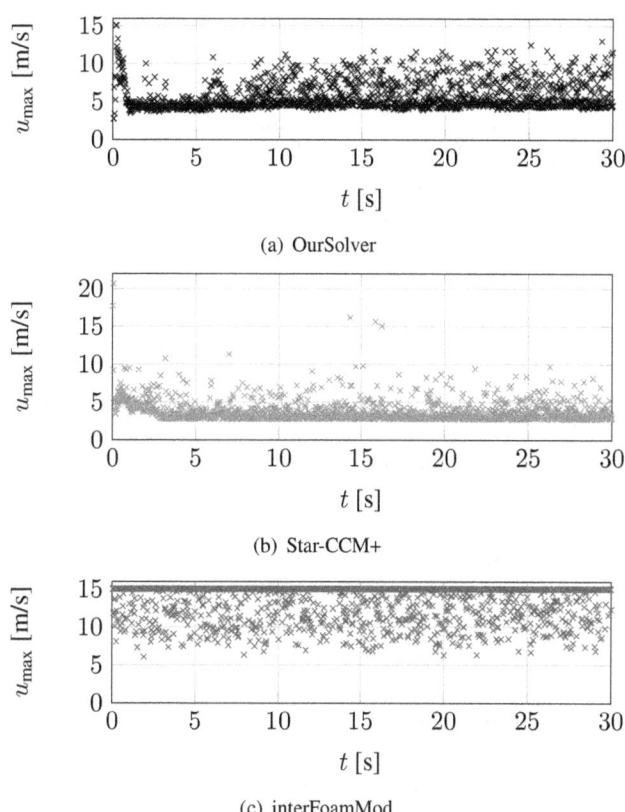

(a) OurSolver

(b) Star-CCM+

(c) interFoamMod

Figure 17: Velocity overshoots

without pressure reconstruction. Without pressure reconstruction the maximal velocities have the same characteristic as the velocities of the interFoamMod simulation. For this test case, besides the velocity overshoots the result looks almost identical to the result of the simulation with reconstruction. Thus, in this case the reconstruction is only necessary for the stability of the solver. It delivers a physically correct solution of the velocity overshoot problem compared to a hard limitation. Additionally it does not require a user-specified value. This simplifies automated set-ups and more complex cases.

Computation Time

For the comparison of the computational time the case was varied in cell size and time step. The computations were done on a node with 16 cores at 2.6Ghz. Only 15 cores were used to have enough power left for background processes of the machine. Different nodes with identical construction have been used for all simulations. Overall, this is not a scientific investigation on the computational time and the time needed for the same simulation might vary about 20%. Still,the results given in Table 1 show a significant trend. The table demonstrates the solver capabilities for real engineering situations where grid or timestep studies are required.

For all grids calculated with the smallest timestep of $\Delta t = 0.02$ OurSolver is about one third slower than Star-CCM+. For the timestep variation OurSolver needs about twice the time of Star-CCM+. The reason could be the additional non-orthogonal correction iteration and the settings for the absolute residual which breaks the inner iterations of the linear solvers. Also, OurSolver is compiled with g++ which might be up to 20% slower for specific tasks than other well known compilers.

To facilitate the calculation with the interFoamMod solver its settings had to be adjusted. These adjustments are the major reason for the significant slower computations. For the grid variation the solver needs two to six times the Star-CCM+ calculation time. For the time step variation it

needs up to 33 times of the calculation time. Increasing the timestep should decrease the calculation time. The cases with $\Delta t = 0.06$ and $\Delta t = 0.08$ show that this is not always the case for interFoamMod just because of stability reasons. Additionally it has to be emphasized that a typical grid or timestep study is based on scripting the simulations and then running all in one chunk. Here a lot of failing runs were necessary to get a setup with acceptable calculation time for each variation. Of course this is an unacceptable workflow. The 8.3 million cells test case forced a decrease in the timestep to $\Delta t = 0.01$. A grid with 10.8 cells has been tested, also. OurSolver and Star-CCM+ run as stable as before without modifying any settings. For interFoamMod the settings were adjusted until a simulation time of 2 weeks was reached, but the simulation still crashed.

Immersed Transom Study

Farr Yacht Design, in association with the *Sailing Yacht Research Foundation*, kindly released the results of a towing tank study on the impact of transom immersion respectively displacement and LCG on the resistance of a canonical VO70 yacht into public domain (see Farr Yacht Design (2005)). Within this study the transom immersion at rest was varied by adding weights to the yacht and sail trimming moment was applied by shifting the weights according to standard towing tank practice. The experiments were car-

Table 1: Simulation time on 15*2.6GHz for different case setups

	Δt [s]	nCells [E6 cells]	θ [ ]	κ [ ]	ξ [ ]	ϵ [ ]	Computation Time [h]	Deviation to Star-CCM+ [%]
Star-CCM+	0.02	2.7	5	1	1	-	8.39	0.0
OurSolver	0.02	2.7	5	1	1	1	10.71	+27
interFoamMod	0.02	2.7	5	3	5	1	27.95	+233
Star-CCM+	0.04	2.7	5	1	1	-	2.68	0.0
OurSolver	0.04	2.7	5	1	1	1	5.92	+121
interFoamMod	0.04	2.7	5	3	10	1	22.53	+772
Star-CCM+	0.06	2.7	5	1	1	-	1.77	0.0
OurSolver	0.06	2.7	5	1	1	1	2.92	+64
interFoamMod	0.06	2.7	10	3	15	1	42.51	+2301
Star-CCM+	0.08	2.7	5	1	1	-	1.24	0.0
OurSolver	0.08	2.7	5	1	1	1	2.41	+94
interFoamMod	0.08	2.7	10	3	20	1	42.07	+3392
Star-CCM+	0.02	4.5	5	1	1	-	14.91	0.0
OurSolver	0.02	4.5	5	1	1	1	18.91	+26
interFoamMod	0.02	4.5	5	3	5	1	48.23	+223
Star-CCM+	0.02	8.3	5	1	1	-	29.94	0.0
OurSolver	0.02	8.3	5	1	1	1	40.21	+34
interFoamMod	0.01	8.3	5	3	5	1	213.12	+611

θ = number of outer iterations, κ = number of Piso corrector iterations,
ξ = number of α subcycles, ϵ = number of non-orthogonal correction iterations

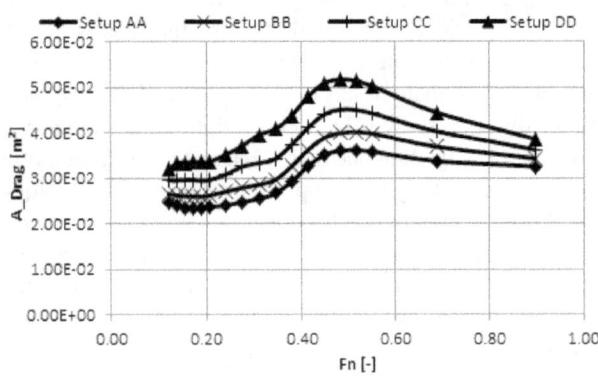

Figure 18: Resistance of canonical VO70 design (Towing tank results)

ried out at a scale of 3.2, and four basic flotations were tested at Froude numbers from 0.121 to 0.898. All tests were run in upright condition without yaw. Experimental results are given in Figure 18, the mid-forward (BB) and mid-aft (CC) LCGs are selected for validation.

Simulation Setup

The flow around the yacht is simulated at selected velocities corresponding to Fn from 0.345 upwards for LCG / displacement combinations denoted BB and CC. As only upright resistance is of interest only one half of the yacht is modeled with a domain size of 1 length ahead, 3 astern, 2 aside, 1.5 below and 0.5 above the yacht. The computational mesh is generated using a combination of snappyHexMesh (SHM) and related utilities, resulting in a domain size of about 1.8E+06 cells. Analogous to the experiments the only appendages considered are keel and bulb. The free surface region perpendicular to the undisturbed free surface is resolved down to a thousandth of length over a total height of 3% of length.

An incident turbulence level of 1% is prescribed with turbulence modeled by the $k\omega$-SST-model using wall functions. Average y^+ is about 50. Velocity is ramped over 5s, body motion is released at 1s with forces and moments ramped over another second.

Results

Figure 19 gives a comparison of drag areas (drag divided by dynamic pressure head) for the evaluated cases. As can

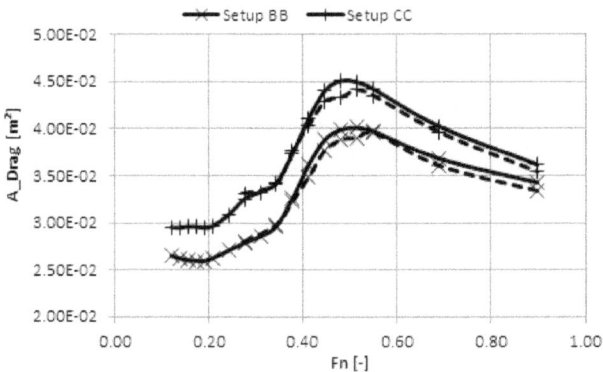

Figure 19: Drag areas from simulation and experiment

be seen, the agreement is very good with maximum deviation of about 3% (turbulence stimulation in experiment is not accounted for in the diagram). In particular, the effect of the LCG and displacement variation is correctly captured. Figures 20 to 22 give a comparison of near hull wave patterns from simulation and tank test photographs at selected velocities. Here as well, good agreement can be observed.

APPLICATION

A good example to show the code's capabilities is the simulation of the flow around a VO65OD yacht (geometry kindly provided by *Farr Yacht Design*, see figure 23). This particular design features an extensive set of appendages, comprising canting keel, twin dagger boards and twin rudders. The keel root and canting axis are located in a recess on the hull. At typical operating conditions the leeward daggerboard is lowered, the windward rudder piercing the free surface and the keel is fully canted to windward with the recess at the free surface. Due to the inclined keel pin the canted keel generates significant amounts of vertical lift, especially at larger Froude numbers, having significant impact on the wave pattern and resulting in strongly dynamic behavior of the boat.

Case Setup

The flow around the yacht is simulated at full scale at a velocity of 17kts (Fn = 0.63), leeway angle of 5°, heel angle of 26°. Domain size is 1 length in front, 3 astern, 2 aside, 1.5 below and 0.5 above the yacht with a body fitted split cartesian mesh of 6.96E+06 cells generated using SHM. The capabilities of SHM and associated utilities, using proper settings and scripting, to provide appropriate free surface resolution and capture complex geometries can be seen in figures 24 and 25. Velocity is ramped over 5s (total simulation time 20s), body motion is released at 1s with force / moment ramping over 2s. Sail forces / moments depending on resistance and side forces are taken into account by an appropriate model.

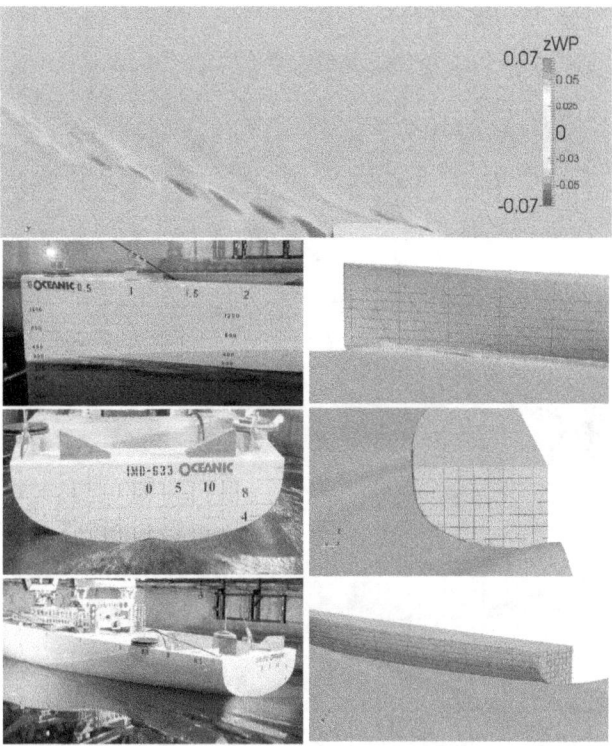

Figure 20: Wave pattern and near hull waves at Fn = 0.345

Results

Figure 26 shows the global wave field generated by the yacht. Even here the impact of the appendages on the hull-generated wave system can already be clearly seen. Figures 27 to 31 give closer views of flow features of particular interest. The breaking bow wave with rising sheet of water as well as the wave systems generated by the keel and recess close to the free surface and the surface-piercing rudder can clearly be seen.

Of particular interest for the correct prediction of resistance and trim is the volume fraction underneath the hull. In many simulations using various codes significant amounts of air can be observed here (aka numerical ventilation or streaking), strongly affecting viscous as well as pressure forces and moments. As can be seen in figure 32, this numerical phenomenon is successfully suppressed in this implementation.

Figures 33 and 34 show the evolution of forces and body motion. The rise of resistance, side force, yawing and righting moment as well as the evolution to a steady flotation state are clearly observable.

CONCLUSION

A new OpenFOAM-based solver for the calculation of free surface flow based on state of the art methods has been described. The important changes to the standard OpenFOAM solvers interFoam / interDymFoam have been presented. A modern discretisation scheme for the convective term of the

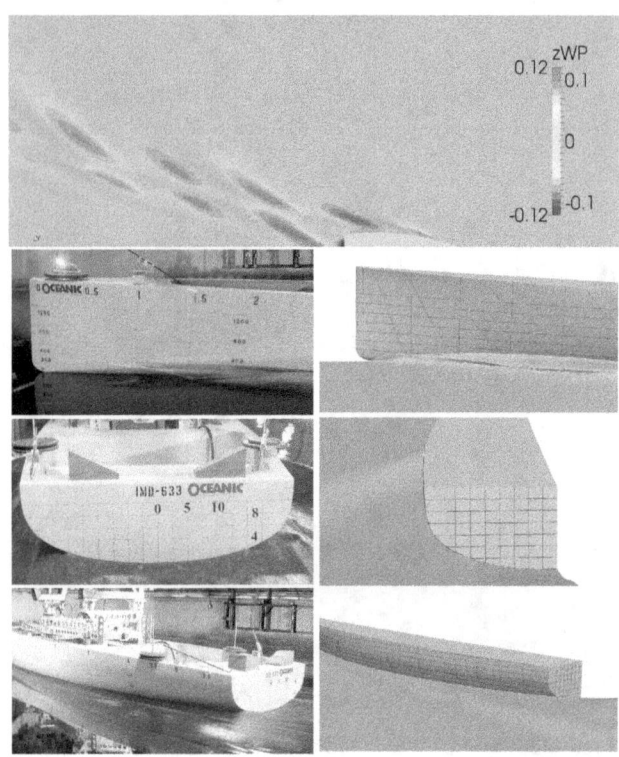

Figure 21: Wave pattern and near hull waves at Fn = 0.484

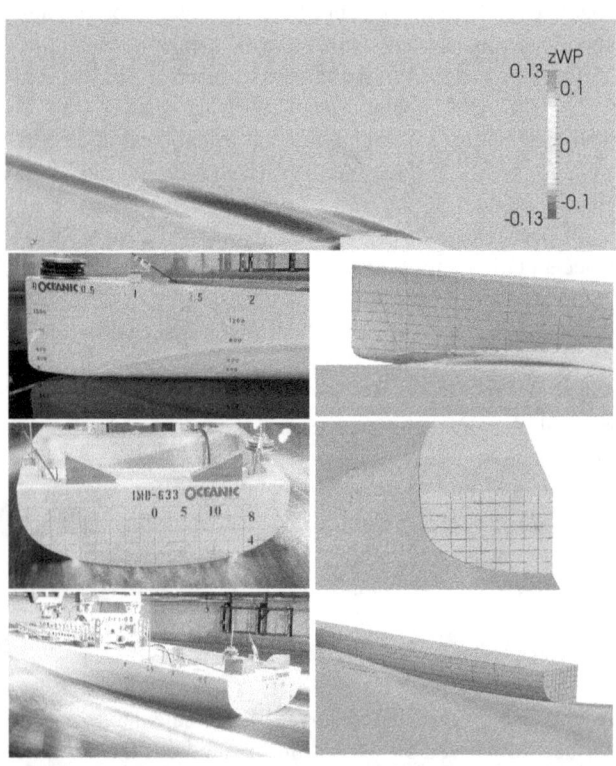

Figure 22: Wave pattern and near hull waves at Fn = 0.898

Figure 23: Rendering of VO65 OD

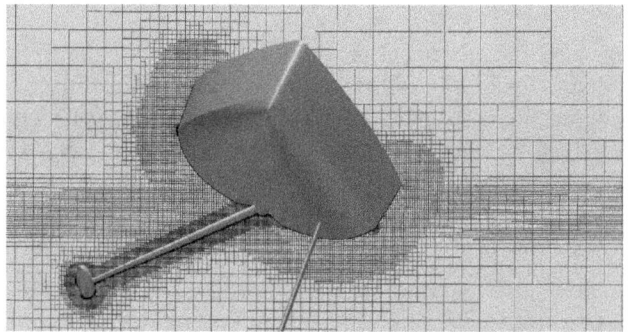

Figure 24: Global and free surface resolution

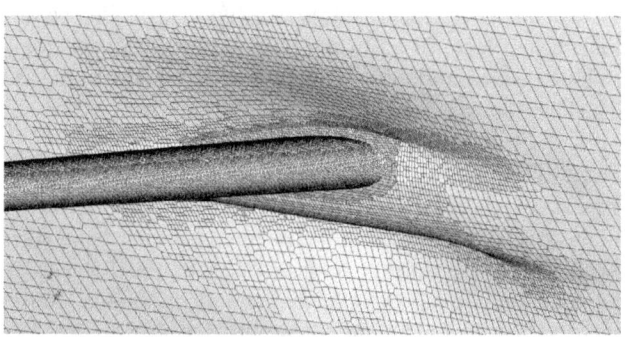

Figure 25: Resolution of keel detail

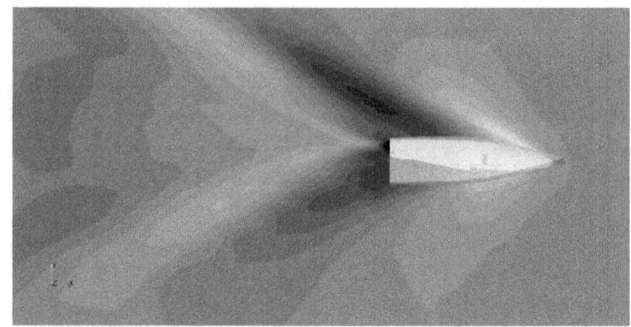

Figure 26: Global wave pattern, weather side at bottom of picture

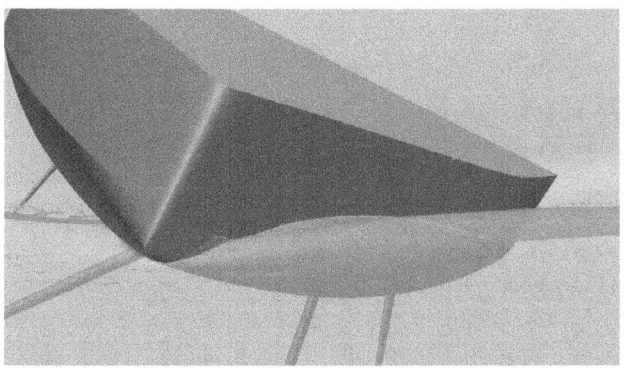

Figure 27: Bow wave system, leeward

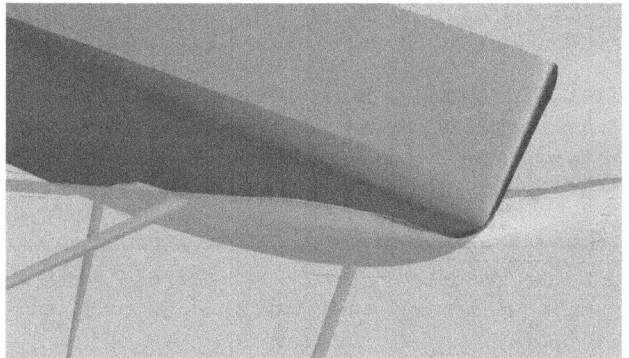

Figure 28: Bow wave system, windward

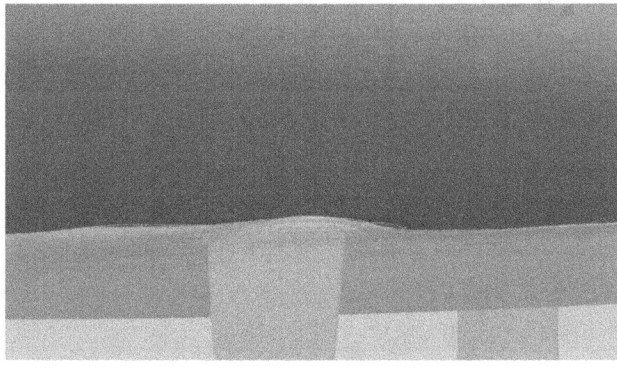

Figure 29: Local free surface at keel root (above)

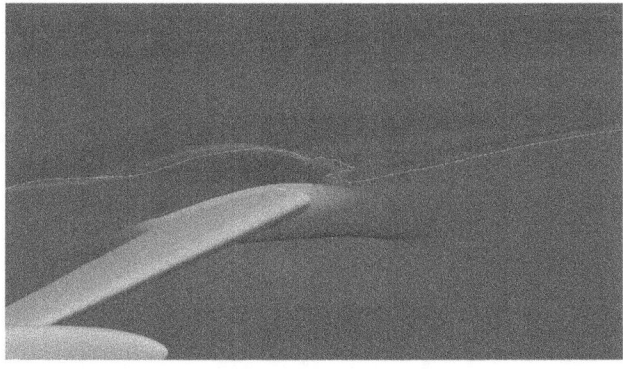

Figure 30: Local free surface at keel root (below)

Figure 31: Local wave system from windward rudder

Figure 32: Volume fraction below hull

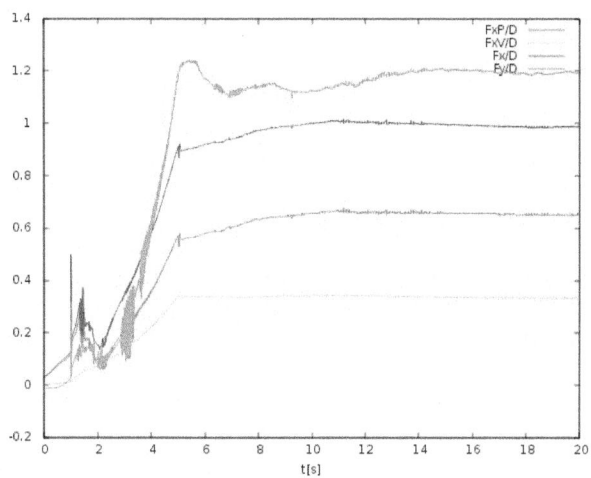

Figure 33: Evolution of resistance and side forces divided by displacement over simulated time

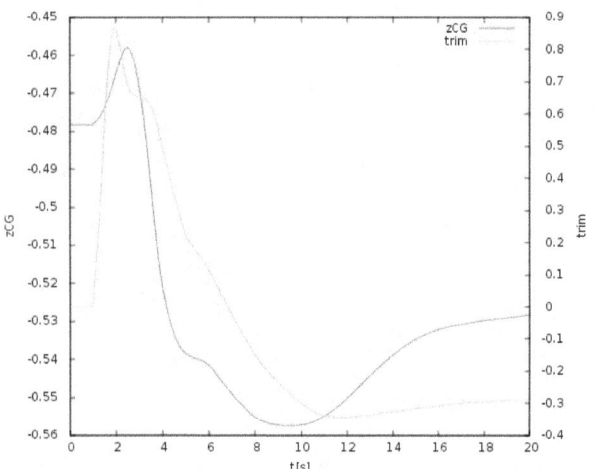

Figure 34: Evolution of trim and sink over simulated time

transport equation of the volume of fluid method has been implemented. Additionally a method for the reconstruction of the pressure discontinuity at the interface has been implemented. A robust body motion solver has been implemented and tested, a development path for further improvements of efficiency is described. First simulations for the validation have been done. The Sysser 60 case of the Delft Systematic Yacht Hull Series showed results of quality and computational time comparable to another commercial free surface code, Star-CCM+. It has been shown that the results of the new solver have better quality than the original interFoam solver. Also the stability of the solver was improved significantly compared to interFoam. On the Volvo 70 validation case good agreement to tank test results has been shown for a range of Froude numbers. The solver's capabilities to capture complex flows of different scales have been shown on the VO65OD case.

REFERENCES

Böhm, C., Graf, K.: *Advancements in free surface simulations for sailing yacht applications*, The Third International Conference on Innovation in High Performance Sailing Yachts, Lorient, France, (2014).

Delft University of Technology: *Delft Systematic Yacht Hull Series*, http://www.dsyhs.tudelft.nl/dsyhs.php, Delft, Netherlands, (retrieved 2013).

Devolder, B., Schmitt, P., Rauwoens, P., Elsaesser, B., Troch, P.: *A Review of the Implicit Motion Solver Algorithm in OpenFOAM to Simulate a Heaving Buoy*, 18th Numerical Towing Tank Symposium, Cortona-Italia, 2015.

Farr Yacht Design: *Farr Volvo70 Immersed Transom Study*, Sailing Yacht Research Foundation [distributor], http://sailyachtresearch.org/19-library/44-farr-volvo70-immersed-transom-study, 2005

Förster, C., Wall, W., Ramm, E.: *Artificial added mass instabilities in sequential staggered coupling of nonlinear structures and incompressible viscous flows*, Computational Methods in Applied Mechanics and Engineering, **196**, (2007), 1278 –1293.

Hadžić, I., Hennig, J., Perić, M., Xing-Kaeding, Y.: *Computation of flow-induced motion of floating bodies*, Applied Mathematical Modeling, **29**, (2005), 1196 –1210.

Hirt, C.W., Nichols, B.D.: *Volume of Fluid (VOF) Method for the Dynamics of Free Boundaries*, Journal of Computational Physics, **39**, (1981) 201-225.

Jasak, H., Weller, H.G., Gosman, A.D.: *High resolution NVD differencing scheme for arbitrarily unstructured meshes*, International Journal for Numerical Methods and Fluids, **31**, London, (1999), 431-449.

Muzaferija, S., Perić, M.: *Computation of free surface flows using interface-tracking and interface-capturing methods*, Computational Mechanics Publications, Southampton, (1998), 59-100.

Queutey, P., Visonneau, M.: *An interface capturing method for free-surface hydrodynamic flows*, Computers&Fluids, **36**, (2007), 1481-1510.

Soeding, H.: *How to integrate free motion of solids in fluids*, 4th Numerical Towing Tank Symposium, Hamburg-Germany, 2001.

Wackers, J., Koren, B., Raven, H.C., van der Ploeg, A., Starke, A.R., Deng, G.B., Queutey, P., Visonneau, M., Hino, T., Ohashi, K.: *Free-surface viscous flow solution methods for ship hydrodynamics*, Archives of Computational Methods in Engineering, **18**, (2010), 1-41.

Insights from the Load Monitoring Program for the 2014-2015 Volvo Ocean Race

Suzy Russell, Gurit Composite Components, UK
Gaspar Vanhollebeke, Gurit Composite Components, UK
Paolo Manganelli, Gurit Composite Components, France

ABSTRACT

This paper describes insights into keel and rigging loads obtained through a data acquisition system fitted on the fleet of Volvo 65 yachts during the 2014-2015 Volvo Ocean Race. In the first part, keel fin stress spectra are derived from traces of canting keel ram pressures and keel angle; these are reviewed and compared against equivalent spectra obtained by applying methods proposed by Det Norske Veritas - Germanischer Lloyd ("DNVGL") guidelines and the ISO 12215 standard. The differences between stress spectra and their validity are discussed, considering two types of keel: milled from a monolithic cast of steel, and fabricated from welded metal sheets. The second part discusses predicted and actual rigging working loads for the Volvo 65 yachts, and considers how safety factors vary between design loads proposed by DNVGL and actual recorded loads.

NOTATION

ρ	Density of water (kg/m^3)
a_{vw}	Vertical surface acceleration (m/s^2)
CDR	Cumulative damage ratio
DAQ	Data Acquisition
FAT	Classification reference to S-N curve
GPS	Global Positioning System
H	Wave height (m)
IMU	Inertial Measuring Unit
LCG	Location of Centre of Gravity
LJB	Lightweight Junction Box
LLB	Lightweight Logger Box
n_i	number of cycles at given stress level "i"
N_i	average number of cycles to failure at given stress level "i"
TWA	True wind angle (degrees)
TWS	True wind speed (knots)
$VO65$	Volvo 65 yacht
ω_e	Wave encounter frequency (rad/s)

1. INTRODUCTION

For the first time in history, the 2014-2015 edition of the Volvo Ocean Race has been competed in with a fleet of one-design offshore racing yachts (the "VO65"). Green Marine, a specialist of high performance composite structures based in the United Kingdom, has coordinated the work of a consortium of composite builders including Decision S.A., Multiplast, and Persico S.P.A, while Farr Yacht Design (FYD) has been responsible for the design of the yachts, including the composite structure engineering. Gurit's engineering team has been engaged by Farr Yacht Design to provide finite element analysis (FEA) to support the VO65 design.

Green Marine has also been responsible for assembling and fitting out the VO65s and providing overall quality control to guarantee that all the yachts would meet the strict "one-design" rule.

As part of their endeavour, Green Marine has taken the initiative to have bespoke data acquisition (DAQ) systems fitted to all seven Volvo 65 yachts that have entered the 2014-15 Volvo Ocean Race. The systems have recorded rig loads, keel ram pressure, keel angle, acceleration and sailing parameters obtained from the navigation instruments. Gurit has been responsible for specifying the data acquisition equipment, assisting Green Marine through the installation and commissioning stages, creating an algorithm to program the DAQ system, and for analysing all the recorded data and reporting the results to Green Marine and to the shore crew before and during the race.

The main purposes of the load monitoring system were:

- to monitor (and record continuously whilst sailing) the loads on all the main items of standing rigging, canting keel and rudder stocks, through-out the life of the yachts

- to provide the racing crew with real-time feedback on the load and slamming acceleration levels experienced while sailing

- to allow in depth analysis of loads at the end of each leg of the race to inform the decisions of the maintenance teams.

- to provide additional knowledge on the loads that VO65s are subject to and verify the validity of the design assumptions, with an aim to improving the reliability of offshore racing yacht structures

In total, approximately 14030 hours of continuous recording were produced during the race and many more hours during the lead up including during the 2014 Round-Britain-and-Ireland race and multiple transatlantic crossings. This is believed to be the largest and most comprehensive set of data collected on offshore racing yachts to date (1) (2). It is a rare event for yacht builders and designers to have access to such an array of data. Green Marine has worked with Gurit to convert this wealth of information into useful insight about offshore racing yacht structures, so far focusing particularly on two subjects: fatigue loading of keel structures and maximum loads of rigging components measured while underway.

While there is a significant body of scientific literature and design guidelines covering the subject of fatigue loading for large commercial vessels (e.g. (3) and (4)), very little information has been published that applies to offshore sailing yachts. Fatigue in racing yachts is typically associated with failure of keels. Indeed, there are several reported cases of yachts capsizing due to keel loss or having to complete races under reduced sail as they lost their keel before reaching the finish line. Often, fatigue is quoted among the likely causes of these accidents (5); however historical records of the loads endured by keel structures are rarely available.

The first part of this paper focuses on the characteristics of stress spectra applicable to the keel structures of offshore racing yachts like the VO65s and delves into the different implications on the design of milled and fabricated metal keel fins. It compares a fatigue assessment achieved following existing guidelines and a fatigue assessment based on the fatigue loading recorded while sailing. In order to compare those methods, two hypothetical keels are considered, a milled and a fabricated keel, under the hypothesis that both are designed to meet the minimum static strength requirements to withstand the DNVGL static heeled loadcase (6).

Rig failure is another key factor that forces high performance sailing yachts to retire from racing. As well as disappointing and expensive, rig failure is a primary concern for the safety of the crew. Gaining a better understanding of the rig loads experienced during sailing is critical for increasing the reliability of racing yachts while continuing to improve their performance.

The second part of this study focuses on the maximum working rig loads measured during the 2014-2015 Volvo Ocean Race and presents a comparison between these, the typical design loads that Gurit would have used to design the structure supporting the rigging and the design loads recommended by the DNVGL (7).

2. VOLVO 65 DATA ACQUISITION SYSTEM

The primary function of the load monitoring system was to log selected data from force and motion sensors whenever the yacht is sailing. The data acquisition system was set up to start logging either when the boat speed sensor measured more than 1 knot or when the heel angle was greater than $20°$; no intervention from the sailors was required to start or stop the acquisition. All of the yachts were fitted with the same set of sensors, shown in Figure 1, including:

- One triaxial accelerometer in way of the bow, measuring the three components of linear acceleration

- Inertial measurement unit (IMU), fitted close to the LCG, measuring heel and trim angle, roll, pitch and yaw angular velocity and the three components of linear acceleration

- Two pressure transducers fitted on the keel canting hydraulic system measuring pressures acting inside the keel rams with a pressure range from 0-700 bar and ±1% accuracy

- One sensor measuring the cant angle of the keel

- Strain gauges fitted on both rudder stocks measuring strains induced by lateral bending

- Eleven load cells measuring tension, each with an accuracy of ±2%, in way of the following rigging components:

  - 'J1' stay
  - 'J2' stay
  - 'J3' stay
  - 'V1' shrouds (port and starboard)
  - 'D1' shrouds (port and starboard)
  - Mainsheet
  - Runners (port and starboard)
  - Bobstay

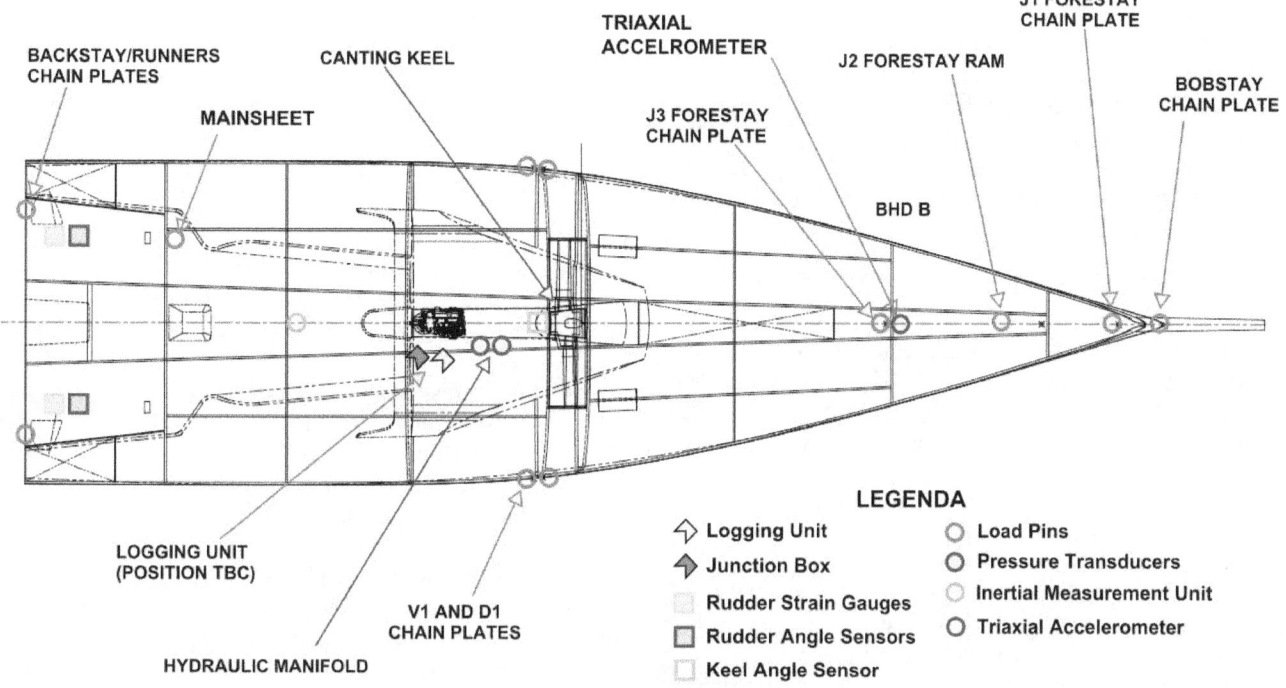

BACKSTAY/RUNNERS CHAIN PLATES

CANTING KEEL

MAINSHEET

TRIAXIAL ACCELROMETER

J3 FORESTAY CHAIN PLATE

J2 FORESTAY RAM

J1 FORESTAY CHAIN PLATE

BOBSTAY CHAIN PLATE

BHD B

LOGGING UNIT (POSITION TBC)

V1 AND D1 CHAIN PLATES

HYDRAULIC MANIFOLD

LEGENDA

Logging Unit
Junction Box
Rudder Strain Gauges
Rudder Angle Sensors
Keel Angle Sensor

Load Pins
Pressure Transducers
Inertial Measurement Unit
Triaxial Accelerometer

Figure 1: Volvo 65 load monitoring system

A Cosworth lightweight logger box (LLB) logging unit was used, in combination with a Cosworth lightweight junction box (LJB), to process and store the data from all of the sensors; the unit had a storage capacity of 1Gb which allowed for the following variables to be logged continuously throughout the race:

- Sailing performance Data (Boat speed, TWS, TWA, GPS Latitude and Longitude, Rudder angle, Deflector settings), logged at a frequency of 1Hz
- Rigging loads, logged at a frequency of 10Hz
- Keel ram pressures, logged at a frequency of 2Hz

Besides the continuous logging, the LLB unit was programmed to perform "burst" logging when either accelerations, keel loads or rudder deflections exceeded set thresholds. Burst logs typically included all motion variables logged at 100Hz and keel ram and rudder loads logged at 50 Hz. The burst logs covered a period of time spanning between 4 seconds prior to the trigger event and 10 seconds after. The data was offloaded from the logging units at the end of every leg. In particular, the data was logged into buffer memory at the burst frequency and every minute a check was made to assess whether any value exceeded the threshold. If this was the case, the data points spanning 4 seconds before the event and 10 seconds after were kept while the other data points were decimated to the continuous acquisition frequency. If the threshold value was not reached, the entire minute worth of data was decimated to the continuous acquisition frequency. In the

case of acceleration measurements, the data was entirely erased if the threshold value was not met. This was done in order to obtain a record of all significant loads and events at high sampling frequency and for the entirety of each leg without exceeding the storage capacity of the logging unit.

3. KEEL LOADS ASSESMENT

Some of the first insights that have been obtained from the complete data set are about the magnitude of the maximum loads experienced by the keel structures during the race and about the characteristics of fatigue loading that affects keels and their supporting structures.

3.1. Maximum Dynamic Keel Loads

The maximum dynamic load on the keel was calculated from the keel ram pressure. If a keel was designed to the maximum allowable of DNVGL static loadcase it would be designed for a 1g vertical acceleration with the yacht heeled 30 degrees and the keel fully canted (40 degrees) and material properties reduced with a partial material factor, with an additional partial load factor ("c_d") of 1.4 to account for canting keel configuration (6),.

The partial material factor ("g_m") on steel properties is dependent on the yield strength of the steel used. It ranges from 1.96 to 2.74 for steel with yield stress ranging respectively from 235MPa to 900MPa. Thus for high strength steel (e.g. Weldox 900 with yield strength of 900 MPa) the overall static reserve factor relative to yield strength would be 3.842 (product of both partial material factor and partial laod factor). The maximum force recorded in the canting keel rams would have corresponded to a keel lateral acceleration of 1.897g.

Thus, based on the keel forces measured across the entire fleet and for the duration of the whole race, the actual reserve factor between the minimum yield strength of the fin required by DNVGL and the largest estimated transient stress would have been of the order of 1.90. This number was obtained by dividing the product of partial material and load factors mentioned above by the maximum keel lateral acceleration.

$$RF = c_d * \gamma_m * \sin(40° + 30°) \frac{1}{1.897} = 1.90$$

Note that the keels of the VO65s were designed by Farr Yacht Design for higher transverse bending strength requirements than those of the DNVGL rules; hence, their actual reserve factors relative to the highest measured transverse load were higher than the number quoted above.

3.2. Existing Standards for Keel Fatigue Assessment

3.2.1. DNVGL

Det Norske Veritas - Germanischer Lloyd (DNVGL) has published guidelines for the fatigue assessment of racing yacht keels (6). The DNVGL fatigue assessment is based on the assumption of a 5 year design life during which the yacht is expected to sail 15% of the time. During this design life the structure should not endure fatigue degradation leading to premature failure.

DNVGL considers two types of cyclic loads: inertial loads due to cyclic vertical accelerations when the yacht sails in a seaway heaving up and down, and full reversal cyclic loads due to tacks, gybes and "knock-downs". In order to establish a stress spectrum to be used for a fatigue assessment, DNVGL applies the following first principles approach:

- Stress cycles are distinguished between cycles due to motion through the waves and tacks/gybes.

- It is assumed that the yacht will spend a given percentage of its design life sailing at 4 different TWA: 45°, 90°, 135° and 180°.

- In each of these headings, a given percentage of the time sailed is spent in different sea conditions associated with a given wave length and amplitude, and at a given heel angle.

- The number of stress cycles due to waves is calculated from the encounter frequency and the time spent in each condition.

- The stress range is calculated based on the vertical acceleration, heel angle, keel cant angle and the nominal stress in the keel in the static "90 degree" load case.

- Vertical acceleration ("a_{vw}") is calculated based on the relationship between wave height ("H") and encounter frequency ("ω_e") given below:

$$a_{vw} = \frac{H * \omega_e^2}{2}$$

- The yacht is considered to tack or gybe 30 times each day.

- The stress range due to tacks is calculated based on the nominal stress in the "90 degree" load case and the angle of the keel to vertical.

- All stress cycles are considered to be tension-tension, no consideration is made over the mean stress of each cycle. This remains a conservative assumption for metals and is reflected in the method by the choice of "S-N" curve.

3.2.2. ISO Standard 12215 PART 9

ISO standard 12215 part 9 (8) contains a simplified method for assessing keel fatigue strength. The method essentially stipulates a stress spectrum to be used for the fatigue assessment. It considers that over its operational life, the yacht will endure 8 million stress cycles which should cover tacking and gybing, rigid body motions and flutter or vibration related phenomena. The standard assumes that 8 million cycles should cover an operational life of 25-30 years of moderate to high usage recreational sailing or 5 years of very extensive ocean racing. The design life considered by ISO is thus very similar to that considered by DNVGL.

Similarly to the DNVGL guidelines, the stress range is established on the basis of a nominal stress obtained from the static transverse load case. The peak stress range is considered as 1.5 times the nominal stress and also includes a factor to take into account specific characteristics of the keel (i.e. with/without flange, design category, canting or fixed, fabricated or solid). The factor of 1.5 is explained as such: the largest cycle will have a peak stress equal to the nominal stress multiplied by the keel type factor and a trough equal to minus half the nominal stress multiplied by the keel type factor.

Hence, it considers cycles to be an intermediate case between full-reversal and unidirectional stress (no load reversal). This is reflected by the "S-N" curve specified by ISO (8).

3.3. Fatigue assessment based on actual recorded data

On the basis of the data recorded from the race, a fatigue assessment has been performed with the following approach:

- An actual stress spectrum was derived for the keel based on the assumption that it was designed to meet the DNVGL static design criteria (6) and subjected to the same number and amplitude of cycles as recorded during the race.

- The set of cycles recorded during the race was extrapolated to provide an equivalent design life as used by DNVGL and ISO. That is, the numbers of cycles recorded during the race were multiplied by 274/141 where 141 was the number of days of recording during the race and 274 days is the DNVGL fatigue design life.

- The partial damage factors corresponding to the nominal load spectra obtained by applying the DNVGL and ISO methods were compared with those obtained from the extrapolation of measured data.

- The cumulative damage ratios (*CDR*) obtained with the DNVGL and ISO methods for both a fabricated (welded) and a milled keel design were compared with the *CDR* based on the measured data.

- Considerations have been made on how the DNVGL and ISO assumptions compare with actual measured data and how this affects the results of the fatigue assessment for different types of keels

In order to compare the findings from the measured data with the predictions of the DNVGL and ISO methods, a measure of the stress cycles in the keel fin had to be derived from the data acquired during the Volvo Ocean Race. The method and its assumptions are described below.

3.3.1. From Ram Pressures to Accelerations
The force perpendicular to the head of the keel was obtained on the basis of the known keel structure geometry and from the logged keel cant angle and keel ram pressures. For the purpose of this investigation, the keel was considered as a rigid body and the acceleration perpendicular to the fin at its centre of gravity was derived from the force at the head of the keel (inferred from keel ram pressures) and the known weight of the keel fin and bulb.

3.3.2. From Acceleration to Stress
In order to translate acceleration amplitude into stress amplitude the following assumptions were made:

- When the keel is lying horizontal with 1g acceleration applied at its centre of gravity, the highest stress in the keel structure is equal to DNVGL's allowable stress and is referred to as "nominal stress". This assumption is legitimate for a racing yacht where the designer would typically

optimise the structure to achieve the smallest acceptable reserve factor over the requirements of the design rules.

- The stress amplitude is equal to the nominal stress multiplied by the acceleration amplitude perpendicular to the keel expressed in "g".

- Only stresses due to lateral acceleration are accounted for and the contribution of longitudinal acceleration is neglected, in-line with DNVGL provisions.

3.3.3. Stress Spectra
The number of stress cycles was obtained by applying a "rainflow counting" algorithm (9) on the measured accelerations. To allow direct comparison with the DNVGL method, a stress spectrum was derived from the measured data by counting stress cycles on the basis of stress amplitudes and independently from mean stress levels. In practice this means that the method did not make a distinction between tension-tension and tension-compression cycles.

3.3.4. Extrapolating Complete Design Spectra
The complete design spectrum that covers a design life corresponding to 15% usage during 5 years (equivalent to 274 days of continuous sailing) was obtained by scaling the number of cycles counted during the race by the ratio of design life over time logged. For reference, the total logged time over the nine VOR legs for one boat amounts to 141 days of continuous sailing, which is equivalent to 51% of DNVGL design life. The assumption is made that the data gathered during 141 days of continuous logging contain a representative sample of the stress cycles that the yacht will experience throughout its design life.

Only the data logged during offshore legs was used for the fatigue assessment. In comparison with inshore racing, the rate at which yachts tack and gybe when sailing offshore is lower. Tacks and gybes generate large amplitude stress cycles. Thus it could be argued that, by considering only the data recorded when sailing offshore, one would miss a high proportion of large amplitude stress cycles. However, Figure 6 demonstrates that the cycles with ranges of 400 MPa to 600 MPa (which are typically caused by tacks and gybes) only contribute to a *CDR* of 1%. Thus doubling the number of tacks would increase the *CDR* by 1%.

3.3.5. Reference S-N curves
To establish whether a keel is critical in fatigue rather than in static strength, the designer must compare the stress spectrum chosen with a reference "S-N" curve. "S-N" curves can be obtained from various sources (e.g. (10) and (8)) and are specific to a material and the detail under consideration (for example edge of a plate or butt weld).

For the present study, a fatigue assessment based on the three different stress spectra obtained as described in sections 3.2.1, 3.2.2 and 3.3.3 was performed for a hypothetical fabricated (welded) keel and a milled keel.

Four relevant S-N curves along with the three mentioned stress spectra are represented in Figure 2.
The reference "S-N" curves were sourced from the International Institute of Welding (10) and from ISO standard 12215-9 (8). As "S-N" curves vary depending on the type of structural detail and alloy being considered (e.g. welded or solid plate, high or low yield strength, etc.), a curve for a "FAT" class of 140 was chosen for a milled keel fin (i.e. milled from a monolithic metal casting), and a curve for a "FAT" class of 90 was used for the assessment of a hypothetical fabricated keel. FAT 90 class corresponds to a continuous transverse butt weld which has been inspected with non-destructive testing (see (10) for a more detailed illustration of FAT classes).

3.3.6. Cumulative Damage Ratio

In a structure subjected to variable amplitude cyclic loading, the fatigue life is usually assessed using the "Palmgren-Miner" rule, which derives the "cumulative damage ratio" from the summation of the "partial damage factors".

Partial damage factors are calculated as the ratio of the number of cycles ("n_i") at a given stress range versus the allowable number of cycles ("N_i") at that stress range before fatigue failure occurs. "k" is the number of sets of cycles at given stress range. This allowable number of cycles is typically sourced from the S-N curve that is relevant to the material and structural item under consideration.

Figure 3,Figure 4 and Figure 5 show plots of partial damage factors and their corresponding stress ranges, obtained for a milled keel respectively from DNVGL, ISO and measured VO65 stress spectra. Figure 6,Figure 7 andFigure 8 show similar plots, but for a fabricated keel. They are useful for assessing which type of cyclic loading has the greatest effect on the fatigue life of the structure. In the case of DNVGL, it is apparent that the high number of tacks and gybes assumed and the corresponding large amplitude stress cycles would be the primary contributor to fatigue of the milled keel fin.

However, Figure 5 shows that large amplitude stress cycles measured on the VO65 are not the main cause of fatigue, as these racing yachts complete much fewer

Figure 2: "S-N" plot

tacks/gybes while racing offshore than the DNVGL method

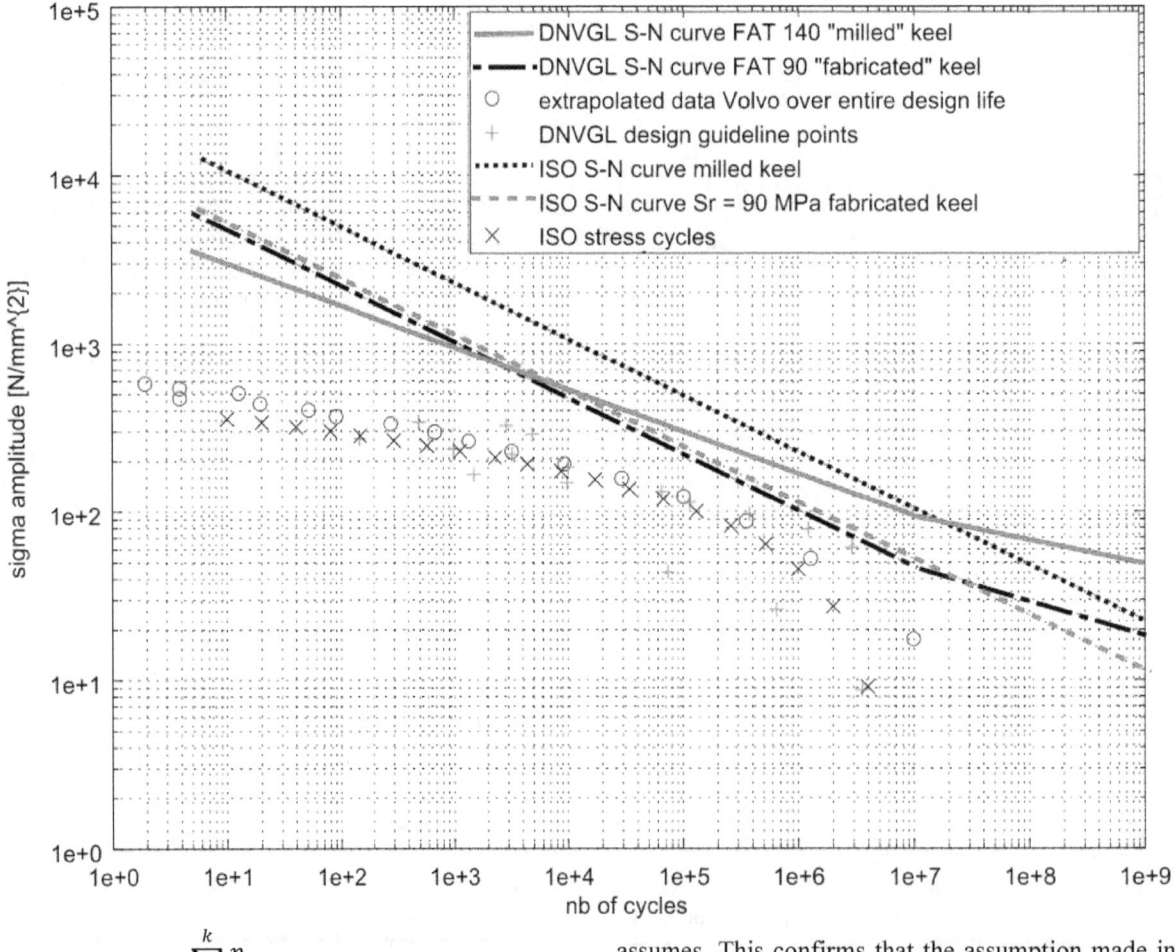

$$\sum_{i=1}^{k} \frac{n_i}{N_i} = CDR$$

assumes. This confirms that the assumption made in 3.3.4 of scaling the number of cycles by the ratio of design life to logged time is suitable: any very large amplitude cycle with small probability of occurrence that could have been missed would not have greatly influenced the *CDR*.

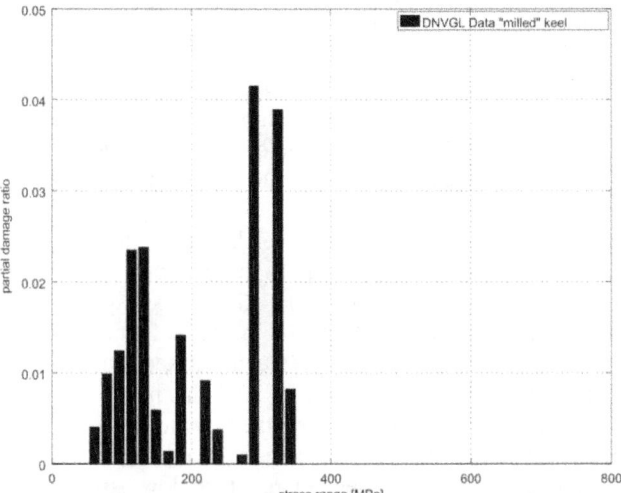

Figure 3: partial damage factor vs stress ranges obtained from DNVGL guidelines for a milled keel

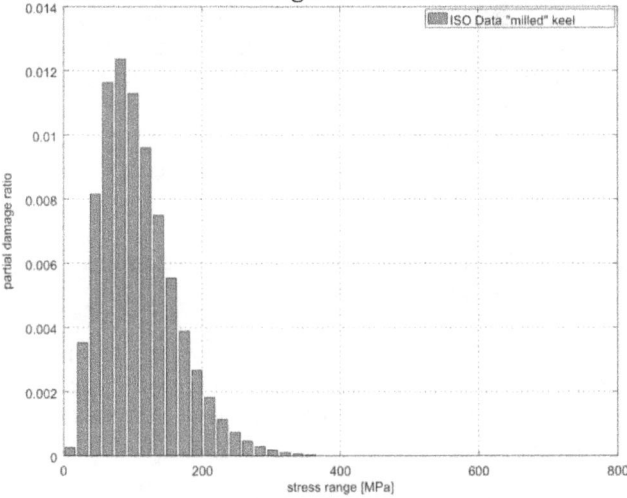

Figure 4: partial damage factor vs stress ranges for ISO standard for a milled keel

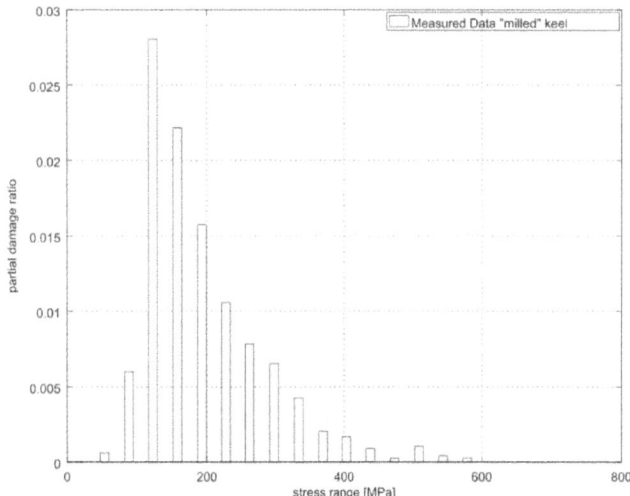

Figure 5: partial damage factor vs stress ranges for extrapolated Volvo 65 measured data for a milled keel

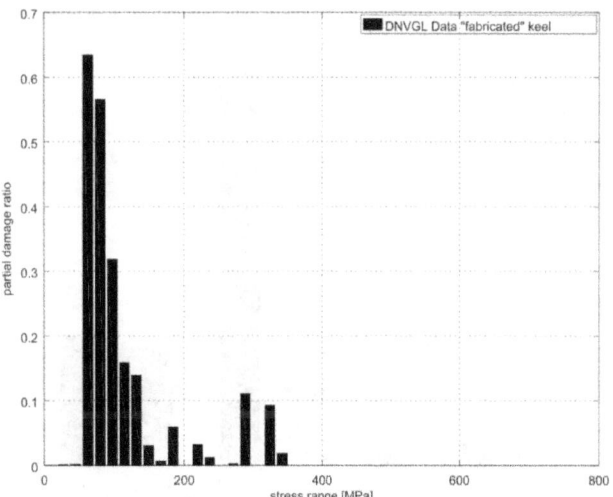

Figure 6: partial damage factor vs stress ranges obtained from DNVGL guidelines for a fabricated keel

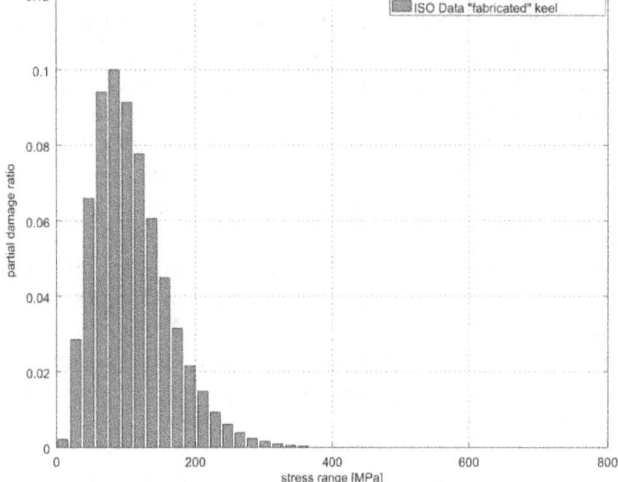

Figure 7: partial damage factor vs stress ranges for ISO standard for a fabricated keel

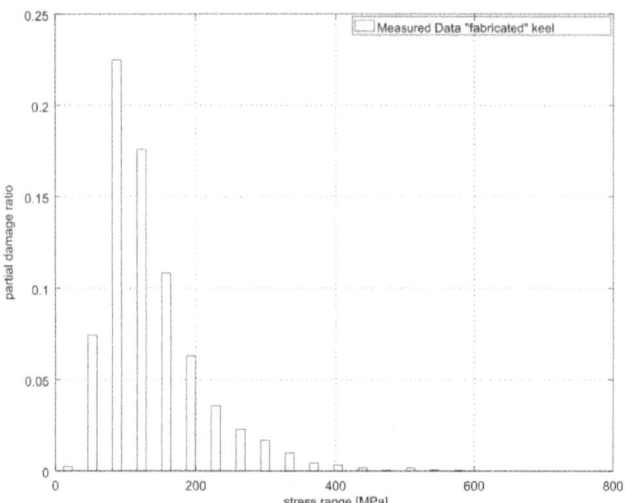

Figure 8: partial damage factor vs stress ranges for extrapolated Volvo 65 measured data for a fabricated keel

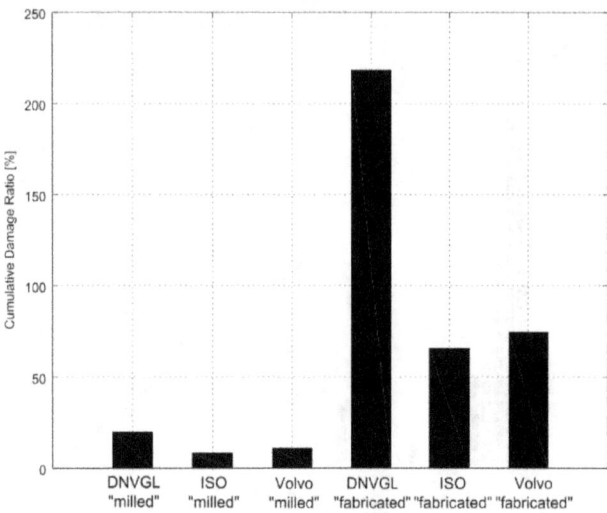

Figure 9: Cumulative damage ratios

Figure 9 displays the cumulative damage ratio (*CDR*) expressed as a percentage obtained for both a milled (solid) and a fabricated (welded) keel using the three different stress spectra described previously. A *CDR* of above 50% indicates a high likelihood of premature fatigue failure and indicates that further investigation is required and *CDR* of 100% indicates that fatigue failure will occur (6). It is interesting to note how the fabricated keel *CDR* exceeds the 50% limit, with both "nominal" (DNVGL and ISO) and "actual" (from VO65 data) stress spectra. Conversely the *CDR* of a milled fin is consistently below 25%, irrespective of which stress spectrum is used for the fatigue assessment.

From the data represented in Figure 9, one may also understand that:

- By applying the DNVGL fatigue assessment method, a milled keel designed for DNVGL static strength criteria would be appear to be adequate to perform 2 round-the-world races without experiencing fatigue failure.

- Based on the stress cycles actually recorded during the Volvo race, the same keel would be considered to be adequate for more than 4 round-the-world races

- Based on actual Volvo race data and using a "S-N" curve for FAT class 90 for reference, a fabricated keel fin would not be considered to be suitable for more than one round-the-world race

The *CDR* of the welded keel based on the DNVGL stress spectrum is significantly higher than the *CDRs* obtained from the VO65 and ISO stress spectra. An explanation for this can be found in Figure 2 and Figure 6: the reference S-N curves for the fabricated fins lie well below the ones for milled fins, and the number of low stress range cycles predicted by the GL method is close to the allowable level and contributes to a very large *CDR*.

Based on the recorded VO65 data, low stress range cycles appear to be less numerous than predicted by the DNVGL method and thus make a lower contribution to fatigue degradation. This shows how the assumption of the number of cycles due to wave encounters (usually associated with lower stress ranges) is critical in the assessment of keel fatigue life, particularly for fabricated keels, and highlights the importance of using stress spectra derived from experiments to improve the quality of fatigue assessments.

3.4. Discussion of DNVGL and ISO standard assumptions

It is of interest to consider how the assumptions contained in both the GL and ISO fatigue assessment methods compare with the data recorded on the VO65s and how this is likely to affect the results of these methods.

3.4.1. Discussion of DNVGL Assumptions

In comparison with the results obtained from the Volvo 65 data, the DNVGL method appears to lead to a rather conservative estimate of the cumulative damage factor, particularly for fabricated (welded) fins.

For the milled keel, this is due to the large contribution of the number of tacks predicted by DNVGL on the *CDR*. If *CDR* was recalculated following the DNVGL approach but using the actual rate of tacks or gybes per day, one would obtain 12.0%, which is close to what is obtained when using the Volvo stress spectrum (10.8%).

The fabricated (welded) keel *CDR* as calculated per DNVGL method is mainly driven by low stress range cycles as demonstrated in Figure 6. In the DNVGL method, low stress cycles are calculated based on the assumed percentage of time spent sailing at different true wind angles, design boat speed and nominal wave length. This method is conservative but the impact of those assumptions (6) on *CDR* compared with *CDR* obtained from test data is very large and again this highlights the importance of using experimental data to refine the analysis. Without experimental data the method could be refined if the boat speed used to calculate the encounter frequency with the waves was a function of boat heading and sea state rather than being one single value. Similarly, the percentage of time spent at different TWA could be a function of the design: for instance a VO65 yacht is clearly likely to spend more time sailing at TWA of 135° than 45°.

3.4.2. Discussion of ISO Assumptions

The ISO fatigue assessment appears to lead to a slightly optimistic conclusion when compared with the assessment conducted with the stress spectrum for the acquired VO65 data. The stress spectrum in the simple fatigue assessment proposed in ISO 12215 part 9 (8) is issued from a typical stress spectrum distribution for ships. The difference between the spectra reflects the fundamental differences in the motion of a ship and of a racing yacht in waves: there is less high amplitude cycles compared with the experimental data, typically caused by tacking or gybing. However, fatigue prediction is in good correlation with experimental data as it has been shown that tacks and gybes have a small impact on the fatigue life prediction according to the experimental data.

4. RIGGING LOAD ASSESMENT

4.1. Nature of Rig Loads

The rig of a sailing yacht experiences continuously varying loads during sailing. A set of nomenclature to describe the different load states was proposed by McEwen and Belgrano (11) as described below:

W1:   Maximum steady-state load
W2:   Peak dynamic load
LIMIT: Elastic limit of supporting composite
            structure
ULTIMATE: Break load of supporting composite
            structure

W1 and W2 represent real or anticipated working loads. These are used for stiffness calculations and as an input into calculating LIMIT and ULTIMATE design loads. At the LIMIT load the supporting structure should continue to perform as designed without any sign of degradation. It is expected that beyond the LIMIT load the structure may begin to yield, , crack etc. but should not fail catastrophically until the ULTIMATE load is surpassed. (11)

4.2. Analysis of Rig Loads

Unfortunately there were a relatively high number of instances throughout the race when particular load cells failed either intermittently or permanently due to loss of water-tight integrity by cable connectors that were exposed on deck, thus the data has been sanity checked and filtered

to remove false readings prior to the analysis.

4.2.1. Working loads

The designer of the VO65 rigs has supplied the race teams with a table detailing allowable sailing load cases. The maximum load specified for each rigging component is referred to as the "nominal W1".

The rigging data acquired during the Volvo Ocean Race has been post-processed to determine "VO65 W1" and "VO65 W2" loads. A moving average over 25 second windows (typically equivalent to at least 5-10 wave encounters) was calculated for each load cell and the maximum value for this moving average for each load cell across the whole fleet was determined as the "VO65 W1" load for that load cell. The maximum load for a given load cell across the whole fleet and throughout the entire race was determined as the "VO65 W2" load. Thus, it must be noted that "VO65 W1" and "VO65 W2" loads do not necessarily occur on the same boat or the same leg of the race.

Furthermore, transient peak loads were calculated as the "moving maximum" over 25 seconds for each load cell; these were used to calculate the ratio of transient peak to steady state load for "VO65 W1" loads greater than 80% of the nominal W1 load. Generally, the ratio of transient peak to steady state is expected to be larger at lower loads as the increase in load makes up a larger proportion of the initial load.

Figure 10 shows (1)the ratio of "VO65 W2" over "VO65 W1", (2)the maximum calculated ratio of transient peak over steady state load, and (3)the ratio of "VO65 W2" over nominal W1.

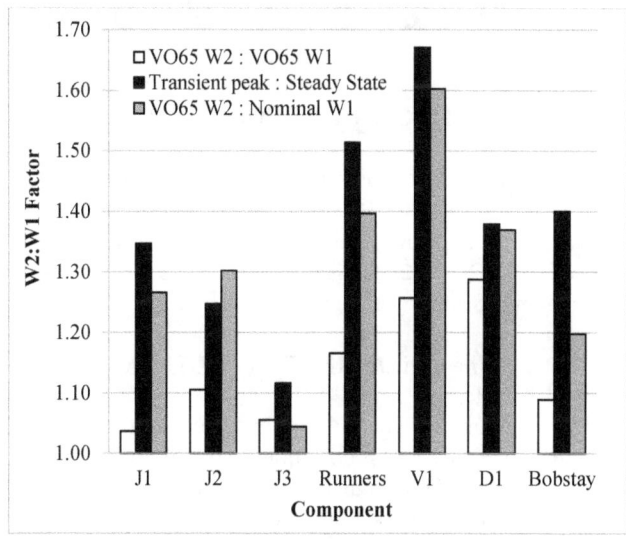

Figure 10: Working load ratios

With the exception of the J3 forestay, the "VO65 W2":

"VO 65 W1" is the smallest ratio. It should also be noted that the "VO65" loads often exceeded the nominal W1 loads. It is important to observe that the maximum W1 and W2 loads in each data set never occurred within the same 25 second window. This explains why the ratio of "VO65 W2" over "VO65 W1" show in Figure 10 above, is smaller than the largest observed ratio of transient peak over steady state.

In other words, the largest dynamic responses did not occur when the rigging experienced the highest "steady state" loads and all observed values of transient peak/steady state ratio were in line with what is typically observed for structures subject to short transient excitations with low damping.

Figure 11 shows a comparison of the W2 loads predicted according to the standard Gurit design approach (referred to as "Gurit W2'") and the actual measured "VO65 W2" loads. The Gurit design approach involves applying a safety factor over the nominal W1 load of between 1.15 and 1.6 depending on the rigging component; these factors are based on historical data records.

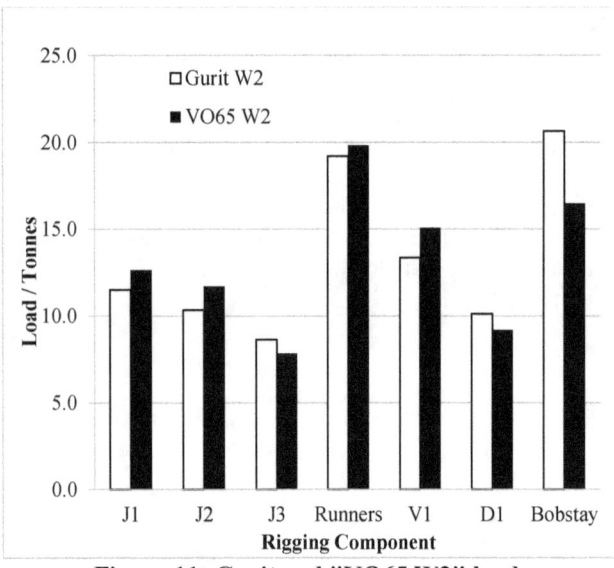

Figure 11: Gurit and "VO65 W2" loads

It can be seen from Figure 11 that the Gurit W2 loads are consistently close to the "VO65 W2" loads. The prediction of all of the rigging loads is within ±12% of the measured "VO65 W2" loads except for the bobstay tension which would seem to be overestimated by 25%.

4.2.2. Theoretical LIMIT and ULTIMATE design loads

It is important when designing the rigging support structure to ensure that it does not break before the rigging itself. The LIMIT design load of the composite support structure, which must be greater than the W2 load by some safety factor, is typically determined by the break load of the rigging. An increasing number of modern high performance sailing yachts, such as the VO65, are equipped with composite rigging. As the specification of

these rigging cables tends to be driven by stiffness rather than by strength requirements, composite rigging cables end up being significantly stronger than their "stiffness-equivalent" metal rods or cables.

This means that when designing composite support structure for composite rigging, the design load would typically be higher than for metal rigging by a significant amount. Hence, to avoid being overly conservative, Gurit typically takes the break load of the stiffness equivalent metal rigging as the LIMIT load for the supporting composite structure and applies a suitable safety factor between ULTIMATE and LIMIT loads to guarantee that the ULTIMATE load for the supporting structure is greater than the break load for the composite rigging by at least 10%. The reserve factor between the Gurit determined LIMIT design load, the break load of the cable and the Gurit determined ULTIMATE design load over the "VO65 W2" loads is shown by Figure 12.

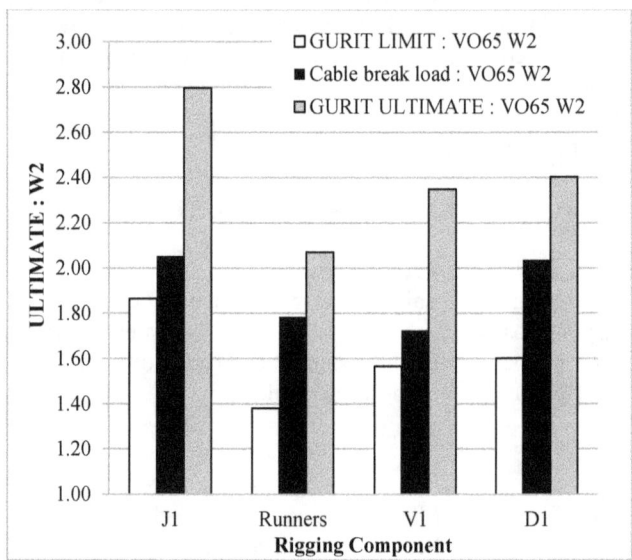

Figure 12: Reserve factor above "VO65 W2" for Gurit LIMIT, cable break and Gurit ULTIMATE design loads

4.3. DNVGL Guidelines

DNVGL guidelines (12) recommend that the rig is analysed using finite element analysis (FEA) with load cases accounting for four sailing conditions upwind and one spinnaker case, typically at 30° of heel.

For "Nitronic 50" rod rigging, provided the working loads have been calculated by static, geometric non-linear analysis, GL recommend a minimum safety factor of 2.5 for the transverse rigging and 2.0 for the fore and aft rigging between this load and the cable break load (7).

For composite rigging, again provided the working loads have been calculated by static, geometric non-linear analysis, GL recommend that the design working load should not exceed the cable's maximum working load as specified by the cable supplier. This maximum cable working load must either be a GL-certified value as determined by testing (13) or, in lack of such proof, a minimum safety factor of 4.5 is applied between cable break load and design working load for transverse rigging and of 3.6 for fore/aft rigging. (7) This highlights the importance of using approved test values when designing to GL guidelines.

Furthermore, DNVGL recommends a safety factor of 1.6 between the break load of the cable (6) (12) and the design load of the composite supporting structure (here referred to as "GL ULTIMATE" load). Figure 13 shows that the factor between the ULTIMATE load as determined by Gurit over "VO65 W2" is less conservative than the factor between "GL ULTIMATE" load and "VO65 W2", but the minimum factor is still fairly conservative at 2.07. The minimum factor between "GL ULTIMATE" load and "VO65 W2" load is 2.76. This is due to the approach of Gurit effectively using a lower safety margin over the break load of the cable for composite rigging relative to metal rigging.

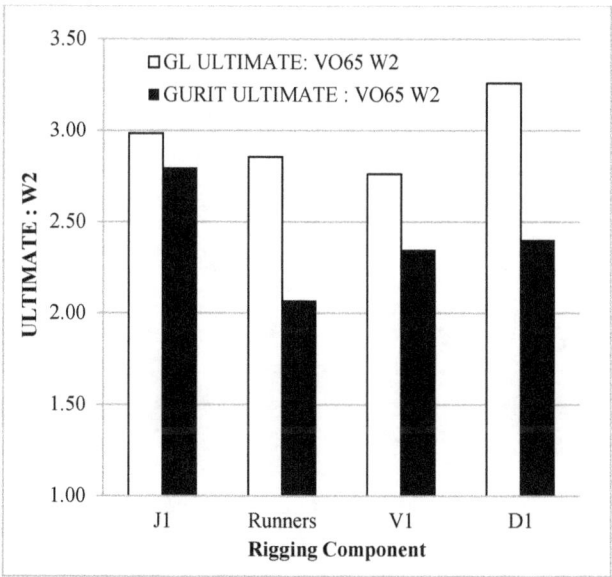

Figure 13: Factors of GL ULTIMATE and GURIT ULTIMATE over VO65 W2

5. CONCLUSIONS

Valuable insights could be obtained from data collected on the VO65 fleet during the 2014-2015 Volvo Ocean Race. This paper focused on some of the findings obtained through the analysis of keel and rigging load recordings, in particular:

- On the basis of the comparison between cumulative damage ratios (*CDR*) obtained from actual data with *CDR*s from the DNVGL fatigue assessment method, the latter appears to give a conservative result when designing for fatigue. With respect to an offshore racing yacht, a correct estimate of the number of tacks and gybes is critical to improving the accuracy of the prediction of fatigue life, particularly for solid keels. Equally, the DNVGL estimate of the

number of cycles due to wave encounters (small stress amplitude) may be improved in order to provide more reliable indication of fatigue life for fabricated (welded) keels.

- The stress spectrum used by ISO for the keel simple fatigue assessment leads to results of CDR close to those obtained from using a stress spectrum derived from actual measurements on a VO65, and to a slightly optimistic prediction of fatigue life for the same type of yacht.

- Based on the results of the DNVGL fatigue assessment method and on the data recorded on the VO65s, one may conclude that designing a fabricated canting keel fin only to meet DNVGL static "knock-down" loadcase, would result in a significant probability of fatigue failure after less than 2 round-the-world races. To avoid this type of failure, extreme care should be taken by the structural designer in:
  a) carrying out adequate fatigue analysis for the keel structure and using an appropriate stress spectrum
  b) designing the welds with higher safety factors with respect to static loads than used for the continuous parts of the keel fin
  c) designing welds to be appropriately far away from high stress zones

- The approach used by Gurit to predict peak dynamic load (W2) from maximum steady-state load (W1) for rigging components shows good correlation with the data from the Volvo 65s, but in 4/7 cases underpredicts the maximum.

- For composite rigging, Gurit use the approach of taking the break load for metal stiffness equivalent rigging as the LIMIT design load and applying a safety factor to this to obtain the ULTIMATE design load, which must be greater than the composite cable break load. The minimum reserve factor above the peak dynamic load from the Volvo 65 data for Gurit LIMIT design load is 1.38, for cable break load is 1.73 and for Gurit ULTIMATE design load is 2.07.

- The reserve factor used by DNVGL between peak dynamic loads (W2) and minimum design break load of supporting composite structure (ULTIMATE) when using composite rigging is higher than using the approach taken by Gurit to determine the ULTIMATE design load. Both methods have a reserve factor greater than 2 between ULTIMATE and cable break load.

ACKNOWLEDGEMENTS

We would like to acknowledge:

- Green Marine for allowing us to use the data gathered from the Volvo Ocean Race 2014-2015 to write this paper.
- Det Norske Veritas - Germanischer Lloyds ("DNVGL") for their support in the investigation on appropriate stress spectrum for keel fatigue assessment
- Diverse Yachts and Cosworth for supplying information on load cells and DAQ equipment

REFERENCES

1. **MANGANELLI, P.** Experimental investigation of dynamic loads on offshore racing yachts. 2006.

2. **MANGANELLI P. & HOBBS M.** Measurement of acceleration and keel loads on canting keel race yachts. *RINA Modern Yacht Conference.* 2007.

3. **PETERSHAGEN, H.** Fatigue Problems in Ship Structures. *Advances in Marine Structures Conference.* 1986.

4. **DET NORSKE VERITAS.** Fatigue Assesment of Ship Structures. *Classification Notes No 30.7.* April 2014.

5. **VAN GORKOM, G.** Keels Falling Off. *IBEX.* 2013.

6. **GERMANISCHER LLOYD.** Guidelines for the Structural Design of Racing Yachts ≥ 24m. *Rules for Classification and Construction Ship Technology.* 2012.

7. —. Guidelines for Design and Construction of Large Modern Yacht Rigs. *Rules for Classification and Construction Ship Technology.* 2009.

8. **ISO.** ISO 12215 Part 9: Sailing craft appendages. *Small craft - Hull construction and scantling.* 2011.

9. **ASTM E1049 - 85.** Standard Practises for Cycle Counting in Fatigue Analysis. 2011.

10. **HOBBACHER, A.** Recommendation for Fatigue Design of Welded Joints and Components. *document XIII-2151-07/XV-1254-07.* s.l. : International Institute of Welding, 2008.

11. **BELGRANO, G. & MCEWEN, L.** Working load to break load: Safety factors in composite yacht structure. *RINA High Performance Yacht Design Conference.* December 2012.

12. **GERMANISCHER LLOYD.** Yachts and Boats up to 24m. *Rules for Classification and Construction Ship Technology (Chapter 3, Section 2).* 2003.

13. —. Guidelines for the Type Approval of Carbon Strand and PBO Cable Rigging for Sailing Yachts. *Rules for Classification and Construction Ship Technology.* 2008.

DISCLAIMER

The influence of sailor position and motion on the performance prediction of racing dinghies

Joshua C Taylor, Performance Sport Engineering Lab, University of Southampton, UK
Joseph Banks, Performance Sport Engineering Lab, University of Southampton, UK
Stephen R Turnock, Performance Sport Engineering Lab, University of Southampton, UK

ABSTRACT

The time-varying influence of a sailor's position is typically neglected in dinghy velocity prediction programs (VPPs). When applied to the assessment of dinghy race performance, the position and motions of the crew become significant but are practically hard to measure as they interact with the motions of the sailboat. As an initial stage in developing a time accurate dinghy VPP this work develops an on-water system capably of measuring the applied hiking moment due to the sailor's pose and compares this with the resultant dinghy motion. The sailor's kinematics are captured using a network of inertial motion sensors (IMS) synchronised to a video camera and dinghy motion sensor. The hiking moment is evaluated using a 'stick man' body representation with the mass and inertial terms associated with the main body segments appropriately scaled for the representative sailor. The accuracy of the pose captured is validated using laboratory based pose measurements. The completed work will provide a platform to model how sailor generated forces interact with the sailboat to affect boat speed. This will be used alongside realistic modelling of the wind and wave loadings to extend an existing time-domain dynamic velocity prediction program (DVPP). The results are demonstrated using a single handed Laser and demonstrate an acceptable level of accuracy.

NOTATION

υ	Kinematic viscosity (N s m^{-2})
ρ	Density of water (kg m^{-3})
P	Pressure (N m^{-2})
LOA	Length over all (m)
BWL	Beam waterline (m)
SA	Sail Area (m^2)
Δ	Total mass displacement (kg)
VPP	Velocity Prediction Programme
L^{seg}	Body segment length (m)
C_g	Segmental centre of gravity (m)
IMU	Inertial Measurement Unit

RMS	Root mean squared
λ	Relative segment centre of gravity
μ	IMU positon
a_{IMU}	Acceleration in IMU axis
a_{SEG}	Acceleration in segment axis
m_i	Mass of segment i (kg)

(Tensor, $\bar{\bar{T}}$, and vector $\bar{T}$ notation, quaternions are 1x4 vectors which are identified by a bold capital $\boldsymbol{Q}$)

INTRODUCTION

The rapid rise in the competitive use of foiling yachts either in classes such as the International Moth, the America's Cup or in ocean races requires an ever deeper understanding of the physics of yacht performance. At the smaller scale the behaviour of the sailor, where and how they move, can have a marked effect on overall boat performance. Understanding how the influence of the sailor can be captured during the evaluation of alternative hull-foil design combinations and for overall race prediction analysis motivates the work described.

The performance of sailing boats is commonly assessed by the time required for them to complete a mile-long racecourse. This is calculated using a velocity prediction programme (VPP). The resultant race time is derived by balancing the resistive and propulsive forces acting on the vessel at different points of sail to determine the maximum velocity made good (VMG) around the course. For this method to accurately predict course times sophisticated force modelling is required. This must include the predominant sail and hydrodynamic forces but also the effect of perturbations caused by the naturally varying wind and wave environment as well as the motions of the sailor.

Typically, VPP's assume that the forces are varying in a quasi-steady manner and that the sailor remains in a fixed position to generate a hiking moment. A logical step to assess the significance of this is to extend the VPP to include realistic loadings including the effects of sailors' motions. To achieve this it is important to understand what the sailor's motions are whilst out sailing and how these effect the righting moment that they are generating. Therefore, this research focuses on the development of a system that is able to estimate the on-water sailor loadings by measuring the dynamic pose of the athlete. The paper layout looks first at the background to dynamics velocity prediction programs (DVPP's) followed by the development of a motion capture system suitable for on-water measurements.

BACKGROUND

Single handed dinghies provide convenient platforms to study the significance of sailor motions. Sailors' essentially have three controls to react to their environment: (1.) Changing course; (2.) Trimming the sail; or (3.) Altering the loading they exert on the boat. The most significant of these loadings is the hiking moment. A change in loading can be in response to a varying wind vector, wave loadings or tactical decisions to events in the race.

A typical response to a long duration wind increase would be to increase the hiking moment by leaning out further. However short duration responses to waves or gusts can be made using rapid body motions which can spill wind from the leech or ease the hull through a wave by controlling trim. Although it is standard practice for the sailor to constantly adapt their hiking moment most VPPs, including unsteady versions, assume the hiking moment to be either constant or simply equal to the wind heeling moment. The current paper presents a technique to capture sailor-induced loadings exerted on the boat using an array of inertial sensors.

Dynamic sailor motions are most significant in tacking and gybing manoeuvres. The tack and gybe manoeuvres have been modelled in detail for yachts (Masayuma et al, 1995; Breschan et al, 2013; Spenkuch et al, 2010; Keuning et al, 2005; De Ridder et al, 2004) but so far has been neglected for dinghies.

The sailor can affect the forces acting on the boat by adjusting the rudder, mainsail and their own positon in the boat. Human tactical models based on either rules or decision algorithms have been developed for yacht VPP's (Spenkuch, 2014). While these still assume constant sailor loadings, they have also been extended to include wave induced motions (Keuning, 2005). Spenkuch and Scarponi

(2010) developed the tacking simulations of Matsuyama (1995) to investigate how the effect of human decision making impacts on course time. A Bayesian belief-based human decision engine was developed to include the effects of tactics in a sailing race (Spenkuch, 2011).

Although dynamic body motions are acknowledged in sailing literature to react to short term changes in wind speed and to promote foiling (Findlay and Turnock, 2008), no VPPs or studies into the sailors motions have yet been published. This is despite international sailing rules being developed to prevent sailors from over using such methods (ISAF, 2013).

FULL SCALE ANALYSIS OF MOTIONS INCLUDING DYNAMIC POSE CAPTURE

It is challenging to directly measure a sailor's loading whilst on the water. This is due to the generated hiking moment being transmitted to the hull through a combination of the deck, toe-straps and the mainsheet. Therefore, a method based on wireless inertial sensors is presented to estimate the athlete loadings exerted on the boat. Alternative approaches to capturing motion can include automated video capture. Phillips et al (2014) compared the use of inertial sensors and video motion capture for a swimmers underwater fly-kick and demonstrated comparable accuracy. The advantage of sensors is that they do not rely on high quality images at all times. Disadvantages of using the inertial sensors are associated with sensor drift and ensuring they are operating all the time. For a multi-sensor system the latest sensors are now much more reliable, which offers the opportunity to capture on-water motions as long as issues with waterproofing can be addressed. Similar challenges have been overcome for use in capturing model ship motions (Bennett et al, 2014).

The method we adopt to control the influence of drift is to use our knowledge of the fixed geometry of the sailor. From this the sailor's motion is based on using a kinematic chain. This deliberate simplification allows the model to be calibrated to match the subject's own body segment weights using an anthropometric model as a starting point. This allows subject specific loadings which are required to reverse engineer the sailor's dynamic pose. Figure 1 illustrates the main body segments and key dimensions of the sailor. For this work the body is assumed to be symmetric.

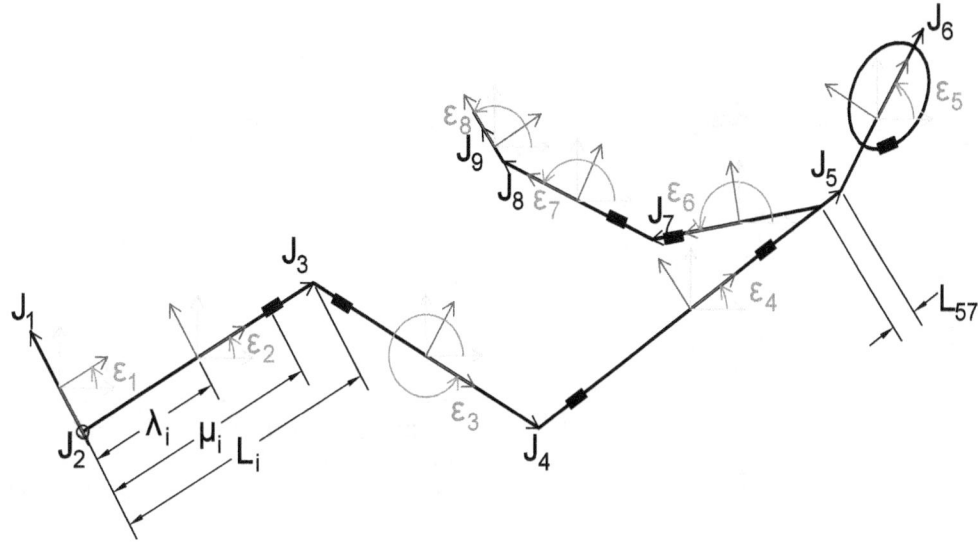

Figure 1 - Hiking Pose Schematic showing key dimensions and approximate sensor locations

The sailor's position is represented by a stick figure, as shown in Figure 1, where each line represents a body segment. The segments are linked using joints constrained at their endpoints in position but free to rotate. Each segment has a constant length, L_i^seg, measured for a given subject, as described in Figure 2, Table 1 and the vector set in equation (1).

static moment taken about a specific subject's ankles this equates to an error of only $\pm$ 0.6% (0.4 Nm) for a 75 kg, 1.8 m tall individual. The curvature of the back is approximated as the interpolated orientation of the small of the back and the sternum using equation (3). For simplicity, only the left arm and leg are instrumented and limbs on the right side are assumed to mirror the left.

$$L^{seg} = \left\{ \begin{array}{c} L_1^{foot} \\ L_2^{shank} \\ L_3^{thigh} \\ L_4^{back} \\ L_5^{head \& neck} \\ L_6^{U-arm} \\ L_7^{L-arm} \\ L_8^{Hand} \end{array} \right\} \quad (1)$$

$$\epsilon^{seg} = \left\{ \begin{array}{c} \epsilon_1^{foot} \\ \epsilon_2^{shank} \\ \epsilon_3^{thigh} \\ \epsilon_4^{\overline{back}} \\ \epsilon_5^{head \& neck} \\ \epsilon_6^{U-arm} \\ \epsilon_7^{L-arm} \\ \epsilon_8^{Hand} \end{array} \right\} \quad (2)$$

$$\epsilon_4^{\overline{back}} = \frac{\epsilon_{41}^{L-back} + \epsilon_{42}^{sturnum}}{2} \quad (3)$$

Table 1: Representative body segment details

Index	Segment	Origin	End Point	Length	Mass	COG	Rg x	Rg y	Rg z
1	Foot	HEEL	TTIP	258.1	1.4	44.2	25.7	24.5	12.4
2	Shank	KJC	LMAL	434.0	4.3	44.6	25.5	24.9	10.3
3	Thigh	HJC	KJC	422.2	14.2	41.0	32.9	32.9	14.9
4	Trunk	CERV	MIDH	603.3	43.5	51.4	32.8	30.6	16.9
5	Head	VERT	CERV	242.9	6.9	50.0	30.3	31.5	26.1
6	Upper arm	SJC	EJC	281.7	2.7	57.7	28.5	26.9	15.8
7	Forearm	EJC	WJC	268.9	1.6	45.7	27.6	26.5	12.1
8	Hand	WJC	MET3	86.2	0.6	79.0	62.8	51.3	40.1

The orientation and acceleration of each body segment is measured using Xsens' MTw inertial sensors (Xsens, 2013). These sensors provide an orientation expressed as a unit quaternion (a three dimensional rotation) dynamically stable to 2 degrees RMS. The highest impact of this error is when the subject is horizontal with their arms over their head and each body segment is rotated by 2 deg. For a

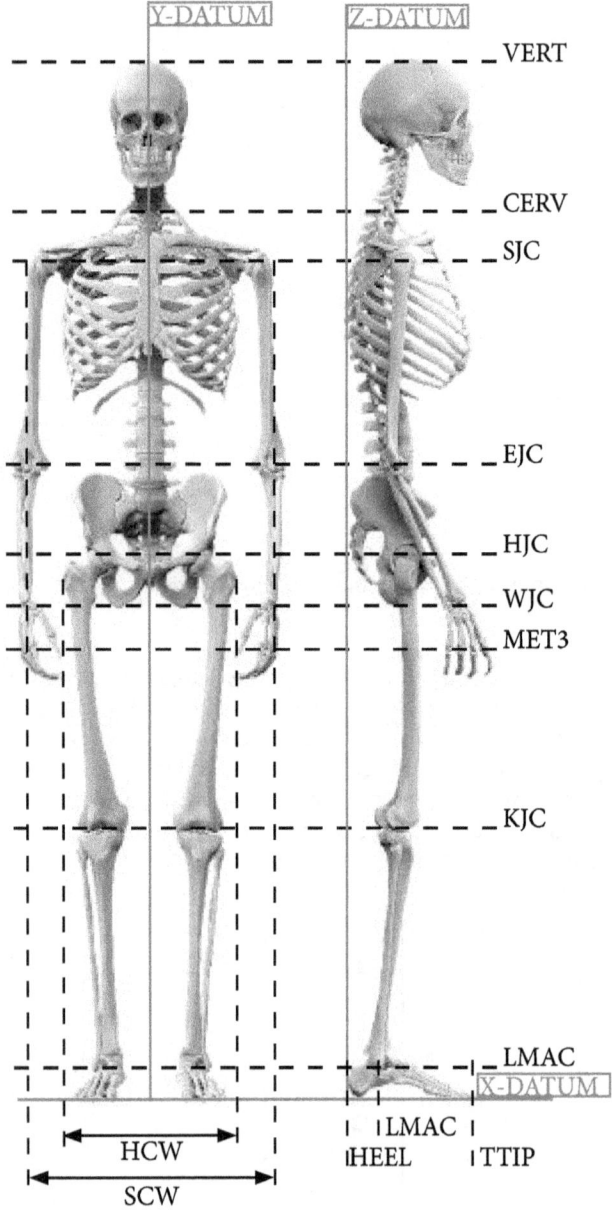

Figure 2 - Subject joint distances and anthropometric data

Seluyanov, 1983). A regression based mass model of Russian PE students is presented by Zatsiorsky (1983). This was later adjusted to reference more convenient segment boundaries located at joint centres and extended to include male and female data in a proportional mass model (Leva, 1996). For the purpose of this paper the later model will be used to define the mass distribution. An example of the resultant mass distribution for a 75kg, 1.8m tall male is presented in Table 1. The masses are given in kilograms, the centre of gravities (COG) are provided as a percentage of the segment lengths.

The following section uses quaternion rotations to switch frames of references. This provides a method which does not suffer from a mathematical singularity when pitch orientation approaches 90 degrees as Euler angles suffer from. The rotation notion used **Q** to represent a quaternion and a superscript to represent the order of the rotation, where B represents the body segment, Y the yacht, G gravitational and S the sensor frame of reference. E.g. Q_i^{BG} represents the quaternion to rotate from body to gravity frames of reference for segment i.

The orientation of the body segment to the gravity vector $\mathbf{Q}^{BG}$ is calculated by finding the difference of the sensor orientation relative to gravity $\mathbf{Q}^{SG}$ and the sensor to body orientation $\mathbf{Q}^{SB}$, as described in equation 4. $\mathbf{Q}^{SB}$ is found by capturing a pose when the subject is lying down so the joint angles are zero.

$$\mathbf{Q}_i^{BG} = \mathbf{Q}_i^{SG} \times inv(\mathbf{Q}_i^{SB}) \qquad (4)$$

A separate sensor mounted on the boat captures the yacht orientation and is zeroed when the boat is flat. This allows the orientation of the body segments relative to the yacht to be calculated.

A body segment is rotated to the sailors pose using the body-yacht rotation $\mathbf{Q}^{BY}$, starting from the fixed point at the ankle. The body segment orientations therefore provide the coordinates for each joint, J_i, moving up the body.

The linked chain pose is given by (5): -

$$J_{i+1} = \left\{ Q_i^{BY} \begin{bmatrix} L_i^{seg} \\ 0 \\ 0 \end{bmatrix} \right\} + J_i \qquad (5)$$

for $1 \leq i \leq 5$ and $7 \leq i$ where: -

$$Q_i^{SB} = Q_i^{SG} \, for \, zero \, pose$$

The shoulder joint is not located at the end of the back segment therefore a non-dimensional distance is used to locate it between the neck and palpaple point of the 5th vertibre.

The mass and location of the centre of mass for each segment is required to convert the pose position into the generated force. There are several body-segment data sources in the literature. Models can be based on cadaver data which tend to have an elderly population. More recent studies use radiation to measure living subject's mass distribution. It is important for the studies' mean population to be close to the target population. There are two main methods for determining the mass distribution. The first assumes each body segment has uniform density. This relies on many measurements to define the geometry of each segment. The second method uses either proportional or regression based mass distribution methods using the subject height and mass (Zatsiorsky and

The entire pose is then translated so the ankle joint, J_2 becomes the origin, as it will be used as the datum in experiments.

$$J = J - J_2$$

(6)

Position vectors for segment centre of gravity, C_G and sensor positions S_{IMU} are defined for each segment in terms of COG coefficient, λ and sensor position coefficient, μ using equations (7) and (9) respectively. C_G is derived from published data (Zatsiorsky and Seluyanov, 1983; Leva, 1996; Dumas et al, 2007), gven in table 1.

$$C_G = J_i + \lambda_i(J_i + J_{i+1})$$

(7)

Where: -

$$\lambda_i = \frac{L_i^{COG}}{L_i^{seg}}$$

(8)

$$S_{IMU} = J_i + \mu(J_i + J_{i+1})$$

(9)

Where: -

$$\mu_i = \frac{L_i^{IMU}}{L_i^{seg}}$$

(10)

For the on-water experiments the pose is offset by a measured vector found from the scaled video, providing the distance between the centreline of the dinghy at deck level, amidships and the midpoint between the sailor's ankles. The pose is finally rotated about the centreline of the dinghy to coincide with the roll angle (θ) of the boat:

$$C_{G_i}{}' = Q^{BG} C_G$$

(11)

The 3D acceleration, a' at $C_{G_i}{}'$ is then used to to calculate the generated hiking moment:

$$HM = CG_{xi}' m_{seg} a_z'$$

(12)

In summary, the following modelling assumptions have been made: -

- The head and neck are rigid and pivot about the palpable point of the 7th vertebrate.
- The back is straight and its curvature can be approximated by taking the mean of the upper and lower back orientations.

- The left limbs mirror the right limbs.
- The anthropometric model accurately predicts the segment COG.
- The mass distribution (i.e. the relative mass of each segment) can be adjusted to tune the model for a given subject if required.

ESTIMATED POSE LOAD VALIDATION

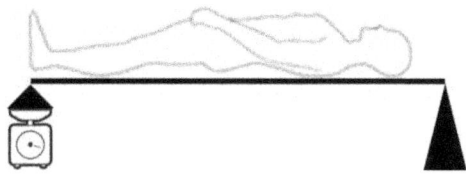

Figure 3 - Schematic of direct measurement of hiking moment

The pose algorithm is validated using a board with one end resting on a set of scales, as depicted in Figure 3. As the total mass of the subject and the length of the board is known the position of the centre of gravity, and hence the hiking moment about the ankles, can be calculated. The subject is asked to hold a set of poses chosen to isolate the effect each body segment has on hiking moment. Each of these poses are held for 30 seconds.

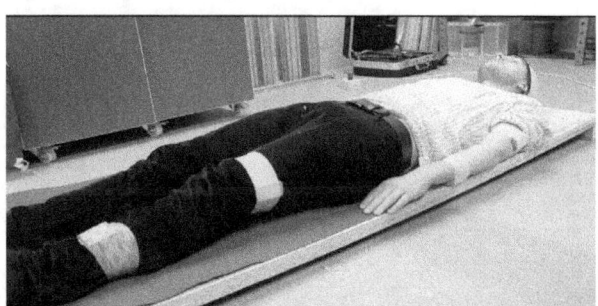

Figure 4 - Direct measurement of Hiking Moment to validate Pose capture system.

Each segment's centre of gravity given by Leva (1996) is assumed to be correct and the mass is redistributed from the initial assumed values using equation (13). The estimated mass of the feet is assumed correct (relative mass 2.8% m_{aser}) since the feet location is close to the centreline.

$$m_i^{cal} = m_i^{seg} \left(\frac{m_{slr}}{\sum m_i^{seg}} \right)$$

(13)

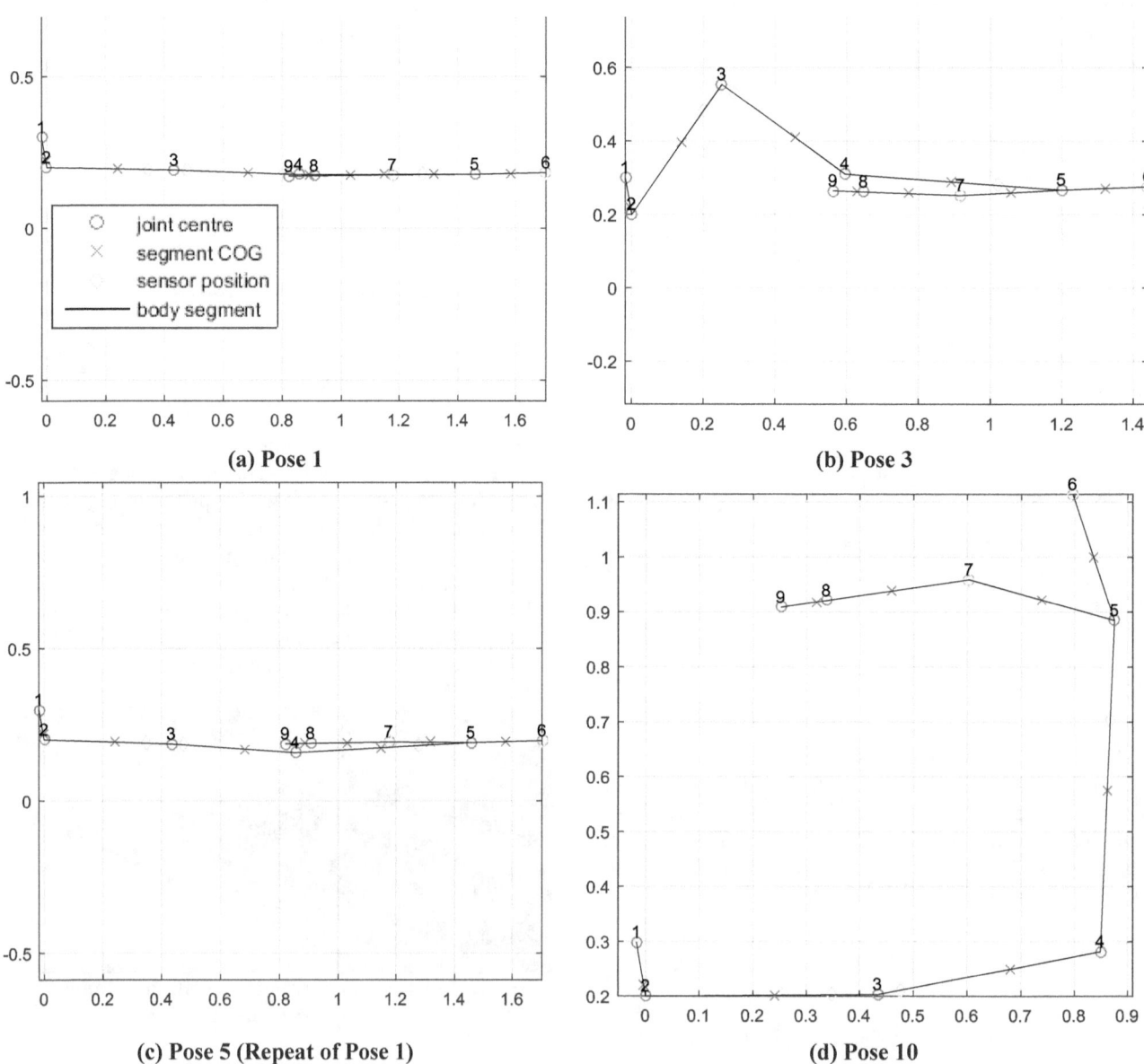

(a) Pose 1

(b) Pose 3

(c) Pose 5 (Repeat of Pose 1)

(d) Pose 10

Figure 5 Four examples of the pose captured by the inertial sensor system using the flat board. Axes are in m relative to ankle joint

A series of 13 body positions were conducted during the pose validation process. The comparison between the direct measurement of the hiking moment and the estimate provided by the pose capture system is provided in Table 2. A number of the captured poses are provided in Figure 5 for visual reference.

The baseline position depicted in Figure 4 was conducted several times to check the repeatability of the system. It can be seen in Table 4 that, although the hiking moment calculated from the pose capture system repeatedly estimated the same value, the measured value from the board and scales varies slightly. This is most likely due to a slight change in the position of the ankle (i.e. a longitudinal change in the position of the subject on the measurement board). Comparing pose 5 with pose 1, in Figure 5, it is evident that either the physical sensor position has moved or the sensors are drifting as there are slight changes in the captured pose for the repeated position. Many factors could be contributing to this, including the movement and stretching of clothing, the presence of high voltage/current electrical cables present in the test venue or slightly different physical positions being adopted by the subject.

Table 2 – Comparison of measured Hiking Moment (HM) compared to the estimate from the pose capture system. (Poses 2-4 did not maintain the ankle at the end of the board, therefore had to have the measured HM corrected for comparison)

Pose#	HM Measured (Nm)	HM from Pose (Nm)	Error
1	713	707	-1%
2	57	54	-5%
3	525	534	2%
4	513	547	7%
5	710	707	0%
6	574	551	-4%
7	567	543	-4%
8	551	531	-4%
9	563	533	-5%
10	539	531	-1%
11	547	537	-2%
12	562	535	-5%
13	692	707	2%

On average the pose capture system was observed to correctly estimate the hiking moment generated in a range of different positions within 5% accuracy. Therefore, it was concluded that the system would provide useful measurements of the on water hiking position of sailors. However, it must be accepted that this as yet does not represent a detailed study of the sailing population. The results relate to a single subject and neglect dynamic loads. The validation procedure should be extended to account for this by imposing motion and comparing pose data with load cell data on an instrumented platform.

ON THE WATER METHODOLOGY

To establish if the developed pose-capture methodology would work out on the water an initial case study was conducted using a Laser Radial sailing dinghy. The Laser is a small solo planing dinghy class. The craft is LWL = 3.81m, BWL = 1.37m Δ > 58.9kg. SA = 7.06m² (Laser, 2012).

To enable the boat motions to be captured the same model Xsens IMU was attached to the right of the daggerboard's leading edge. A 10Hz GPS receiver is used to track the boat's speed and course over ground. The inertial sensors used to capture the athlete's pose wirelessly stream the data to a tablet that logs the data. The estimated combined weight of the instrumentation was approaching 10kg. All of this mass was carried forward of the mast which significantly affected boat motions and acts as a weight penalty. The equipment can be observed in Figure 6.

Figure 6 – On water equipment

The experiments were conducted close to slack water to minimise tidal effects. The ankles are taken as the pose datum so the mounted video can be used to measure their offset away from the centreline in post-processing using video footage captured from a tiller-mounted camera, as shown in Figure 7. The sensor positions on both the sailor and boat are chosen for the best compromise between being close to the segment centre of gravity and located on a rigid part of the body (i.e. close to the bone) to reduce skin movement artefacts.

Figure 7 Using video to offset pose

The GPS track of the on-water data recording session is provided in Figure 8. A representative plot of the unsteady Hiking moment acquired by the pose capture system is provided in Figure 9. This initial on the water study shows that the developed system can work in a real sailing situation enabling valuable dynamic hiking moment data to be obtained. The presented data is from a lit wind condition where the boats roll angle can be seen to respond to changes in generated hiking moment.

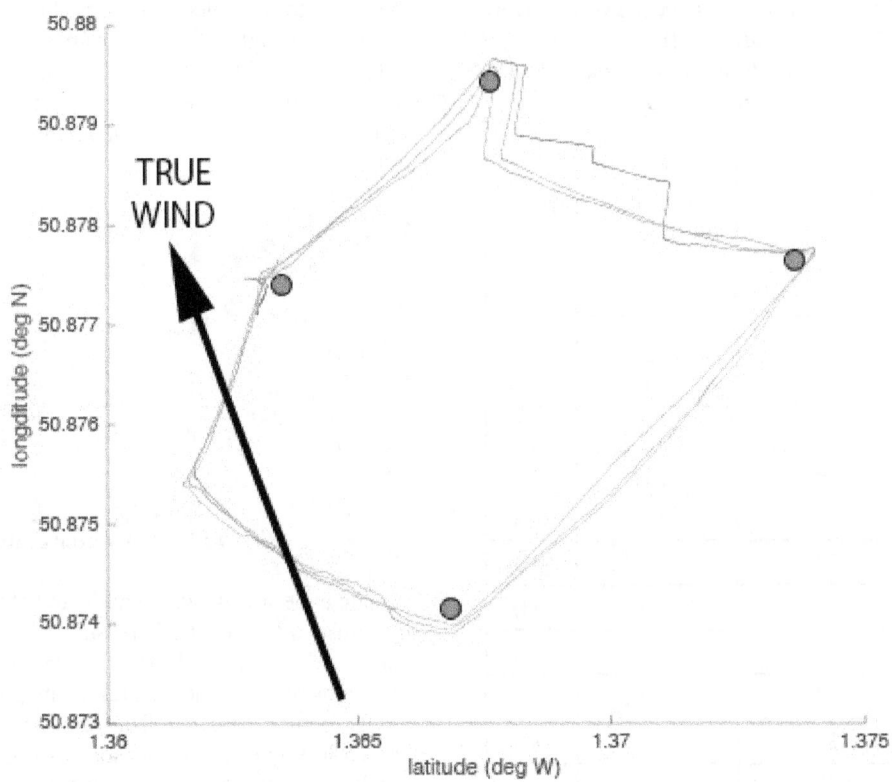

Figure 8 GPS track of Race, with line colour representing boat speed.

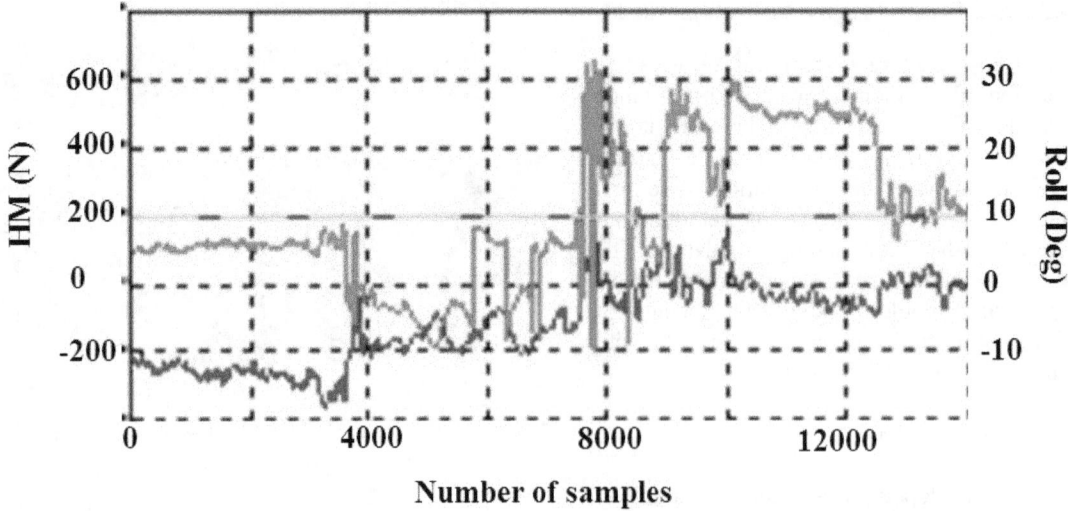

Figure 9 Light wind example of measured Unsteady Hiking Moment (Red) compared to roll angle (Blue) from on the water tests.

FUTURE APPLICATION TO A DYNAMIC VPP

The ability to acquire unsteady hiking moments relative to boat motions allows these loadings to be characterised and fed back into a fully dynamic VPP algorithm in the future. This could be coupled with unsteady wind, sail and foil models to provide a more realistic evaluation tool for sailing dinghy performance. The next stage in the work is to classify the sailor position and adjustment during typical race phases. From these a model can be developed to use within the dynamic VPP.

The single-handed sailor is the human-in-the-loop controller for the selection of the 'best' performance of the yacht through adjustment of position, helm angle and trim of sails. Previous work (Spenkuch, 2014; 2011, 2010) demonstrates that understanding the tactical and strategic behaviour of competitor sailors requires the unsteady physics of the yacht performance to be captured realistically enough.

CONCLUSION

The work has demonstrated that a network of appropriately located inertial motions sensors can provide assessment of both pose and motion. The ability to estimate the developed hiking moment provides an opportunity to better understand the on-water performance of dinghies in the stochastic environmental challenge of wind and waves.

The developed system still requires some further development to reduce its mass and on-board footprint. One area requiring further work is in ensuring automated operation with minimal interaction required of the sailor.

The development work has used a limited set of sailors and in the next phase a wider group will be studied for pose and motion calibration within a laboratory environment. On-water studies will move onto the International Moth once remaining mass issues are resolved.

ACKNOWLEDGEMENTS

The author's gratefully acknowledge the funding support of the EPSRC iCASE Studentship which support Josh Taylor's PhD from the English Institute of Sport (Research and Innovation). All work requiring ethical approval has been registered through the Faculty of Engineering and Environment ethics review process.

REFERENCES

L.-M. Breschan, A. Lidtke, L. M. Giovannetti, A. Sampson, and M. Vitti, "America's Cup Catamaran Tacking Simulator," Southampton University Group Design Project Thesis for MEng Ship Science degree course, 2013.

S.S. Bennett,C.J. Brooks,B. Winden,D.J. Taunton,A.I.J. Forrester,S.R. Turnock,D.A. Hudson, "Measurement of ship hydroelastic response using multiple wireless sensor nodes", Ocean Eng., Vol. 79, 67–80. 2014.

R. Dumas, L. Chèze, and J.-P. Verriest, "Adjustments to McConville et al. and Young et al. body segment inertial parameters," J. Biomech., vol. 40, no. 3, pp. 543–553, Jan. 2007.

E. J. De Ridder, K. J. Vermeulen, and J. A. Keuning, "A Mathematical Model for the Tacking Maneuver of a Sailing Yacht," Int. HISWA Symp. Yacht Des. Yacht Constr., pp. 1–34, 2004.

M. W. Findlay and S. R. Turnock, "Development and use of a Velocity Prediction Program to compare the effects of changes to foil arrangement on a hydro-foiling Moth dinghy," Proccedings Int. Conf. Innov. High Perform. Sail. Yachts., 2008.

ISAF, INTERPRETATIONS OF RULE 42 , PROPULSION, May. 2013.

J. Keuning and E. De Ridder, "A Generic Mathematical Model for the Maneuvering and Tacking of a Sailing Yacht," Chesap. Sail. Yacht Symp., no. 1458, pp. 143–163, 2005.

Laser Class Association, "Laser Class Rules," 2012.

P. de Leva, "Adjustments to Zatsiorsky-Seluyanov's segment inertia parameters," J. Biomech., vol. 29, no. 9, pp. 1223–1230, Sep. 1996.

Y. Masayuma, T. Fukasawa, and H. Sasagawa, "Tacking Simulation of Sailing Yachts – Numerical Integration of Equations of Motion and Application of Neural Network Technique," 12th Chesap. Sail. Yacht Symp., 1995.

C. Phillips, A.Forrester, D.Hudson, S.Turnock, "Comparison of Kinematic Acquisition Methods for Musculoskeletal Analysis of Underwater Flykick", Procedia Eng., Vol 72, 2014, 56–61.

T. Spenkuch, S. Turnock, M. Scarponi, and A. Shenoi, "Real time simulation of tacking yachts: how best to counter the advantage of an upwind yacht," Procedia Eng., vol. 2, no. 2, pp. 3305–3310, Jun. 2010.

T. B. Spenkuch, "UNIVERSITY OF SOUTHAMPTON A Bayesian Belief Network Approach for Modelling Tactical Decision-Making in a Multiple Yacht Race Simulator By," no. April, 2014.

T. Spenkuch, S. R. Turnock, M. Scarponi, and R. A. Shenoi, "Modelling multiple yacht sailing interactions between upwind sailing yachts," J. Mar. Sci. Technol., vol. 16, pp. 115–128, 2011.

T. Spenkuch, S. Turnock, M. Scarponi, and A. Shenoi, "Real time simulation of tacking yachts: How best to counter the advantage of an upwind yacht," Procedia Eng., vol. 2, no. 2, pp. 3305–3310, 2010.

Xsens, "MTw User Manual", Document MW0502P, Revision H, 2013. (https://www.xsens.com/download/usermanual/MTw_usermanual.pdf)

Zatsiorsky and Seluyanov, "The mass and inertia characteristics of the main segments of the human body," Biomechanics, 1983.

AUTHORS BIOGRAPHY

Josh Taylor is a BEng graduate of the University of Newcastle-upon-Tyne. He is a keen sailor and is now pursuing his Doctoral studies in the Performance Sports Engineering Lab at the University of Southampton.

Dr Joe Banks is a New Frontiers Fellow at the University

of Southampton where he also completed his Master of Engineering and PhD. In this he worked closely with UKSport (now EIS) Research and Innovation in the run upto London 2012. His current research focusses on the development of experimental techniques to capture time accurate coupled fluid structure interaction alongside Computational Fluid Dynamics.

Professor Stephen Turnock directs the Performance Sports Engineering Lab at the University of Southampton. He is a graduate of Cambridge(MA), MIT (SM) and Southampton (PhD). He leads the Maritime Engineering group in the Faculty of Engineering and Environment and in recent years has led the fitout of the new wave/towing tank (138m x 6m x 3.5m, max speed 12m/s). His research interests ship hydordynamcs, renewable energy, autonomy alongside his interests in performance sport.

DEVELOPMENT OF A ROUTING SOFTWARE FOR INSHORE MATCH RACE

Francesca Tagliaferri, Newcastle University, Newcastle upon Tyne, UK
Ignazio Maria Viola, The University of Edinburgh, Edinburgh, UK

ABSTRACT

Yacht races are won by good sailors racing fast boats. A good skipper takes decisions at key moments of the race based on the anticipated wind behaviour and on his position on the racing area and with respect to the competitors. His aim is generally to complete the race before all his opponents, or, when this is not possible, to perform better than some of them. In the past two decades some methods have been proposed to compute optimal strategies for a yacht race. Those strategies are aimed at minimizing the expected time needed to complete the race and are based on the assumption that the faster a yacht, the higher the number of races that it will win (and opponents that it will defeat). In a match race, however, only two yachts are competing. A skipper's aim is therefore to complete the race before his opponent rather than completing the race in the shortest possible time. This means that being on average faster may not necessarily mean winning the majority of races. This paper presents the development of software to compute a sailing strategy for a match race that can defeat an opponent who is following a fixed strategy that minimises the expected time of completion of the race. The proposed method includes two novel aspects in the strategy computation:

- A short-term wind forecast, based on an Artificial Neural Network (ANN) model, is performed in real time during the race using the wind measurements collected on board.

- Depending on the relative position with respect to the opponent, decisions with different levels of risk aversion are computed. The risk attitude is modeled using Coherent Risk Measures.

The software is tested in a number of simulated races. The results confirm that maximising the probability of winning a match race does not necessarily correspond to minimising the expected time needed to complete the race.

NOMENCLATURE

Acronyms	
ANN	artificial neural network
BS	boat speed
DP	dynamic programming
RMP	race modelling program
TWA	true wind angle
VMG	velocity made good
Symbols	
$\mathbb{E}$	expected value
$C(U,\omega)$	cost function
G_x, G_y	grid point matrices
t_k	sailing time at step k
$U = u_0, \ldots, u_{N-1}$	policy
U^{opt}	optimal policy
$\mathcal{U}$	set of admissible policies
(x,y)	position coordinates
(x_0, y_0)	initial position
$(x_{1L}, y_{1L}), (x_{1R}, y_{1R})$	reachable nodes coordinates
w	wind vector
ω, ω_k	random variables

INTRODUCTION

A yacht race is a competition where two or more boats race each other to complete a certain course in the shortest time. Traditionally, the problem that a sailor has to solve is addessed as an optimisation problem consisting in going from point A to point B in the shortest possible time, under certain constraints given by the dynamics of the yacht and racing rules.

This approach however doesn't really capture the competitive aspect of a race. In fact, the real aim of a sailor is not to get to the finish line as fast as possible, but rather to get there before their opponent(s). Moreover, the speed of a sailing yacht is highly dependent on the behaviour of the wind. A sailor doesn't have perfect knowledge of the future wind patterns, and therefore the problem must be adddessed as a stochastic problem, based upon probability distributions of the wind behaviour, as done, for instance, by **?**.

In previous studies (Tagliaferri et al., 2015) the authors have shown how the accuracy of a wind forecast can improve the chances of winning a race. The use of artificial neural networks (ANN), compared to other forecasting techniques, was identified as suitable for very-short-term wind prediction (order of seconds/minutes in advance). It was also shown (Tagliaferri et al., 2014) that strategies with different risk tolerance can be computed, but that the strategies that aim at minimising the time needed to complete a race are not necessarily the ones that lead to a higher chances of winning.

This paper focuses on the development of a methodology that allows the computation of a strategy for a sailor, combining the ANN wind forecast and optimal risk modelling. For the first time the presence of a moving opponent is included in the computation of a strategy, and the opponent is not only seen as a moving obstacle, but also as an element of influence in the yacht's speed. The computation of the optimal strategy is based on dynamic programming (DP) over a time-dependent lattice, which is generated according to an ANN-based wind forecast.

Background on yacht racing

Yacht races are held in many different formats and levels: in the case of a *match race* only two boats face each other, while in a *fleet race* the number of participants can be very high. One of the most prestigious sailing competition (and by far the oldest and most expensive) is the America's Cup, which includes match races between various teams fighting for the chance of challenging the Cup defender, i.e. the winner of the previous edition.

Usually, a race course includes several turns around an upwind and a downwind mark, where the marks are aligned with the wind. The course is designed to present some challenges to the skippers. In fact, the speed of a sailing yacht depends on the wind speed and the True Wind Angle (TWA, the supplementary angle between the wind velocity and the boat heading). Figure 1 presents an example of boat speed (BS) as a function of the TWA for a given wind speed in a polar diagram. A polar plot of this kind, which may include different curves associated to different wind speeds, is the conventional way of presenting the boat speed, and although the actual BS can depend on other factors (such as waves and crew), it is considered as a characteristic of a yacht.

As shown in the plot, the highest values for the BS are achieved when sailing at a TWA of approximately 90° (on a *beam reach*). Conversely, when the TWA tends to zero, BS tends to zero. Therefore, when sailing upwind (for instance, from a downwind mark to an upwind mark), the most effective route consists in a zig-zag in the wind direction, sailing at a TWA of 35°-50° (*close hauled*). In this case, a skipper's aim is to maximise the speed in the

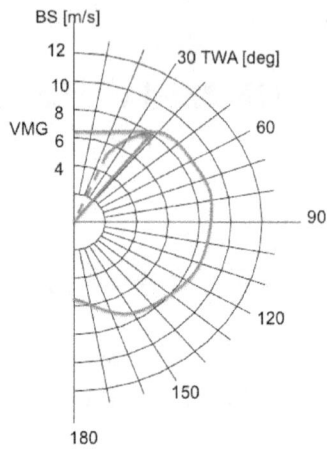

Figure 1: Example of polar diagram.

upwind direction, which means to find the TWA such that the projection of the boat velocity on the upwind direction is a maximum. The corresponding velocity is referred to as Velocity Made Good (VMG) and is shown in red in Figure 1. Similarly, a VMG can be defined for downwind sailing as the projection of the boat velocity on the downwind direction.

The VMG can be defined also for downwind sailing. In

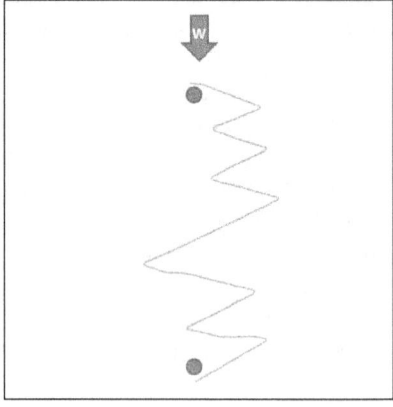

Figure 2: Example of upwind leg.

fact, as shown in Figure 1, even if the velocity is not null when the TWA is 180°, the maximum projection on the downwind direction for this example is obtained at angles of approximately 150°. However, the optimal angle for downwind sailing can have significant variations depending on the yacht geometry.

Yacht racing strategy

Initial research related to competitive sailing was mainly focussed on understanding a yacht's dynamics, in order to improve the design process and create faster and more efficient boats. The outcome of this research was the development of Velocity Prediction Programs (VPPs), computer programs that solve the equations of motion for a sailing yacht, and determine its velocity for given wind conditions. Kerwin and Newman (1979) present one of the earliest studies detailing the development of a VPP. The introduction of such tools determined a big step forward for competitions such as the America's Cup. In fact, The United States had always successfully defended the trophy since its creation, in 1851, but in 1983, for the first time, an Australian team managed to win, and this success was mostly due to the radical innovation in keel design for the yacht *Australia II*, (Oossanen and Joubert, 1986). Americans learned from this defeat, and the following campaign saw a massive effort in applying state-of-the-art technology to the design of the yacht that would challenge the Australians. The resulting yacht, *Stars and Stripes*, succeeded in the mission of bringing the Cup back to the US in 1987, after a campaign that was the first to see a competition not only between sailors but also between the engineering teams of the different countries. This aspect is passionately described in the paper *Stars and Stripes* by Letcher et al. (1987) which focuses on the advantages brought by computer technology.

Since then, the competition has evolved along those lines, and today it is still a fierce battle between engineering teams besides sailing teams, and in sports journalism America's Cup races are often compared to Formula One GP. The development of VPPs allowed design teams to compare different design choices at an early stage, and techniques used for determining forces acting on the boats are determined using a variety of techniques, both experimental and numerical.

The evolution of VPP led to *Race Modelling Programs* (RMP), computer programs aimed at simulating an entire race between two yachts. The *Stars & Stripes* campaign involved one of the very first RMP to analyse the probabilities of win/loss of a yacht.

The subsequent America's Cup saw the development of a RMP which included a statistical weather model based on site-specific environmental data for San Diego. This RMP was developed by the Partnership for America's Cup Technology, and details are described by Gretzky and Marshall (1993).

In those models, the tactical decision process is modeled as a set of fixed decision rules. The tactical and physical interactions between the yachts are not adequately modeled, and this limitation is reflected in the definition of win in Letcher et al. (1987), where a yacht has to win by a certain time margin to be certain of a win.

An important contribution to RMP came from the studies

carried out at The University of Auckland in collaboration with the New Zealand challenger team. Philpott and Mason (2001) investigated the decision-making process, focussing on the development of a strategy. In this work, which constitutes a fundamental basis for the present study, dynamic programming is used to generate a *policy*, that can be computed before the race, and can then be used during the race. Later, Philpott et al. (2004) developed a model to predict the outcome of a match race between two competing designs, still assuming a set of fixed decision rules but taking into account some interactions between yachts (for instance, when crossing).

In these two studies the tactics and strategy modeling was aimed at obtaining a simulation tool that could replicate as closely as possible the situations that can arise during a yacht race, with the ultimate objective of assessing competing designs. Other studies not directly related to the America's Cup have tackled the problem of decision-making for sailors. Ferguson and Elinas (2011) propose a simple Markov decision model, where at all times the sailor has only two options, "do nothing" or "tack". The work developed in this study, focussed on inshore racing, includes a VPP and a model for wind flow around landmasses. The importance of the tacking penalty is investigated by comparing routes produced by assuming different penalty factors associated to tacks.

Recently, the University of Southampton has developed a sailing simulator called "Robo-Race", a tool to model both the physical behaviour of a yacht and the interaction with the crew (Scarponi et al., 2007a,b). The tool is designed so that human sailors can interact with it, racing against a computer in an artificial environment. A VPP using four degrees of freedom is implemented, including the tacking model based on the studies of Masuyama (1995). Improvements on the first implementation are focused on physical interactions between racing yachts (Spenkuch et al., 2008, 2011), and the dynamics of the yacht during manoeuvring (Banks et al., 2010, Spenkuch et al., 2010), both for upwind and downwind sailing. An important contribution of this work is recognising the existence of conflicts between strategy and tactics, for instance when a yacht decides to tack to avoid the blanketing effect from another yacht, but doing so it incurs in an unfavourable wind.

METHOD

Figure 3 shows the boundaries of the race area used in this work. The dimensions used are inspired by the typical length of a 35th America's Cup race area. The distance between the starting point and the upwind mark is of 5000 m, and the width of the area is 3000 m. The course is assumed to be aligned with an average initial wind direction which is kept constant for the entire race. The area as shown in the Figure is delimited by ideal laylines, but in some cases the actual routes goes beyond those lines. There is a limited tolerance (100 m) on the side boundaries for ease of grid computation.

The DP algorithm is based on a shortest path problem de-

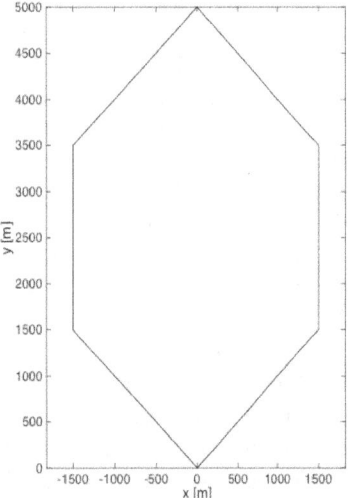

Figure 3: Racing area

fined on a set of nodes connected in a lattice. The set of nodes is not fixed, but their position depends on the wind forecast. Before formally describing the process of grid definition, an example to motivate this choice will be shown.

Let us consider the final phase of the upwind leg when the boat is reaching the mark in the case of a gradual wind shift towards the left. Figure 4 shows the optimal route towards the mark with two different underlying grids. In the left grid, the optimal route does not go through the nodes defined by the grid, therefore a certain approximation in the DP algorithm is needed. Conversely, the grid on the right shows an

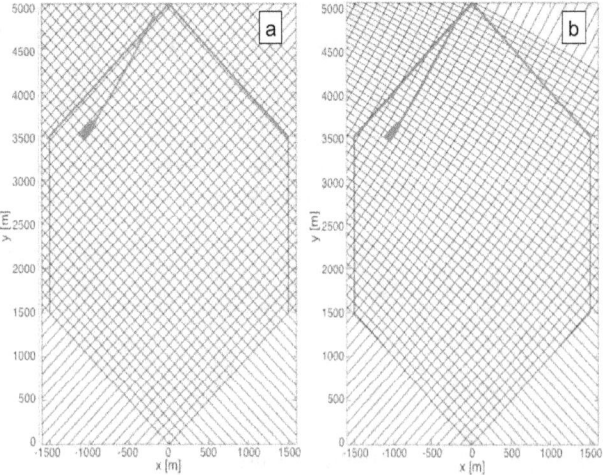

Figure 4: Comparison between grid with fixed spatial steps (a) and with wind-dependent steps (b).

exact superposition of the route and one of the lines consti-

tuting the grid. Ideally, the nodes defined by the grid should correspond to the reachable points on the racing area. Of course the racing area is a continuum, so every point within the race boundaries is always reachable, but the discretisation should be developed so that, if the yacht is in a given node belonging to the set of nodes defined by the grid, then the neighbour nodes should be reachable from that node. This property is not satisfied by the left grid in Fig. 4, but it is satisfied by the right grid, as shown by the red path followed by the yacht to reach the upwind mark. A curvilinear grid that matches the optimal route can be drawn if the future wind evolution is known.

The grid defining the lattice used in this work is therefore based on the wind forecast, in order to predict the possible reachable points. The grid is then recomputed every time step. The main assumption underlying the construction of the grid is that, in the absence of tactical interactions due to the presence of a competitor, a skipper will always sail at maximum VMG. Figure 5 shows how to build the subse-

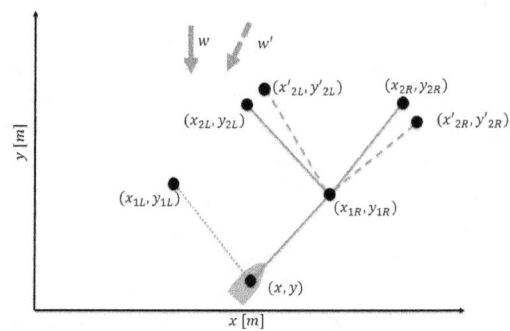

Figure 5: Construction of grid points.

quent grid points given an initial node of coordinates (x, y). If the forecast wind when reaching the point (x, y) is represented by the wind w, then the possible reachable points in a given time step dt have coordinates $(x_{1L}, y_{1L}), (x_{1R}, y_{1R})$ depending on the current tack. Let us assume that the boat is on a port tack. Then in a period of time of $2dt$ the points $(x_{2L}, y_{2L}), (x_{2R}, y_{2R})$ can be reached. w is the wind which is *expected* at the moment when the grid is generated. A subsequent forecast could predict a different wind (e.g. w' in Figure 5), in which case the reachable points become $(x'_{2L}, y'_{2L}), (x'_{2R}, y'_{2R})$. This is why the grid construction is updated at every step.

If every node generated two subsequent nodes, the size of the grid would grow exponentially at each iteration. Rather than building the grid point by point, the grid is therefore built by defining a set of lines and then considering their intersections as the nodes constituting the graph underlying the DP algorithm. Figure 6 shows the construction of the initial grid. A set of M_0 evenly spaced points is defined on the x axis, where M_0 depend on the desired grid resolution. At step one of the computation the following operations are performed:

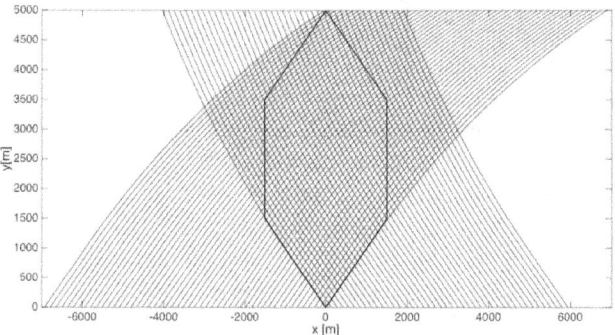

Figure 6: Example of grid construction.

1. Yacht position: $(x_0, y_0) = (0, 0)$

2. Generate wind speed and direction forecast

3. Compute grid lines

4. Compute lines intersections. These points constitute the DP nodes

5. Store grid points in matrices G_x, G_y

The distance between grid points depends on the chosen time step. The step used for the simulations for this work is $dt = 5$s. This is the time step used in the computation of the optimal strategy and does not necessarily correspond to the time step used for the wind forecast.

The grid is built starting from the current position of the boat. The objective of the boat is to round the upwind mark clockwise. The mark itself is not necessarily a point of the grid. However, by construction the mark will lay between four grid nodes defining a grid cell. The leftmost node is considered the arrival node at each iteration.

At each step k of the computation the grid is re-computed. The current position of the boat, (x_k, y_k) becomes the initial node. The equivalent of the initial points laying on the x axis are now a set of points evenly spaced (according to wind speed as for the first step) laying on the line of equation

$$y = \tan(TWA_k)(x - x_k) + y_k \tag{1}$$

where TWA_k indicates the TWA at time step k.

Dynamic programming algorithm

Let us consider a dynamic system evolving according to the following Equation 2:

$$x_{k+1} = f_k(x_k, u_k, \omega_k), \ k = 1, \ldots, N - 1 \tag{2}$$

where k represents a discrete step, x_k and x_{k+1} represent the state of the system at steps k and $k + 1$, respectively, u_k represents a decision, also called *control*, and ω_k is a random variable influencing the evolution of the system, characterised by a certain probability function p_k. The step index may refer to an increment over time or space, and the

increment doesn't need to have fixed amplitude. Usually the initial state x_0 is fixed. All the variables defined take values in some determined interval or space; in particular, for a given state of the system x_k, the set of admissible controls $\mathcal{U}_k(x_k)$ is defined as the set containing all the possible decisions that can be taken at that stage. For instance, in financial problems, $\mathcal{U}_k(x_k)$ may be the set of all the possible assets that it is possible to buy or sell. In sailing applications, x_k can represent the state of a yacht on the race area (in this case x_k can be the vector constituted by the yacht's coordinates and the observed wind, assuming values on a limited subset of $\mathbb{R}^n$), u_k the course followed by the skipper ($u_k \in \mathcal{U}_k \subseteq [0, 360)$), and ω_k the unknown wind evolution between step k and step $k + 1$. The position of the yacht at step $k + 1$ is then a function of those three variables.

A control, or a *policy*, is a finite sequence $U = u_0, \cdots, u_{N-1}$, where $u_k = u_k(x_k)$ is a function of the current state of the system, and all the $u_k \in \mathcal{U}_k(x_k)$ for all x_k. In the following, $\mathcal{U}$ will denote the set of the admissible policies.

The aim of DP is to find an admissible policy $U = u_0, \cdots, u_{N-1}$ that minimises a cost function which can assume the generic form as expressed in Equation 3:

$$C(U, \omega) = \sum_{k=0}^{N} c_k(x_k, u_k(x_k), \omega_k) \tag{3}$$

where $\omega = [\omega_0, \ldots \omega_N]$, subject to the system constraint specified in Equation 2. In sailing, this cost corresponds to time:

$$T(U, \omega) = \sum_{k=0}^{N} t_k(x_k, u_k(x_k), \omega_k) \tag{4}$$

where t_k represents the time needed to sail from state x_k to state x_{k+1}.

For this class of problems, the cost function is known at every stage. Unfortunately, in practical applications (including sailing) the cost function is only known in terms of a probability distribution, and rather than minimising a cost the aim is to minimise its expected value. In this case, the stochastic version of dynamic programming is used. Going back to the general description, a solution for the problem is then a policy U^{opt} such that

$$\mathbb{E}(C(U^{opt})) = \min_{U \in \mathcal{U}} \mathbb{E}(C(U, \omega)) \tag{5}$$

We assume that the minimum in Equation 5 is well defined. A discussion of this aspect can be found in Bertsekas (2007). According to the principle of optimality, an optimal solution has the property that, considering the subproblem starting at stage M, then the subpolicy ($U^{opt,M} = (u_M^{opt}, u_{M+1}^{opt}, \ldots, u_N^{opt})$) is optimal for that subproblem. The expected values in Equation 5 are computed by using a Markov model for the distribution of wind speed and direction. The Markov model is derived from wind data as detailed in Tagliaferri et al. (2014). This model is also the basis for the risk model. In fact, by using coherent risk measures, the transition matrix for the Markov process is multiplied by a transformation matrix which has the function of shifting the probabilities of favourable/unfavourable events. A complete description of this procedure can again be found in Tagliaferri et al. (2014). The optimal transformation is found among a set of matrices heuristically selected according to the following principles.

A boat skipper who is losing will seek risk. If she adopts a minimum expected finish time strategy against another skipper who minimises his expected time to finish, then she will tend to make the same decisions (unless the boats see very different winds) and lose the race almost certainly. She will instead seek different wind conditions from the competitor, being optimistic about the possible advantageous wind shifts and assigning a higher probability to these outcomes (i.e. lifting shifts). Being optimistic about random outcomes increases risk, as well as incurring some loss in expected performance.

A sailor who is losing will seek risk. This corresponds to increasing her confidence of a lifting wind shift while discounting the likelihood of a heading wind shift. The transition matrices used to represent the two attitudes are shown in Figure 7. Advantageous shifts (cells below the diagonal when the skipper is to the left of the opposition, and cells above when on the right) happen with higher probability than in the risk-neutral case. The remaining probabilities in each row are reduced to add to one. Following the notation in the aforeentioned paper, the transition matrices are represented by using a gray scale, where darker colours represent higher probabilities.

A set of rules aimed at avoiding collisions between the

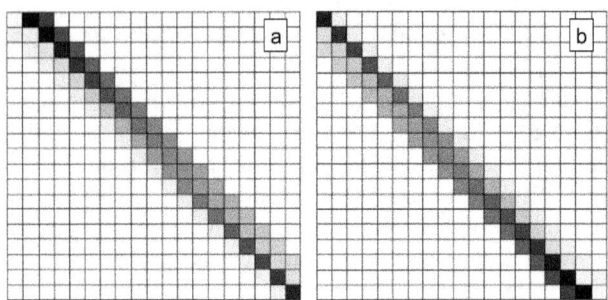

Figure 7: Modified transition matrices for a risk-seeking skipper. Advantageous wind shifts occur with higher probability than disadvantageous ones. (a) Yacht on the left-hand side of competitor and (b) yacht on the right-hand side of competitor.

boats and at respecting the racing rules are implemented. In particular:

1. If the two boats meet, then the boat on a port tack increases the TWA, passing behind the other boat.

2. A boat cannot tack if this leads to its track crossing the opponent's under a certain fixed safety distance.

The safety distance is defined noting that the boats are modelled as points. The longitudinal safety distance is 10 m, the side distance is 5m.

The computations for the manoeuvre of bearing away and passing behind the opponent's boat is carried out by adding a node to the set of reachable nodes. This temporarily modifies the assumption that a boat always sails at maximum VMG. In the example shown in Figure 8, the red yacht expects to meet the opponent at the node indicated by the red dot. The black node is therefore added to the set of the reachable points. The model for physical interactions

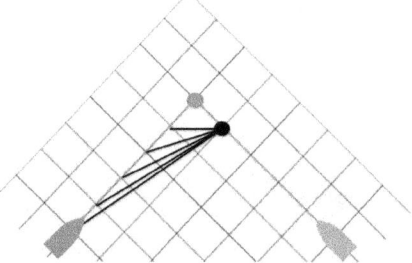

Figure 8: Example of grid modification when two boats meet.

between the two yachts competing in a match race is based on the experimental results of wind tunnel tests presented in Aubin (2013). These results are based on a series of wind tunnel experiments carried out in the Twisted Flow Wind Tunnel facility of Auckland University. The complete set of experiments is described in Aubin (2013).

The opponent is assumed to follow a strategy aimed at minimising the expected time to complete the race. This means that he is expected to follow a reasonable route that depends on the forecast wind. To compute an optimal strategy, it may be possible to forecast the future position of the opponent, to take into account that the two boats might meet further in the future. However, in order to properly take into account such events, the computation of a probability distribution is required. In fact, let's assume that with the wind conditions forecast at the beginning of the race the two boats can compute an optimal strategy which will lead them to meet in proximity of the upwind mark. This event will actually happen with a probability which is equal to the probability that the wind realisation is exactly the one forecast at the beginning and that the boats actually follow the computed strategy. The further this event is in the future, the closer this probability is to zero. In the current software implementation, the future window is set at one minute, which is the time frame at which the wind forecast has an average error lower than 2° (Tagliaferri et al., 2015) and because a yacht is expected to perform not more than one tack in one minute.

An important hypothesis, not necessarily corresponding to reality, is that no yacht will take a decision that leads to a higher expected time with the aim of slowing the other yacht down.

RESULTS

Routing examples from America's Cup data

The algorithm is tested using recorded wind scenarios from the past edition of the America's Cup held in San Francisco. A strategy based on the ANN forecast is compared with a strategy which assumes perfect knowledge of the wind behaviour.

The results of the simulated races are summarised in Table 1. Only 12 upwind legs are simulated using the initial minutes of the last 12 races, as the data relative to the other races was used for training. The average difference be-

Table 1: Simulated races with San Francisco wind dataset.

Race [deg]	1	2	3	4	5	6	
Time (perfect wind knowledge) [s]	603	743	618	743	642	818	
Time (ANN forecast) [s]	608	747	634	781	648	821	
Difference [%]	0.8	0.5	2.5	4.8	0.9	0.4	
Race [deg]	7	8	9	10	11	12	13
Time (perfect wind knowledge) [s]	597	661	654	712	748	684	697
Time (ANN forecast) [s]	608	672	659	715	761	693	712
Difference [%]	1.8	1.6	0.7	0.4	0.3	1.2	2.1

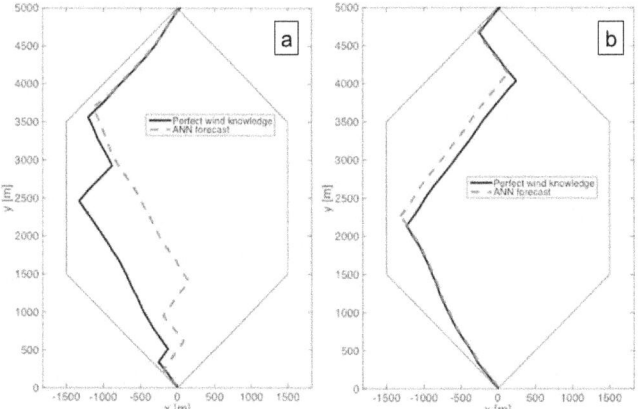

Figure 9: Routes computed using forecast and assuming perfect wind knowledge for Race 2 (a) and Race 4 (b).

tween the different times to completion between a boat with a perfect knowledge of the wind and a boat which uses the ANN forecast is 10.7s. A representative example is given in Figure 9(a), corresponding to Race 2. In this example, the difference between the two strategies is limited to a slight delay of the second tack when the ANN forecast is used.

One of the worst cases is shown in Figure 9(b), corresponding to Race 4. The black trajectory is the one computed by the algorithm having perfect knowledge of the future wind, while the red dashed one is computed by the algorithm using the ANN forecast. The ANN-based algorithm leads to an extra tack at the beginning of the race, due to a wrong forecast for the end of the race. However, the error is soon recovered and in the final part of the race the two strategies become almost indistinguishable.

Optimal risk model

The optimum risk management is investigated considering two boats racing each other. At every step of the simulated race, if A is more than 15 s behind B, she uses the risk-seeking, optimistic matrix for the relevant side of the course. If B is more than 15 s behind A, she uses the risk-averse, pessimistic matrix. For this case, $T_{switch} = 15$ s. The time difference and the matrix transformations are arbitrarily fixed, and the results obtained confirm the results presented in Tagliaferri et al. (2014). Figure 10(a) shows differences between the arrival times of boats A and B. When this time difference is positive, it means that A wins the race. Conversely, if the time difference is negative, B wins the race. This set of results confirms that a risk seeking attitude can constitute an advantage for a skipper who is losing the race. However this advantage can be optimised by changing the amount of risk, i.e. how much the new Markov matrices differ from the original one, and the time at which the attitude is changed. i.e. the time difference between the two boats that triggers the attitude switch.

The amount of risk is investigated by comparing strategies obtained using matrices that have been multiplied for the transformation matrix multiple times. The best outcome, is obtained with the use of the matrix shown in Figure 10(c), and by setting $T_{switch} = 10s$. The optimised risk model leads to the distribution in Figure 10(b), which corresponds to a win for boat A in 74%of the cases, and is obtained by post-multiplying the risk-neutral matrix for the square of the original transformation. This optimisation was carried out over a limited set of possibilities, and it must be the subject of further research.

Example of upwind leg

In this Section a complete race between two boats that follow the optimum course and have an optimum management of risk is presented. The wind which was measured during the last race of the America's Cup is used. Figure 11 shows the wind direction and how this is forecast by the two as shown in Tagliaferri et al. (2015). The red bars highlight the critical points where the forecast error is higher than 3° for two consecutive minutes. Differently from the original America's Cup race where the wind was recorded, in this example the race starts at minute 18, and is made of only one upwind leg. The wind shows a significant shift towards

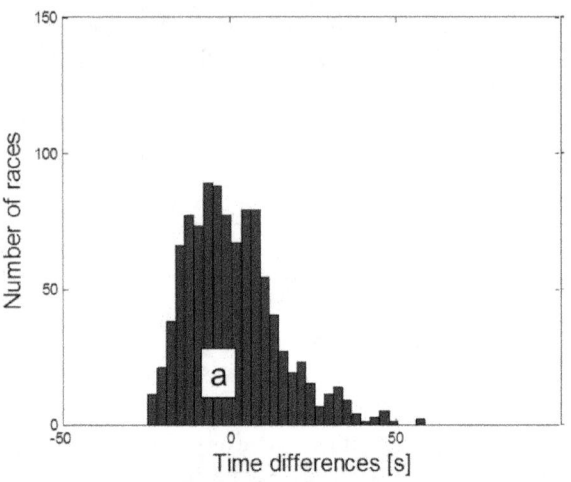

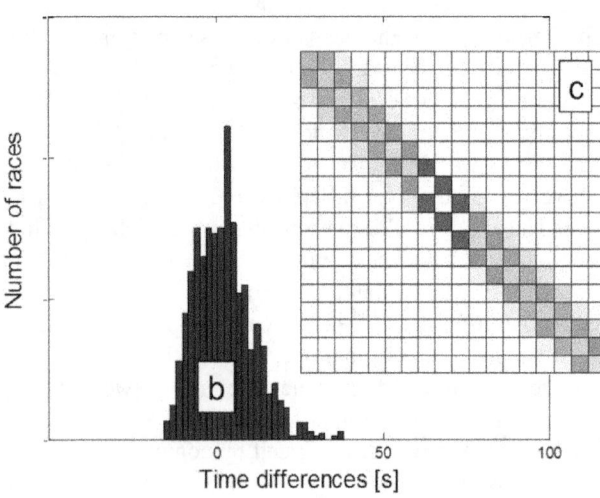

Figure 10: Histograms of time differences for risk-neutral strategy vs optimistic-pessimistic combination (a) and optimal optimistic-pessimistic combination (b) based on an optimal processing matrix (c)

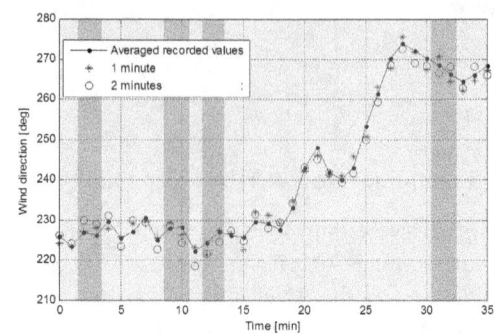

Figure 11: Wind forecast example (Tagliaferri et al., 2015).

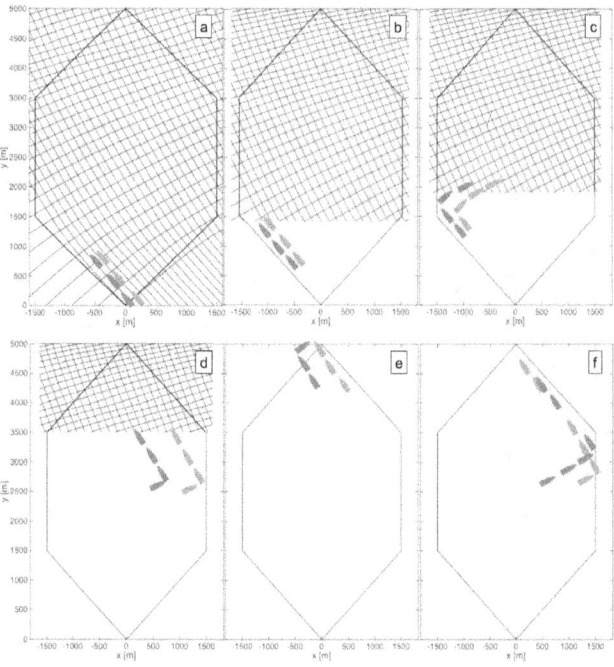

Figure 12: Simulation results for an upwind leg.

the right during the race, while the wind speed variations are negligible. Figure 12(a) shows the grid corresponding to the wind realization. The grid shown is coarser than the one computed by the algorithm for clarity.

Both boats A (blue in Figure 12) and B (red in Figure 12) start on a starboard tack (sailing towards the left) and boat B is on the left of boat A, at a distance of 25 m (distances between the two boats are magnified in the figures for clarity). Figure 12(a) shows the beginning of the race and the grid represents how the wind was forecast at that time. Both boats begin the race sailing on a starboard tack. In fact, as a significant increasing shift towards the right is forecast, the best strategy consists in approaching the mark on a starboard tack. Boat A chooses to sail towards the left

of the race area up to where, with only one tack, she can reach the right-hand-side layline. This would be the optimal choice for B as well in the absence of A. Unfortunately the more the wind shifts towards the right, the more B finds herself in her area of unfavourable aerodynamic influence (Figure 12(b)). B cannot tack until A tacks because the two boats are too close.

When eventually A tacks (Figure 12(c)), B is free to tack as well, but she chooses to wait in order to perform the tack outside of the area where she would still be slowed down because of the presence of A.

In Figure 12(d) both boats are initially sailing on a port tack, and A is leading. A reaches the layline and tacks to sail towards the mark. B adopts a risk-seeking behaviour, and instead of waiting to reach the layline as well she tacks hoping in a favourable wind shift.

Figure 12(e) shows the end of the race. Although B has

207

managed to avoid A to gain more advantage, she is still slightly behind.

Figure 12(f) shows an alternative realization for this example, where the strategy is computed without taking into account the presence of the opponent. In this case, boat B postpones the second tack until she reaches the racing area right boundary, but this then results again in finding herself in an area of negative influence.

CONCLUSIONS

This paper presents a novel methodology for computing an optimal strategy for a sailing match race. This methodology is based on dynamic programming in combination with a forecasting module based on artificial intelligence, which performs a very-short-term forecast of wind speed and direction, and with a risk model that allows a sailor to tune his risk attitude depending on his position with respect to the competitor.

The aim of the method is to compute a strategy that improves the probability of winning the race with respect to strategies aimed at minimising the expected time to complete the race. As an example, for an upwind leg of the 34th America's Cup, the completion time of a boat which uses the proposed forecast is only 10.7 s longer than a boat which has perfect knowledge of the wind. The risk model is based on coherent risk measures in order to investigate whether a change in risk attitude can improve the probabilities of winning a race. An optimistic attitude is associated to a losing skipper, and a pessimistic, conservative one to a winning skipper. The risk model is optimised using three different parameters relative to the distance between boats and the anticipated future wind changes. The results suggest that there is a threshold defining the moment when it is advisable to seek more risk and that not always the risk-seeking and risk-averse behaviours correspond to optimistic and pessimistic anticipations on the wind.

The proposed method is implemented in a computer program with negligible run time and thus capable to compute the optimum route in real-time during a race.

REFERENCES

Nicolas Aubin. Wind tunnel modelling of aerodynamic interference in a fleet race. Master's thesis, Ecole Centrale de Nantes and Universite de Nantes, 2013.

J Banks, A Webb, T Spenkuch, and SR Turnock. Measurement of dynamic forces experienced by an asymmetric yacht during a gybe, for use within sail simulation software. In *Proc. 8th Conf. Int. Sport. Eng. Assoc.*, pages 0–5, Vienna, Austria, 2010. doi: 10.1016/j.proeng.2010.04.024.

Dimitri P Bertsekas. *Dynamic Programming and Optimal Control*, volume I. Athena Scientific Belmont, MA, 2007. ISBN 1886529264. doi: 10.1057/jors.1996.103.

D S S Ferguson and Pantelis Elinas. A markov decision process model for strategic decision making in sailboat racing. *Adv. Artif. Intell.*, 6657 LNAI:110–121, 2011.

J A Gretzky and J K Marshall. The partnership for America's cup technology: An overview. In *Proc. 11th Chesap. Sail. Yacht Symp.*, Annapolis, MD, 1993.

J E Kerwin and J N Newman. A summary of the H. Irving Pratt ocean race handicapping project. In *Proc. 4th Chesap. Sail. Yacht Symp.*, Annapolis, MD, 1979.

John S Letcher, John K Marshall, James C Oliver III, and Nils Salvesen. STARS & STRIPES. *Sci. Am.*, 257(2): 34–40, 1987.

Y Masuyama. Tacking simulation of sailing yachtsnumerical integration of equations of motion and application of neural network technique. In *Proc. 12th Chesap. Sail. Yacht Symp.*, pages 117–132, Annapolis, MD, 1995.

P Van Oossanen and PN Joubert. The development of the winged keel for twelve-metre yachts. *J. Fluid Mech.*, 173:55, 1986. ISSN 1469-7645. doi: 10.1017/S0022112086001076.

Andrew B Philpott, Shane G Henderson, and D Teirney. A simulation model for predicting yacht match race outcomes. *Operations Research*, 52(1):1–16, 2004.

Andy Philpott and Andrew Mason. Optimising yacht routes under uncertainty. In *Proc. 15th Chesap. Sail. Yacht Symp.*, Annapolis, MD, 2001.

M Scarponi, P Conti, R A Shenoi, and S R Turnock. A combined ship science-behavioural science approach to create a winning yacht-sailor combinationl. In *Proc. the18th Chesap. Sail. Yacht Symp.*, pages 143–156, Annapolis, MD, 2007a.

M Scarponi, P Conti, and RA Shenoi. Including human performance in the dynamic model of a sailing yacht: a matlab-simulink based tool. In *Proc. Mod. Yacht Conf.*, pages 143–156, Southampton, UK, 2007b.

T Spenkuch, S Turnock, M Scarponi, and A Shenoi. Lifting line method for modelling covering and blanketing effects for yacht fleet race simulation. In *Proc. 3rd High Perform. Yacht Des. Conf. 2008*, pages 111–120, Auckland, New Zealand, 2008.

Thomas Spenkuch, Stephen Turnock, Matteo Scarponi, and Ajit Shenoi. Real time simulation of tacking yachts: How best to counter the advantage of an upwind yacht. In *Procedia Eng.*, volume 2, pages 3305–3310, Vienna, Austria, 2010. doi: 10.1016/j.proeng.2010.04.149.

Thomas Spenkuch, Stephen Turnock, Matteo Scarponi, and R Shenoi. Modelling multiple yacht sailing interactions between upwind sailing yachts. *J. Mar. Sci. Technol.*, 16(2):115–128, 2011. ISSN 0948-4280. doi: 10.1007/s00773-010-0115-9.

F Tagliaferri, A B Philpott, I M Viola, and R G J Flay. On risk attitude and optimal yacht racing tactics. *Ocean Eng.*, 90:149–154, 2014. ISSN 0029-8018. doi: 10.1016/j.oceaneng.2014.07.020.

F Tagliaferri, I.M. Viola, and R.G.J. Flay. Wind direction forecasting with artificial neural networks and support vector machines. *Ocean Eng.*, 97:65–73, 2015. ISSN 00298018. doi: 10.1016/j.oceaneng.2014.12.026.

Teamwork as Joint Activity in Sailing

Fredrik Forsman, Chalmers University of Technology & Swedish Sea Rescue Society, Göteborg, Sweden

Christian Finnsgård, Chalmers University of Technology & SSPA Sweden AB, Göteborg, Sweden

ABSTRACT

Sailing is a sport and activity that takes a long time both to learn and to master, as much of its competence-based knowledge is acquired through experience. Experience-based learning is very important, time-intensive, and the factors for success are often tacit and hidden. Should these success factors become explicit and salient, learning would occur faster and produce obvious competitive advantages.

This research was conducted by embedding on-going research results into two competitive sailing teams racing in different classes, one offshore keelboat racing with a crew of eight, and a one-design Star-class racing yacht with a crew of two. The data collection consisted of observations, interviews, and video recordings. The results were also verified with the crews to catch biases in the analysis process. A jibe, a specific but common maneuver was analyzed from the perspective of Common Ground within Joint Activity.

Maneuvering a competitive offshore sail racer or a previously Olympic Star-class yacht are tasks that fulfill the requirements for Joint Activity. A high level of Common Ground is required for the effective coordination needed in order to perform at a high level and maintain the safety of the crew and equipment.

Breakdowns in the coordination of maneuvers were observed, although they must be recorded on video for higher analysis reliability. To achieve greater validity, more and different maneuvers should be considered within the analysis.

By better understanding the factors for success, sail racing teams can more quickly gain competence and thus competitive advantages.

The research analyzes the teamwork found in sailing from the perspective of Joint Activity and Common Ground and provides insight into how to achieve performance improvements more efficiently.

NOTATION

CSYS Chesapeake Sailing Yacht Symposium

IMX40 An offshore one-design sail racing yacht manufactured by X-yachts of Denmark, with a total of 99 yachts. The yacht's dimensions are 12.10 meter LOA with 6830 kg displacement. It carries a mainsail of 48.10 m^2 and a symmetrical spinnaker of 121.5 m^2 an a spinnaker pole. For an in-depth description, refer to www.x-yachts.com.

Star An international, and previous Olympic yacht 1932-2012. Designed in 1911, and considered the pinnacle of one design keelboat racing. The yacht's dimensions are 6.92 meter LOA with 671 kg displacement. It carries a mainsail of 24.1 m^2 and a headsail of 4.6 m^2. Downwind, the Star uses a whisker pole to fly the jib downwind. To this date, a total of 8,506 yachts have been manufactured. For an in-depth description refer to either: www.starclass.org, www.sailing.org/classesandequipment/STR.php, or Reynolds and Szabo (2009).

INTRODUCTION

Sail racing is a team activity where many functions must be aligned and synchronized in order to handle the boat efficiently. Boats in offshore sail racing are often bigger then those in inshore racing, and thus these boats have to carry more crew to handle not just heavier gear but more gear. In the Olympic classes, on the other hand, the boats are smaller and have smaller crews. With larger vessels, more options often are available in terms of choice of sails and trimming. Navigation equipment and decision support systems are also present. This paper addresses the problem of teamwork onboard an offshore a sail racer with an 8-person crew and a one-design 2-person keelboat. Success and failure do not stem from individual competence but are the result of team effort. Therefore, it is essential to understand how well a crew operates as a team rather than a group of individuals in order to be competitive. The purpose of this paper is to study a simple and common maneuver through the lens of Joint Activity and Common Ground.

THEORY

Joint Activity

A joint activity is an extended set of behaviors carried out by an ensemble of people who are coordinating with each other (Clark, 1996 p. 3). For activity to be regarded as "joint," the parties have to *intend* to work together and their work has to be *interdependent* (Klein, Feltowich & Woods, 2004). A joint activity doesn't entail questions of either "I do it" or "you do it." Joint activity happens when something genuine is created together that can't be replicated by any single stakeholder.

People engage in joint activity for many reasons. The most salient is the need to accomplish something that is difficult or impossible for any single person to do. Joint activity may occur because multiple competencies and skills are necessary for the achievement. It may also occur because the demands and constraints are so difficult that the task can't be solved without joint activity. Teams who are working jointly often solve highly complicated tasks. The purpose of joint activity is that it enables an ensemble to work in a coordinated and interdependent manner. Joint activity is described from three different aspects: requirements, choreography and criteria (Figure 1).

Interdependence

In a joint activity, party "A's" actions must depend in some significant way on party "B's" and vice versa (Clark, 1996, p. 18). The efforts in joint activity must have mutual influence. If the activities are sequential instead, then joint activity is likely not occurring. When work is done interdependently, any friction within one part of the system will propagate throughout. At a first glance this can be seen as a negative factor, but actually it is not. People working in concert with technology depend on each other to complete a task. Team members simultaneously create the conditions for the next step in the work process. Any frictions and problems within the joint system will be noticed at an early stage and thus can be managed before they create negative consequences.

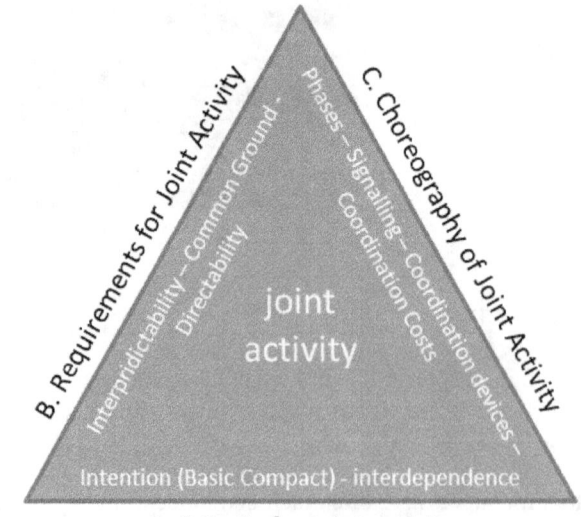

Figure 1 - Joint Activity. Adopted from Klein et al., (2004)

Intention/Basic Compact

Joint activity requires a "basic compact," an agreement between the parties undertaking the joint activity to participate and carry out the required coordinated responsibilities. An aspect of the basic compact is that it often means that the joint activity participants have to relax their personal goals and prioritize the system's global and long-term goals. This implies that the shared goal(s) must be of sufficient importance for the participants to do this. The basic compact also entails a commitment to try to detect and correct any loss of Common Ground that could disrupt the joint activity (Klein et al., 2004). The locus of responsibility in the basic compact is distributed among the team members. This is an essential part in any organization that aims for safety (Rochlin, 1999). The basic compact is not a definite; it is not agreed upon and then set in stone. It needs to be reinforced and confirmed continually, and communication is essential to the process.

Common Ground

Perhaps the most important basis for inter-predictability is Common Ground (Clark & Brennan, 1991), which comprises the pertinent knowledge, mutual beliefs, and mutual assumptions that support interdependent actions in a joint activity. When people share Common Ground, they can use an abbreviated language that carries much information in context that an outside observer would not understand. It is highly advantageous when signals and language can be loaded with dense meaning. Without Common Ground, communication must be very specific and comprehensive and this carries great operation and efficiency costs.

The basic compact also includes the expectation that the parties will repair faulty knowledge, beliefs, and assumptions when detected. Common Ground is what makes joint activity and coordination function. In turn, every action taken will change the Common Ground. Common Ground can be divided into three basic categories (Clark, 1996):

- initial Common Ground
- public events so far
- the current state of the activity

Initial Common Ground refers to the pertinent knowledge and prior knowledge brought to the system. It not only comprises the participants' knowledge of the world but also conventions and norms.

Public events so far describes the understanding of events, actions, and happenings up to the present time. Some such actions exert influence on future possibilities, and a previous action might generate a new option or close a previous one.

The current state of the activity provides cues that help predict subsequent actions and appropriate forms of coordination. In many cases, the physical scene provides sufficient pertinent cues to remove any ambiguity in understanding the next action; the situation is salient. Problems can occur in such a situation when an apparently salient situation fools the participants into framing the situation falsely. This can happen when team members look at the same thing and interpret it differently.

Inter-predictability

To be able to work jointly, a level of inter-predictability must exist among participants. In order to adapt or to give directions, each participant must be able to predict and expect how the other participants will behave and react under certain circumstances. Such a skill is also important for maintaining common ground; it predicts when the common ground is at risk of being corrupted. Shared scripts assist in inter-predictability because they help participants expect how the other participants might behave (Klein et al., 2004).

Directability

Directability is an important aspect of coordination that builds a team's resilience (Christoffersen and Woods, 2002). Interdependence is essential to joint activity. This interdependence also requires participants to adapt to changing situations both internal and external. Within the team, one participant's behavior might signal that the work needs to be directed or that a participant needs direction or must re-direct (adapt) his work and not simply try to direct others.

Signaling

Work as it is conducted is in itself signaling, which is closely linked to predictability. When work is coordinated and performed in concert, participants have the opportunity to respond to variations in work patterns. In the cockpit, this is seen when the crew's communication pattern shifts to contain less information or even becomes silent, often a strong key to act upon. Not all coordination is driven by orders and commands. The crew, as a result of training and experience, recognizes the communication pattern as a signal of threatened Common Ground. An appropriate response to such a situation is sizing up and assessing the situation to repair Common Ground.

METHODOLOGY

The authors have studied the jibes of an IMX40 and a Star by being embedded as an active part in the crew of a vessel of each kind. The authors were given the opportunity to try many but not all of the crew roles over time, which provided a more elaborate understanding of these roles, their tasks, and their constraints then observation alone would have enabled. The maneuvers were also video recorded and post-analyzed to derive who did what in relation to the vessel and other crewmembers. Observations were made and notations taken. Other crewmembers were asked to make notations on their tasks. Three of them accommodated the author's request in the case of the IMX40, and for the Star, two different skippers provided their notations in a corresponding manner. The notations combined with the authors' experiences performing in the different roles have provided guidance interpreting the video recordings. The tasks each crewmember performs in a jibe were mapped in a task analysis, which was later cross-checked with the crew to minimize the effects of biases introduced by the authors and to trap any misconceptions and errors.

Delimitations

The unit of analysis is an IMX40 and a Star and their respective crews during a simple and common maneuver. Other vessels, tactics, and the racing situation as such are not taken into consideration, nor are deviations in the maneuver due to different environmental conditions (i.e.

wind and waves).

RESULTS

Roles

The IMX40's crew consists of eight persons. All are assigned specific roles (Figure 2). Each role has a set of tasks to perform in concert with other crewmembers and the vessel. The Star's crew consists of two persons, each of whom is assigned specific roles (Figure 3). Each role has a set of tasks to perform in concert with the other crewmember and the vessel.

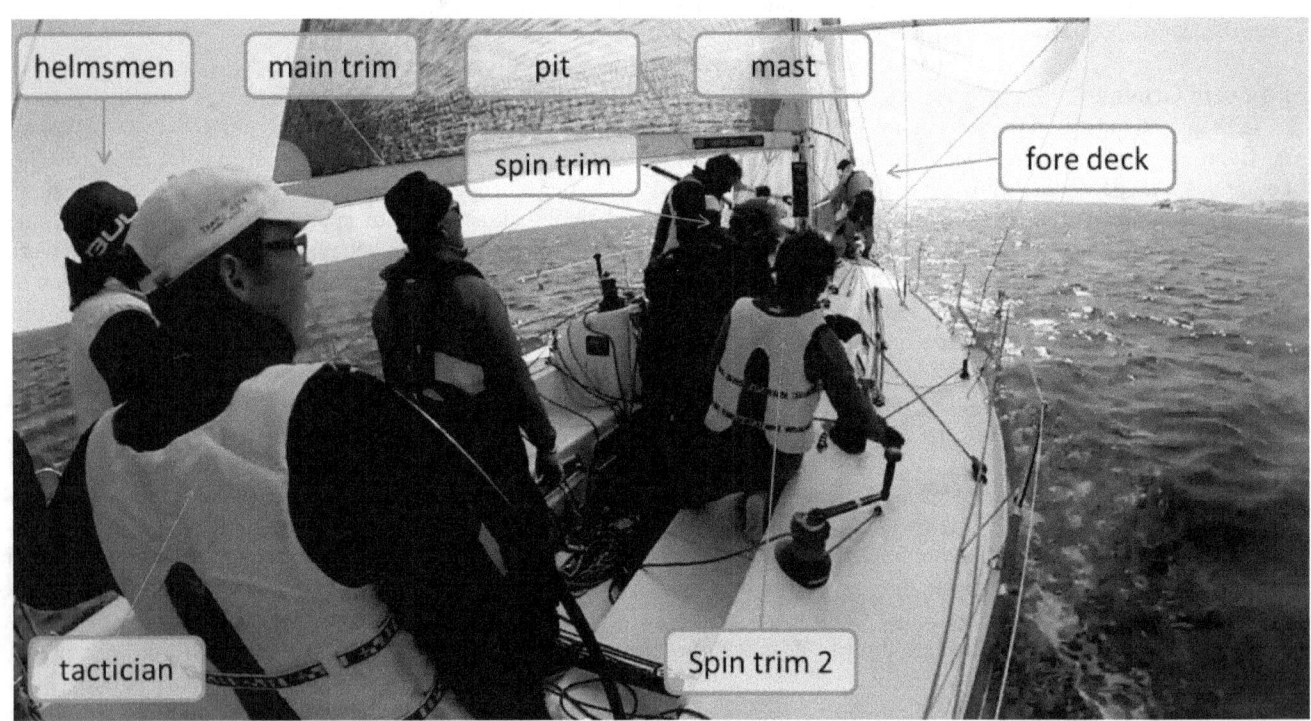

Figure 2 - Roles in an IMX40's crew, one of the authors documenting the crew

Figure 3 - Roles in a Star's crew, the skipper aft and the crew in front

Tasks in a jibe

Jibes on the two described sailing yachts are described in sequence below and thereafter further described.

Sequence of a jibe on an IMX40

The series of tasks on the described IMX40 is described below. Due to the eight different crewmembers, and the tasks' need to be executed simultaneously, describing the tasks in an absolute sequence is not truthful to how work is executed. A truer representation is found in figure 6 and 7.

1. Give instruction to jibe.
2. Communicate "prepare to jibe."
3. First set of tasks by crew to prepare jibe:
    a. Grab new guy
    b. Move in position under boom
    c. Check/untangle uphaul
    d. Load new spin sheet winch

    e.   Feed slack on new guy
    f.   Untangle mainsheet
    g.   Steer steady course

4.   Second set of tasks by crew to prepare jibe:
    a.   Pull slack on new guy
    b.   Grab trip line
    c.   Grab uphaul
    d.   Tension new spin sheet
    e.   Put/check new guy on lee winch
    f.   Grab mainsheet

5.   Third set of tasks by crew to prepare jibe:
    a.   Bring new guy to pulpit
    b.   Open uphaul spin lock half
    c.   Remove winch handle from old spin trim winch
    d.   Loosen windward guy from cleat

6.   Fourth set of tasks by crew to prepare jibe:
    a.   Sit on pulpit
    b.   Loosen downhaul
    c.   Place winch handle in new spin trim winch
    d.   Remove turns and leave 3 turns on windward guy winch

7.   Fifth set of tasks by crew to prepare jibe:
    a.   Give visual OK sign to helmsman
    b.   Grab new and old spin sheet in each hand
    c.   Winch back spinnaker boom

8.   Tasks during turn through wind:
    a.   Hail "Falling"
    b.   Hail "Trip"
    c.   Turn downwind

9.   First set of tasks during jibe:
    a.   Pull the tripline to release the boom
    b.   Release the uphaul spinlock
    c.   Remove winch handle from windward guy winch
    d.   Turn downwind more slowly

10. Second set of tasks during jibe:
    a.   Pull the trip line to help the boom swing over
    b.   Lower the spin boom
    c.   Remove all turns from old guy winch
    d.   Start to pull mainsail over

11. Third set of tasks during jibe:
    a.   Catch the spinnaker boom in the swing
    b.   Duck for main boom
    c.   Move to opposite side of the boat

12. Fourth set of tasks during jibe:
    a.   Clip the new guy in the spinnaker boom
    b.   Put winch handle in new guy winch

13. Fifth set of tasks during jibe:
    a.   Push the spinnaker boom to windward
    b.   Bring home slack on the new guy

14. Sixth set of tasks during jibe:

    a.   Hail "clear"
    b.   Keep eye contact with foredecker

15. Seventh set of tasks during jibe:
    a.   Pull uphaul to get spinnaker boom in position
    b.   Feed out old sheet
    c.   Pull home slack on new guy

16. Eight set of tasks during jibe:
    a.   Tighten new guy on winch for the new guy in the spinnaker boom
    b.   Turn upwind

17. Ninth set of tasks during jibe:
    a.   Lock uphaul spinlock
    b.   Release old sheet

18. Tenth set of tasks during jibe:
    a.   Move back to normal position
    b.   Lock down-haul
    c.   Trim spinnaker
    d.   Winch spinnaker boom back into position
    e.   Finish the turn to the new course

Sequence of a jibe on a Star

The series of the tasks on the Star is described below in sequence. However, a few of the tasks are performed simultaneously, and hence the description is not satisfactory.

1.   Hails "We need to jibe" (Skipper or Crew)
2.   Communicates "prepare to jibe" (Skipper)
3.   Moves aft slowly in sync with waves from standing position forward of the mast (Crew)
4.   Moves down into cock-pit (Crew)
5.   Leans forward and puts hand on whiskerpole (Crew)
6.   Hails "Ready" (Crew)
7.   Hails "Jibing" (Skipper)
8.   Releases windward jibsheet (Skipper)
9.   Removes whiskerpole from mast bracket (Crew)
10. Heels boat to windward (Skipper and crew)
    a.   Boat starts to turn
11. Adjusts (pulls) mainsheet (Skipper)
12. Pulls whiskerpole aft of headstay (Crew)
13. Pushes whiskerpole forward on leeward side (Crew)
14. Sets whiskerpole on mast bracket (Crew)
15. Rough setting (tensioning) of jibsheet on leeward side (Crew)
16. Continuous pulling on mainsheet (Skipper)
17. Turns body aft (180°), (Crew)
18. Checks that boom vang is off (Skipper)
19. Pulls leeward backstay slack to tension (Crew)
20. Last pull on mainsheet (Skipper)
    a.   Boom inside leeward backstay
21. Sets leeward backstay (new windward backstay), (Crew)

a. Boom passes centerline
22. Releases mainsheet (Skipper). Failure to comply will lead to wipeout (depending on wind strength).
23. Releases new leeward backstay (Crew). Failure to comply will lead to wipeout or mast failure (depending on wind strength); refer to Figure 5.
   a. Boom passes leeward backstay fitting (outside side of cockpit)
24. Setts windward jibsheet (Crew)
25. Adjustment of mainsheet (Skipper)
26. Moves forward, in sync with waves (Crew)
27. Stands on deck in front of mast (Crew)
28. Hikes to windward and bears off into new course or on front side of wave (Skipper and crew)

Jibing as Joint Activity

Communication during the jibe

Communication during the analyzed jibes is conducted in different ways. There are several differences between the two boats, easily attributed to crew size, as only two persons have to execute all tasks on the Star. On the IMX 40, verbal communication transpires between the tactician and helmsmen where they come to the decision to jibe. The helmsmen communicate this decision verbally to the rest of the crew by shouting "prepare jibe." This is often unnecessary as the rest of the crew has interpreted the general situation as one where a jibe is probably the next plausible maneuver. The Star has a similar situation, though since the crew is usually standing and has an overview of wind and wave patterns, as well as the competitors, the crew's position is more advantageous compared to the lower situated skipper. Hence the first call for a jibe can easily come from the crew. A depiction of this is provided in Figure 4.

When the crew is preparing for the next maneuver without any specific order, the context requires that cues strong enough for the crew to act upon be provided. This entails both signaling and a coordination device. Signaling can also be seen more explicitly in the crew's work when the foredeck gives the helmsmen the thumbs up to signal that the bow is ready to jibe, while very little verbal communication is needed on the Star if all proceeds according to plan. The crews' different amounts of training together can create a difference in coordination, which can also exert influence. The manner in which events are unfolding or have unfolded up to the present point provides a context that can be seen as a coordination device in itself, as this context guides the understanding of what the next maneuver probably will be.

Coordination of the jibe

Coordination is achieved in different ways. Communication, both verbal or through signals (both tacit and explicit), is used. The presence of a set of phases also helps choreograph or coordinate the maneuver and complements previous communication and signaling. The jibe consists of four phases: decision, preparation, turn through wind, and jibe (Figures 6 and 7). Each phase serves to alter the status of the vessel and to prepare for the next phase in order to arrive at the intended end state, to sail downwind with the wind coming in from the other side of the vessel. Each phase has a desired end state that must be achieved in order to enable continuation to the next phase. This achievement is communicated both explicitly and implicitly. In the case of the IMX40, the foredeck shouts "clear," with explicit communication and signals such as giving the thumbs up. It is also signaled more tacitly or implicitly by the bowman's position on the foredeck and by his activity. The state of the sails is an example of a cue. Changes are also suggested by the coordination devices described in the previous section.

Figure 4 - A Star's crew, standing and having an overview aft

Signaling and coordination devices during the jibe

Figure 5 – Possible outcomes of losing coordination during a jibe on a Star class yacht

Interdependence and Basic Compact in the jibe

Figures 6 and 7 show that the crewmembers' activities in concert with the vessel are highly dependent on each other. Some of the crewmembers' tasks must be conducted sequentially without considering other team members' activities. These comprise only a short set of tasks that eventually need to be coordinated with other crewmembers' activities and the vessel in its present state, and thus, the work is performed interdependently. If the basic compact is compromised, the motivation to execute the task will be impeded, and as the work is highly interdependent, it will cause a disruption that will propagate and cause problems. The basic compact of the crew must be strong and intact for the maneuver's proper execution.

Directability and inter-predictability during the jibe

Figures 6 and 7 show how the jibes' execution is dependent on the crewmembers' ability to receive direction. A certain set of tasks needs to be executed in order to prepare to turn down wind. The helmsmen provide direction by issuing a simple and very abbreviated order when giving the instruction "prepare to jibe." This also requires that the crew is able to predict what effect any action or order will have.

Common Ground

In order to be able to use the context as a coordination device, a significant amount of Common Ground must be established. Common Ground is also required for inter-predictability, directability, and interdependence. As the crew reacts in a certain way and starts to execute all actions necessary to complete the jibe, a certain level of Common Ground must exist for this to take place. Without pertinent knowledge of the domain—an understanding of the events, actions, and happenings up to present and the system's current system state, joint activity will not be conducted.

Team-Task-Role Model

The thin lines connecting the task-boxes in Figure 6 and 7 represents tasks that need to be coordinated by communication or signaling which is essential to choreograph the jibe. Each of the four phases of the jibe (decision, prepare, turn and jibe) needs to be completed before next phase can be initiated. The tasks that each crewmember has to perform in each phase are both sequential and interdependent. Some tasks can be made in sequence without any coordination but al task series end up in a stage where coordination is required; either by the phase coming to an end or by a need of coordination within the phase (represented by the thin lines between the task boxes. The team-Task-Role model is an effort to illustrate joint activity in a more elaborate and descriptive way then the task list presented earlier in this paper.

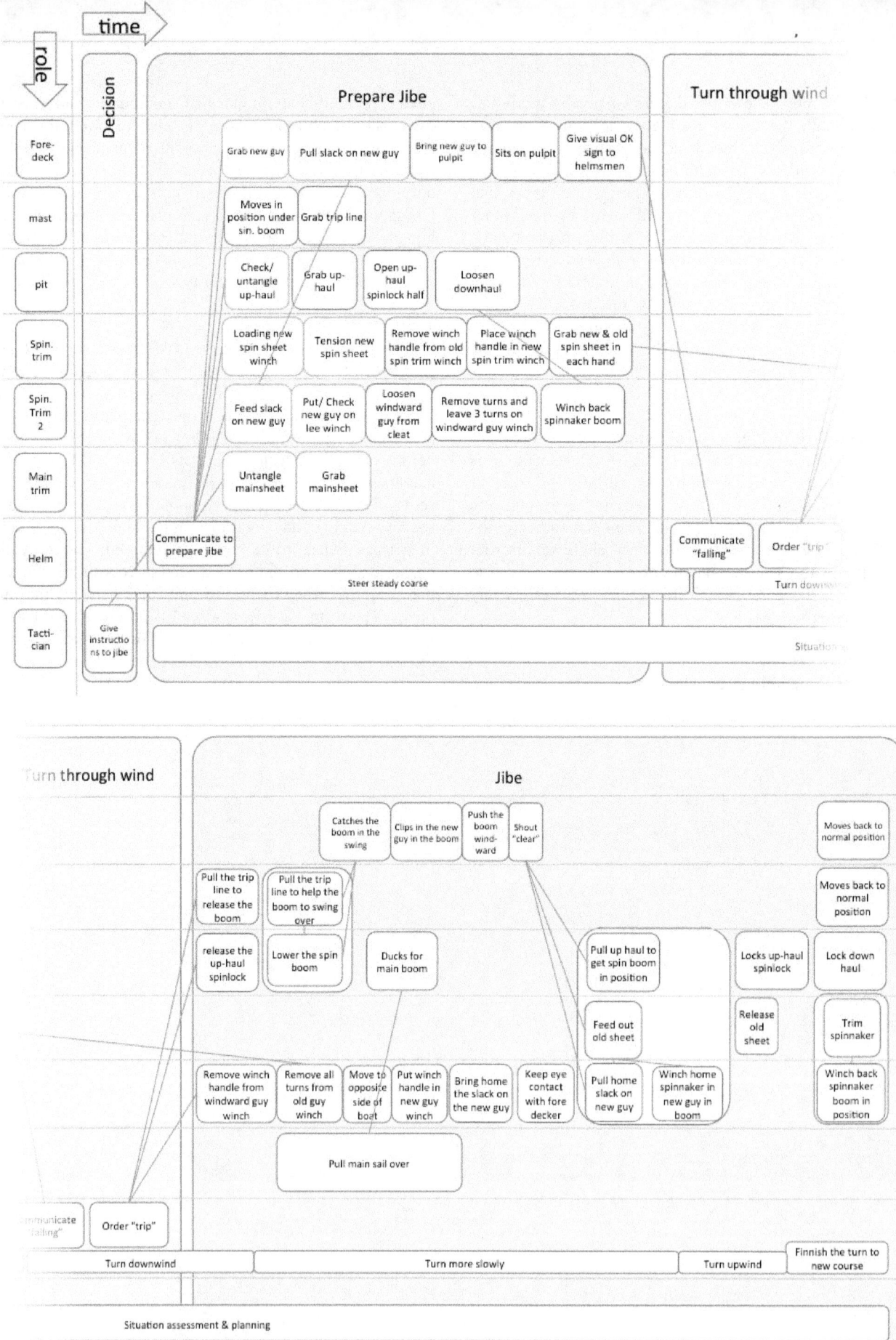

Figure 6. Time-task-role analysis of a jibe with an IMX40

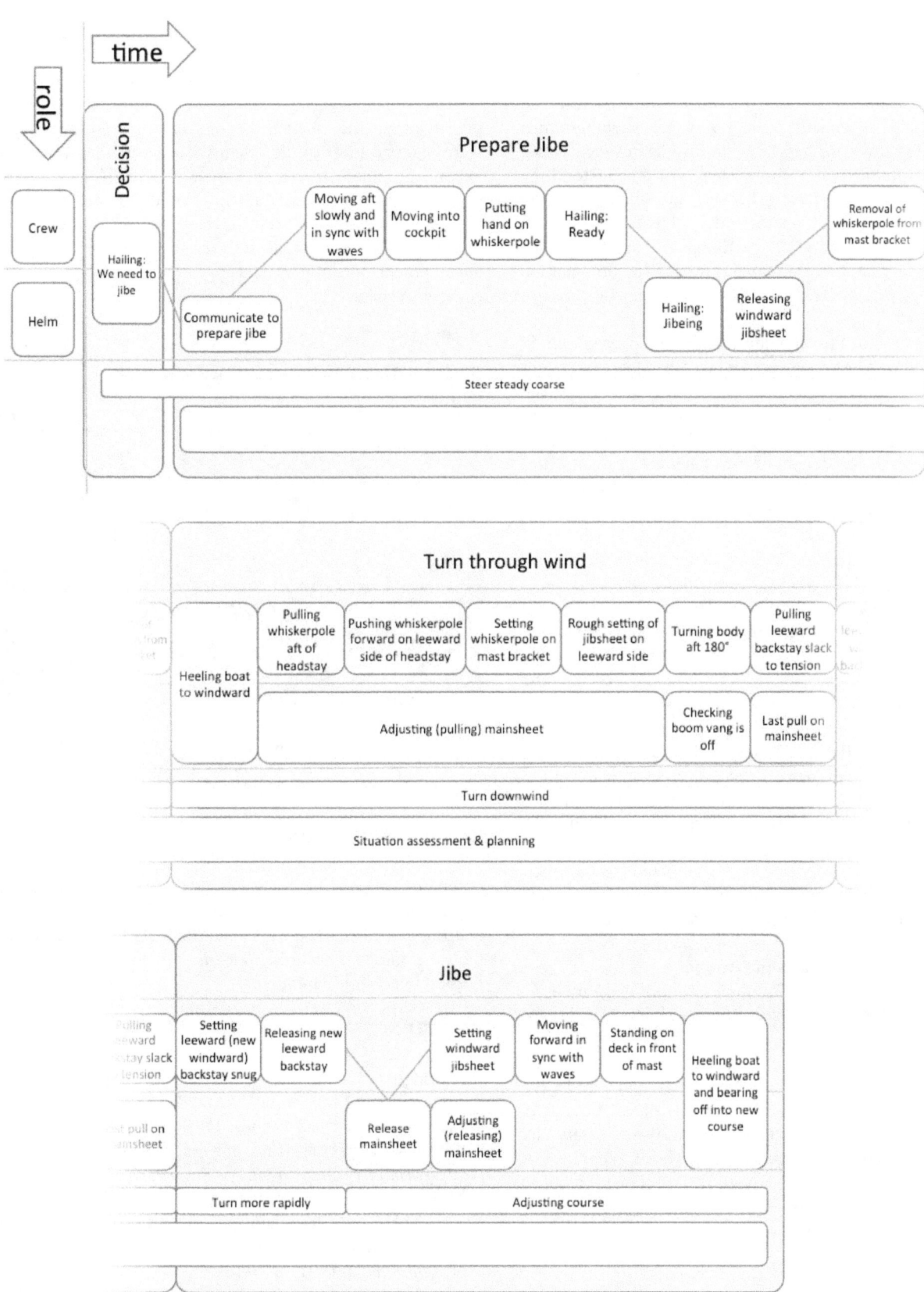

Figure 7. Time-task-role analysis of a jibe with a Star Boat

DISCUSSION

Adaptability of a sailing crew

The roles and distribution of tasks within a crew can create an interdependent crew. Every crewmember must complete his tasks for any other to do his and vice versa. Disruptions in one end of this system will propagate in the other. This is not a bad thing, because any small disruption will likely be caught before they have any serious implications for safety or performance, thanks to the interdependent model of organizing work. The propagation is signaling, which in turn bears on inter-predictability and directability.

An example of this interdependence and propagation in the IMX40's system occurs when the tactician goes unexpectedly silent. The helmsmen or other crew are predicting an action from the tactician that does not occur, which signals to the rest of the crew that the tactician isn't in phase with the procedure of the maneuver or plan. Thus, ,the other crew can notice this and try to compensate by creating time to repair the potentially corrupted/degraded Common Ground.

The idea of organizing work to achieve intrinsic control through Joint Activity also says something about the locus of responsibility. In a hierarchical organization's purest form, initiatives are seen as disobedience (Granström, 2006). In the example above, it is individual initiative that enables reparation of Common Ground. By maintaining Common Ground, the locus of responsibility distributed among the team becomes a key factor for safety and performance (Rochlin, 1999) (Klein et al., 2004).

To a certain extent, heterogeneity is an asset to a team as it enables seeing things from different perspectives and can act as a barrier to counter crewmember bias or false beliefs. On the other hand, when advanced coordination is needed, participants must not be too heterogeneous in their experience and competence so that they can interpret the situation similarly and be directable and predictable.

Loss and Repair of Common Ground

To maintain or repair Common Ground in competitive sailing, the participants must act upon suspicion of threats. Uncertainty in itself is a threat on a meta-level. Uncertainty is a cue to act upon to enable the repair of Common Ground. The loss of Common Ground and the consequences of its loss are what actually must be avoided. Losing Common Ground entails a breakdown in safety and performance and not the consequence itself. The loss of Common Ground can have negative outcomes, such as collisions or groundings, but they should be seen as symptoms rather than threats.

CONCLUSIONS

A maneuver, such as a jibe in sail racing, is a joint activity in its truest sense. A successful maneuver requires

Common Ground, coordination, and interdependence. Success or failure is intimately dependent on how jointly the team is able to work. Knowing the principles of effectively maneuvering a sail racing yacht is explicit. With this in mind, the curriculum for training crews can help them become more skilled then what can be learned from experience alone. If a model of how effective teamwork (joint activity) is constructed and provided to a crew, such a model can help the crew develop its capabilities. This model can act as a goal and set the direction for skill development, thereby increasing the vessel and crew's competitiveness.

Figure 8 – One of the authors crewing and setting the whiskerpole in a Star

ACKNOWLEDGEMENTS

Photos have been used with permission from www.regattaactiveimages.com and Jan Walker (front page and Figures 3, 4, 5 and 8). The authors also wish to extend their gratitude towards Västra Götalandsregionen – Regionutvecklingsnämnden.

REFERENCES

Bainbridge, L. (1983). Ironies of automation. *Automatica, 19*(6), 775-779.

Christoffersen, K., and Woods, D.D. (2002). How to make automated systems team players. *Advances in Human Performance and Cognitive Engineering Research, 2,* 1-12.

Clark, H.H., & Brennan, S.E. (1991). Grounding in communication. *Perspectives on Socially Shared Cognition, 13*(1991), 127-149.

Clark, H. (1996). *Using Language*. Cambridge: Cambridge University Press.

Granström, K. (2006). Dynamik I arbetsgrupper. Lund: Studentlitteratur AB.

Klein, G.A., Feltowich, P.J. & Woods, D.D. (2004)

Common Ground and coordination in joint activity. In W.Rouse & K. Boff (eds), *Organisation Simulation* (pp.139-84). Chichester, UK: John Wiley & Sons.

Osvalder, A. & Ulfengren, P. (2009) Human technology systems. In *Work and Technology on Human Terms* (pp. 377-376). Prevent.

Reynolds, M. & Szabo, G. (2009) Star tuning guide. Quantum Sails, San Diego. http://www.quantumonedesign.com/UserFiles/files/Star%20Tuning%20Guide%202009.pdf

Rochlin, G.I. (1999). Safe operation as a social construct. *Ergonomics, 42*(11), 1549-1560

Bibliography of Prior Chesapeake Sailing Yacht Symposia Papers

The First CSYS- January, 1974
• On the Handicapping of Distance Racing Yachts – A Proposal for IOR IV. Fisher, Bennett
• Handicapping Rules and Performance of Sailing Yachts. Letcher, John S., Jr.
• Analysis of Chesapeake Bay Racing Results 1972 and 1973 Seasons. Peach, Robert W.
• Yacht Rating. Strohmeier, Daniel D.
• Measurement Parameters of the I.O.R. Rule. Stephens, Olin J., II
• Booms Are Obsolete. MacLear, Frank R.
• A Breakthrough - Slotted Headsail Luff Support Systems. MacKenzie, Alan
• Some Observed Effects of Foil Control on Hydrofoil Sailing Vehicle Performance. Bradfield, W. S.
• Scale Experiments with the 5.5 Metre Yacht Antiope. Kirkman, Karl L.
• The Performance of Sailing Yachts in Oblique Seas. Pedrick, David R.
• Directional Stability and Control of Sailing Yachts. Scott, Walter H., Jr.
• Aerodynamics of High Performance Wing Sails. Scherer, Otto

The Second CSYS - January, 1975
• Theory Of Sailing Applied to Ocean Racing Yachts. Myers, Hugo A.
• America's Cup 1974 - An Overview- Organizing to Win. du Moulin, Richard T.
• America's Cup 1974- An Overview- Design and Construction- The First Race. Pedrick, David R.
• A Cruising Boat. Curtze, Charles A., RADM.
• Vane Self-Steerers for Cruising Yachts. Ratcliffe, Gererd
• Extended Cruising - An Overview. Court, Kenneth E.
Seakeeping and the Sailing Yachtsman. Compton, Roger, Johnson, Bruce, and Van Duyne, Carl
• Yacht Keels - An Experimental Study. DeSaix, Pierre
• Kevlar 49 Aramid, A New Material for Boat Hull Construction. Miner, Louis H., Wolffe, Robert A., and Woodrick, James V.
• Flotation for Ballasted Sailing Yachts?. Wilson, Vance O. and Reuter, Wolfgang
• America's Cup 1974 - An Overview- Racing for the Cup. Herreshoff, Halsey Chase

The Third CSYS - January, 1977
• Aluminum Construction. Wyland, Gilbert
• Sailing Yacht Construction in Fiberglass. Goman, William J.
• Surfing: Motions of a Vessel Running in Large Waves. Letcher, John S., Jr.
• The Preservation of Chesapeake Bay Watercraft. Baker, William A.
• Ocean Racing. Strohmeier, Daniel D.
• Principles of Sail Design. Haarstick, Stephen
• Wing Sail Versus Soft Rig: An Analysis of the Successful Little America's Cup Challenge of 1976. Bradfield, W. S. and Madhavan, Suresh
• Analyzing a Yacht for Hydrodynamic Characteristics that Affect What Type of Sails and Rigs Will Work Best. Doyle, Robert E.

The Fourth CSYS - January, 1979
• Evolution of Offshore Ratings - To the Limit. Pedrick, David R.
• A Summary of the H. Irving Pratt Ocean Race Handicapping Project. Kerwin, Justin E., Newman, J. N.
• The Measurement Handicapping System of USRYU. Strohmeier, Daniel D.
• Selecting a Keel Appendage for a Cruising Yacht. Berman, Deborah W.
• Yacht Structural Design for Light Scantlings. Herreshoff, Halsey C.
• Theoretical Estimation of the Influence of Some Main Design Factors on the Performance of International Twelve Meter Class Yachts. van Oossanen, Peter
• Photographic Essay: Ship Training on the Gazella Primeiro. Roewe, George J., Jr.
• A Microcomputer Beats to Windward. Clauser, Milton U.
• A Computer-Based Method for Analyzing the Flow Over Sails. Thrasher, D.F., Mook, D.T., and Nayfeh, A.H.
• The Evolving Role of the Towing Tank. Kirkman, Karl L.

The Fifth CSYS - January, 1981
• Design Development of a 40m Sailing Yacht. Benford,Jay R.
• Yacht Performance Analysis with Computers. Pedrick, David R. and McCurdy, Richard C.
• Geometry of Sailmaking. Andresen, Ted
• Sailing Yacht Capsizing. Stephens, Olin J., II, Kirkman, Karl L., and Peterson, Robert S.
• Kinetics in Small Boat Racing. Smith, Peter G.
• RED HERRING, High Performance Cruising Ketch. Hubbard, David W.

Bibliography of Prior Chesapeake Sailing Yacht Symposia Papers

• Mathematical Hull Design for Sailing Yachts. Letcher,John S., Jr.
• A Design Guide for Estimating Speed Made Good. Berman, Deborah W.

The Sixth CSYS - January, 1983
• Analysis of Steady Flow Over Interacting Sails. Register, David S. and Irey, Richard K.
• Yacht Design With Computers: New Methods for New Tools. Hazen, George S. and Killing, Steve
• Marine Electronic Navigation- A General Overview. Closs, Thomas H., Jr.
• Sailing Yacht Capsizing. Kirkman, Karl L., Nagle, Toby J., and Salsich, Joseph O.
• Rowing and Sailing Craft of the Chesapeake Bay. Tilp, Frederick
• Design and Engineering Aspects of Free-Standing Masts. Sponberg, Eric W.
• FRP Bottom Blistering. Fraser-Harris, A.B.F., COM and Kyle, James H.

The Seventh CSYS - January, 1985
• Experimental Analysis of Five Keel-Hull Combinations. Gerritsma, J. and Keuning, J. A.
• Sailboat Bow Impact Stresses. Ward, Lawrence W.
• Selection Criteria for Plastics Used in Through-Hull Fitting. Fraser-Harris, A.B.F.,COM and Leyden, Jerry J.
• Sailboards, Inventions, Yachts, and Exotic Craft. Russell, Diana
• Extended Cruising the Second Time Around. Court, Kenneth E.
• The Calculation of Sail Panels Using Developed Surfaces. Clemmer, George
• Stress Analysis for Light Alloy 12M Yacht Structures Comparison Between a Transverse and a Longitudinal Structure. Boote, Dario, Ruggiero, Vincenzo, Sironi, Nicola, Vallicelli, Andrea, and Finzi, Bruno
• The Development of the 12 Meter Class Yacht Australia II. van Oossanen, Peter

The Eighth CSYS - March, 1987
• The Application of VPPs to Practical Sailing Problems. Kirkman, Karl L.
• An Assessment of the Progress in Yacht Design Through an Examination of Model Yacht Characteristics. Claughton, Andrew, Howlett, Ian, Stollery, Roger
• Dinghy Design and the International Fourteen. Ames, Robert M. and Weiss, Paul F.

• The Comparison of Potential Driving Force of Various Rig Types Used for Fishing Vessels. Marchaj, C.A.
• Brushfire - An Experience in Building a Masthead Cutter. Williams, John J.
• Keel Design for Low Viscous Drag. Obara, Clifford J. and Van Dam, C.P.
• The Interpretation of Results from Tank Tests on 12m Yachts. Campbell, Ian and Claughton, Andrew
• The Analysis of Wave Resistance in the Design of 12 Meter. Scragg, Carl A., Chance, Britton, Jr., Talcott, John C., and Wyatt, Donald C.
• Stars & Stripes '87; Computational Flow Simulations for Hydrodynamic Design. Boppe, Charles W., Rosen, Bruce S., Laiosa, Joseph P., and Chance, Britton, Jr.
• Data Collection and Analysis for the 1987 Stars & Stripes Campaign. Letcher, John S. and McCurdy, Richard C.

The Ninth CSYS - March, 1989
• Fiber Reinforced Plastic Sailing Yachts - Some Aspects of Structural Design. Curry; Robert
• Guides to the Approximation of Sailing Yacht Performance. Stephens, Olin J.
• Scientific Sail Shape Design. Greeley, David S.,
• Kirkman, Karl L., Drew, Alan L., and Cross-Whiter, John
• The Design and Construction of the Second Pride of Baltimore. Gillmer, Thomas C.
• The Planning, Design and Construction of the 44-foot Offshore Training Craft for the U.S. Naval Academy. McCurdy, Ian and Bonds, John
• The Effect of Counter Length on Hull Resistance. Claughton, Andrew R.
• Performance Prediction Method for Multihull Yachts. Oliver, Clay

The Tenth CSYS - February, 1991
• Gyradius Measurements of Olympic Class Dinghies and Keel Boats. Hinrichsen, Prof. Peter F.
• Structural Design and Construction of America's Cup Class Yachts. Reichard, Prof. Ronnal P.
• The Delft Systematic Yacht Hull (Series II) Experiments. Gerritsma, Prof. ir. J., Keuning, Ir. J., and Onnink, A. R.
• A New Technique for Testing a Sailing Yacht in Waves. Kapsenberg, G. K.
• Magic III - An Old Man's Day Sailer. Miller, Capt. Richards T. and Whitacre, Harold M., III
• A Numerical Approach to the Design of Sailing Yacht Masts. Boote, Dario and Caponetto, Mario

Bibliography of Prior Chesapeake Sailing Yacht Symposia Papers

• Model Test Techniques Developed to Investigate the Wind Heeling Characteristics of Sailing Vessels and Their Response to Gusts. Deakin, Barry
• Sailboat Performance in a Current. Nolan, James P.
• Sailboat Hydrodynamic Drag Source Prediction and Performance Assessment. Boppe, Charles W.
• The Effect of Pitch Gyradius on Added Resistance of Yacht Hulls. Moran, James F.

The Eleventh CSYS - January, 1993
• Applications of Relational Geometric Synthesis in Sailing Yacht Design. Letcher, John S., Jr. and Shook, D. Michael
• Refinements in the Techniques of Tank Testing Sailing Yachts and the Processing of Test Data. Teeters, James R.
• SPLASH Free-Surface Flow Code Methodology for Hydrodynamic Design and Analysis of IACC Yachts. Rosen, Bruce S., Laiosa, Joseph P., Davis, Warren H., and Stavetski, David
• IACC Appendage Studies. Tinoco, E. N., Gentry, A. E.,Bogataj, P., Sevigny, E. G., and Chance, B.
• Modeling IACC Sail Forces by Combining Measurements with CFD. Milgram, Jerome H., Peters, Donald B., and Eckhouse, D. Noah
• Towards a Rational Upwind Sail Force Model for VPPs. Euerle, Steven E. and Greeley, David S.
• Numerical Approach to Aeroelastic Responses of Three-Dimensional Flexible Sails. Fukasawa, Toichi and Katori, Masanobu
• A Review on Il Moro di Venezia Design. Caponnetto, Mario
• The Nippon Challenge America's Cup 1992-Progress in Hull Development. Nagami, Yoshihiro.
• How to Go Cruising. Hays, James O. and Hays, Anne M.
• Notes on Sailing Ship History: Academy Versus Shipyard. Stephens, Olin J., II
• The A-Class Catamaran: Development of Serious Fun. Beadling, Robert G. and Beadling, Walter H.
• Dynamic Performance of Sailing Cruiser by Full-Scale Sea Tests. Masuyama, Yutaka, Nakamura, Ichiro, Tatano, Hisayoshi, and Takagi, Ken
• Hazards and Challenges of Cruising the Northeast Coast of North America. Jordan, Edwin C. and Jordan, Mary K.
• The Partnership for America's Cup Technology: An Overview. Gretzky, James A. and Marshall, John K.
• Stars and Stripes Design Program for the 1992

America's Cup. Todter, Chris, Pedrick, David, Calderon, Alberto, Nelson, Bruce, Debord, Frank, and Dillon, Dave
• Elements of Resistance of IACC Yachts. Milgram, Jerome H. and Frimm, Fernando C.
• Sailing Yacht Performance in Calm Water and in Waves. Gerritsma, Prof. ir J., Keuning, Ir. J. A., and Versluis, A.
• Seakeeping and Added Resistance of IACC Yachts by a Three-Dimensional Panel Method. Sclavounos, P. D. and Nakos, D. E.
• The Effects of Flare and Overhangs on the Motions of a Yacht in Head Seas. Kuhn, John C. and Schlageter, Eric C.
• Analysis of Lift and Drag on a Surface-Piercing Foil. Kuhn, John C. and Scragg, Carl A.
• Performance Prediction Software for IACC Yachts. Schlageter, Eric C. and Teeters, James R.

The Twelfth CSYS - January, 1995
• Scoring IMS Regattas - An Empirical Study of Alternative Methods. Cane, John W.
• Drawing with Performance Prediction. Schwenn, Peter and Hazen, George
• Design Criteria for Composite Masts. Miller, Paul
• The Development of the B&R Rig, Structural Space Frame and Tripod Support System with Integrated Boom. Bergstrom, Lars and Ridder, Sven O.
• The Alexandria Class Dinghy - A Design For Change. Hunley, William H.
• Design, Construction, and Performance of a 27' MORC Boat. Jones, Brian A.
• Imagine- an Open Class 60 BOC Racer- Design and Program Management- Lessons Learned. Court, Kenneth E. and Kaufman, F. Michael III and Whitacre, Harold M III
• The Design of Yacht Sailplans for Maximal Upwind Speed. Day, Dr. Sandy
• Tacking Simulation of Sailing Yachts - Numerical Integration of Equations of Motion and Application of Neural Network Technique. Masayuma, Yutaka and Fukasawa, Toichi and Sasagawa, Hiroshi
• Wing - Body Interaction on a Sailing Yacht. Keuning, Prof. ir. J. A. and Kapsenberg, Ir. G. K.
• Improvement of Sailing Yacht Performance Prediction by Including Force-Moment Equilibrium for the Calculation of Helm Angle in a Velocity Prediction Program. van Oossanen, Dr. Peter
• YACHT97: A Fully Viscous Nonlinear Free-Surface

Bibliography of Prior Chesapeake Sailing Yacht Symposia Papers

Analysis Tool for IACC Yacht Design. Farmer, J. and Martinelli, L. and Jameson A.

The Thirteenth CSYS - January, 1997
• On Test Measurements in Full Scale Sailing Test Programs. Howard P. Grant and Olin J. Stephens
• Full Scale Measurement of Sail Force and the Validation of Numerical Calculation Method. Yutaka Masuyama, and Toichi Fukasawa
• An Investigation of Full Scale Forces Produced by a Sail. Nathan Bossett and Ian Mutnick
• Optimisation of a Sailing Rig using Wind Tunnel Data. IMC Campbell
• Model Tests in Support of the Design of a 50 Meter Barque. Barry Deakin
• The Restoration of AVEL. Clark Poston
• BATOPERF, A Performance Prediction Software and Its Influence on Modern Yacht Design. Sylvain Fargeas and Juan Kouyoumdjian
• Development of Proposed ISO 12217 Single Stability Index for Mono-Hull Sailing Craft. Dr. Peter van Oossanen
• The Cogito Project: Design and Development of an International C-Class Catamaran and Her Successful Challenge to Regain the Little America's Cup. Duncan T. MacLane
• The Institute for Marine Dynamics Model Yacht Dynamometer. B. L. Parsons and R. Pallard
• Model Tests of the PACT Base America's Cup Hull in Following Seas, Jesse Falsone
• Appendage Resistance of a Sailing Yacht Hull. J. A. Keuning and B-J. Binkhorst
• Hull - Appendage Interaction of a Sailing Yacht, Investigated with Wave Cut Techniques. Jonathan R. Binns, Kim Klaka and Andrew Dovell
• SPLASH Nonlinear and Unsteady Free-Surface Analysis Code for Grand Prix Yacht Design. Bruce S. Rosen and Joseph P. Laiosa
• The Effect of Pitch Moment of Inertia in Body Axes on the Performance of a Yacht in Waves. C.J. Sutcliffe and A. Millward
• Experimental Determination of Sail Performance and Blockage Corrections. Dr. Robert Ranzenbach and Chris Mairs

The Fourteenth CSYS- January 1999
• Developments in the IMS VPP Formulations. Andrew Claughton
• Experimental Technique for the Determination of Forces Acting on Sailboat Rigging. F. Fossati, G. Moschini, and D. Vitalone

• Fullscale Hydrodynamic Force Measurement on the Berlin Sailing Dynamometer. Karsten Hochkirch and Hartmut Brandt
• USS Constitution Preparations for Sail 200. Howard Chatterton
• In Search of Power, Pace and Windward Performance In Square Rigged Sailing Ships. Philip Goode
• The Windward Performance of Yachts in Rough Water. Jonathan Binns, Bruce McRae, and Giles Thomas
• A 1997-1998 Whitbread Sail Program - Lessons Learned. Robert C. Ranzenbach, Per Andersson and David Flynn
• Sailing Yacht Design Using Advanced Numerical Flow Techniques. Caponnetto et al
• Parametric Design and Optimization of Sailing Yachts. Stefan Harries and Claus Abt
• An Investigation of the Structural Dynamics of a Racing Yacht. Frederic Louarn and Pandeli Temarel
• Use of CFD Techniques in the Preliminary Design of Upwind Sails. Patrick Couser and Norm Deane
• On the Application of RANS Simulation for Downwind Sail Aerodynamics. William Lasher
• Wind Tunnel Testing of Offwind Sails. Robert Ranzenbach and Chris Mairs
• Approximation of the Calm Water Resistance on a Sailing Yacht based on the "Delft Systematic Yacht Hull Series." J.A. Keuning and U.B. Sonnenberg

The Fifteenth CSYS - January 2001
• The Re-Righting of Sailing Yachts in Waves- A Comparison of Different Hull Forms. Martin Renilson, Jonathan R. Binns and Andrew Tuite
• An Improved Upwind Sail Model For VPP's. Peter Jackson
• On-the-water Measurement of Laminar to Turbulent Boundary Layer Transition on Sailboat Appendages. E.A. Lurie
• International America's Cup Class Yacht Design Using Viscous Flow CFD. Paul Jones and Rich Korpus
• Hydrodynamic Modeling of Sailing Yachts. Stefan Harries, Claus Abt and Karsten Hochkirch
• The Effect of Bow Steepness and Flare on the Resistance of Sailing Yachts in Calm Water and Waves. J.A. Keuning, R. Onnink and A. Damman
• A Time-Domain Simulation for Predicting the Downwind Performance of Yachts in Waves. Dougal Harris, Giles Thomas and Martin Renilson

Bibliography of Prior Chesapeake Sailing Yacht Symposia Papers

• An Experimental Investigation of Slamming on Ocean Racing Yachts. Paolo Manganelli and Philip A. Wilson
• Optimising Yacht Routes Under Uncertainty. Andy Philpott and Andrew Mason
• PCSAIL, A Velocity Prediction Program for a Home Computer. David E. Martin and Robert F. Beck
• Schooner *Brilliant* Sail Coefficients and Speed Polars. Howard Grant, Walter Stubner, Walter Alwang, Charles Henry, John Baird and Paul Spens
• Sailing Performance of "Naniwa-maru" - A Full Scale Reconstruction of a Sailing Trader of Japanese Heritage. Kensaku Nomoto, Yutaka Masuyama and Akira Sakurai
• The *Basiliscus* Project - Return of the Cruising Hydrofoil Sailboat. Thomas E. Speer
• Model Tests to Study Capsize and Stability of Sailing Multihulls. Barry Deakin
• Aerodynamic Performance of Offwind Sails Attached to Sprits. Robert Ranzenbach and Jim Teeters

The Sixteenth CSYS – March 2003
• The Yaw Balance of Sailing Yachts Upright and Heeled. J. A. Keuning and K. J. Vermeulen
• Computational Fluid Dynamics for Downwind Sails. Horst J. Richter, Kevin C. Horrigan and J.B. Braun
• Downwind Load Model for Rigs of modern Sailing Yachts for Use in FEA. Guenter Grabe
• Analysis of Hull Shape Effects on Hydrodynamic Drag in Offshore Handicap Racing Rules. Jim Teeters, Rob Pallard and Caroline Muselet
• Changes to Sail Aerodynamics in the IMS Rule Jim Teeters, Robert Ranzenbach and Martyn Prince
• On the Use of CFD to Assist with Sail Design. Andrea Schneider, Andrea Arnone, Marco Savelli, Andrea Ballico and Paolo Scutellaro
• Numerical Simulation using RANS-based Tools for America's Cup Design. Geoff Cowles, Nicola Parolini and Mark L. Sawley
• Sailing Yacht Design for Maximum Speed. Bob Dill
• Composite Sail Batten Design . Audrey Sery and Jean Paul Charles
• Analysis of 2D Coupled Sails: Use of an Optimization Technique Based on Turbulent Viscous Flows. Giovanni Lombardi, Francois Beux and Mattia de. Michieli Vitturi
• Experimental Study of a Directionally Stable Sailing Vehicle with a Free-Raking Rig and a Self-

Trimming Sail. Akira Sakurai, Takeshi Nakamura and Yuya Nakamoto
.• Student Research Projects for the New Navy 44 Sail Training Craft. Paul H. Miller and NAOE Naval Architecture and Ocean Engineering Department
• Experimental Force Coefficients for a Parametric Series of Spinnakers. William C. Lasher, James R. Sonnenmeier, David R. Forsman, Cheng Zhang and Kenton White
• The Rise of the Hydrofoil and the Displacement of the Hull: The Design, Construction and Performance Measurement of a 6m Flying Catamaran. Edward Chapman and George Chapman
• Numerical Simulation of Maneuvering of "Naniwamaru," A Full-scale Reconstruction of Sailing Trader of Japanese Heritage. Yutaka Masuyama, Kensaku Nomoto and Akira Sakurai

The Seventeenth CSYS – March 2005
• Toward Numerical VPP with the Full Coupling of Aerodynamic and Hydrodynamic Solvers for ACC Yachts. Erwan Jacquin, Yann Roux, Bertrand Allessandrini
• Time Domain Simulation of a Yacht Sailing Upwind in Waves. D. H. Harris
• Geometry and Resistance of the IACC Systematic Series "Il Moro di Venezia" D. Battistin, D. Peri, E. Campana
• Sailing Yacht Rig Improvements Through Viscous Computational Fluid Dynamics. Vincent G. Chapin, Romaric Neyhousser, Stephane Jamme, Guillaume Dulliand, Patrick Chassaing
• A New Velocity Prediction Method for Post-Processing of Towing Tank Test Results. Kai Graf, Christoph Bohm
• Hull Form Optimization of Performance Characteristics of Turkish Gulets for Charter. Mark Gammon, Abdi Kukner, Ahmet Alkan
• The Development of an Integrated Ship Design Environment for the Naval Architect on The Linux Operating System. H. James Parker
• Multiobjective Design Optimization of an IACC Sailing Yacht by Means of CFD High- Fidelity Solvers. Daniele Peri, Fabrizio Mandolesi
• Comparison of Tacking and Wearing Performance Between a Japanese Traditional Square Rig and a Chinese Lug Rig. Yutaka Masuyama, Akira Sakurai, Toichi Fukasawa, Kazunori Aoki
• Relative Performance of Conventional Versus Movable-Ballast Racing Yachts. Frank DeBord, Harry Dunning

Bibliography of Prior Chesapeake Sailing Yacht Symposia Papers

• The Effect of Mast Height and Centre of Gravity on the Re-Righting of Yachts. Jonathan R. Binns, Paul Brandner
• A Generic Mathematical Model for the Maneuvering and Tacking of a Sailing Yacht. J. A. Keuning, K. J. Vermeulen, E. J. de Ridder
• Experimental Methods to Evaluate Underwater Appendages. Robert Ranzenbach, Mathew Zahn
• A Velocity Prediction Program for a Planing Dinghy. Todd Carrico
• Sail Aero-Structures: Studying Primary Load Paths and Distortion. Robert Ranzenbach, Zhenlong Xu Light Weight Sandwich Panels for Yacht Hull Structures. M. C. Rice, C. A. Fleischer, D. D. R. Cartie, Marc Zupan

The Eighteenth CSYS – March 2007
• A Combined Ship Science-Behavioural Science Approach To Create a Winning Yacht-Sailor Combination. Matteo Scarponi, R Ajit Shenoi, Stephen R Turnock, and Paolo Conti
• Database of Sail Shapes vs. Sail Performance and Validation of Numerical Calculation for Upwind Condition. Yutaka Masuyama, Yusuke Tahara, Toichi Fukasawa, and Naotoshi Maeda
• Enhanced Wind Tunnel and Full-Scale Sail Force Comparison. Heikki Hansen, Peter J. Richards, and Peter S. Jackson
• Further Analysis of the Forces on Keel and Rudder of a Sailing Yacht. J. A. Keuning, M. Katgert, and K. J. Vermeulen
• RANSE Calculation of Laminar-to-Turbulent Transition-Flow around Sailing Yacht Appendages. Christoph Böhm and Kai Graf
• Performance Prediction without Empiricism: A RANS-Based VPP and Design Optimization Capability. Richard Korpus
• A Tool for Time Dependent Performance Prediction and Optimization of Sailing Yachts. D. Battistin and M. Ledri
• Hydrodynamic advice of Sailing Yachts through Seakeeping Study. Guilhem Gaillarde, Erik-Jan de Ridder, Frans van Walree and Jos Koning
• Slamming of Composite Yacht Hull Panels. Susan Lake, Michael Eaglen, Brian Jones, and Mark Battley
• ARPRO®: A New Structural Core Material for the Yacht Industry. Corrado Labriola and Vito Tagarielli
• PCLINES, A Parametric Lines Development Program for the Home Computer. David E. Martin

• "That Peculiar Property:" Model Yachting and the Analysis of Balance in Sailing Hulls. Earl Boebert
• SNAME's Stability Letter Improvement Project (SLIP) for Passenger Sailing Vessels. Jan C. Miles, Bruce Johnson, John Womack, and Iver Franzen
• An Aerodynamic Analysis of the U.S. Brig *Niagara*. William C. Lasher, Terrence D. Musho and Kent C. McKee and Walter Rybka,
• Analysis of Wave Making Resistance And Optimization of Canting Keel Bulbs. Karsten Hochkirch and Claudio Fassardi
• Added Resistance in Seaways and its Impact on Yacht Performance, Kai Graf, Marcus Pelz, Volker Bertram, and H. Söding
• Dynamic Lift Coefficients for Spade Rudders on Yachts. Paul H. Miller

The Nineteenth CSYS – March 2009
• CFD and VPP Challenges in the Design of the New AC90 Americas Cup Yacht. Kai Graf, Christoph Boehm and Hannes Renzsch
• A New Method for the Prediction of the Side Force on Keel and Rudder of a Sailing Yacht Based on the Results of the Delft Systematic Yacht Hull Series. J. A. Keuning and B. Verwerft
• CFD-Based Hydrodynamic Analysis of High Performance Racing Yachts. Len Imas, Bryan Baker, Britton Ward and Gregory Buley
• On the Choice of CFD Codes in the Design Process of Planing Sailing Yachts. Jérémie Raymond, Jean-Marie Finot, Jean-Michel Kobus, Gérard Delhommeau, Patrick Queutey and Aurélien Drouet
• Systematic Series of the IACC yacht "Il Moro di Venezia": Heel and Yaw Analysis. D. Peri, F. Di Ci'o,and M. Roccaldo
• Yacht Design Software 2.0: The Open Source Movement. Mathew Bird, William F. Cook, George S. Hazen and Britton Ward
• Upwind Sail Performance Prediction for a VPP Including "Flying Shape" Analysis. Brian Maskew and Frank DeBord
• Photogrammetric Investigation of the Flying Shape of Spinnakers in a Twisted Flow Wind Tunnel. Kai Graf and Olaf Müller
• Sails Aerodynamic Behavior in Dynamic Conditions. Fabio Fossati and Sara Muggiasca
• Assessing the Wind-Heel Angle Relationship of Traditionally-Rigged Sailing Vessels. William C. Lasher, Diana R. Tinlin, Bruce Johnson , John Womack, Jan C. Miles, Walter Rybka and Wes Heerssen

Bibliography of Prior Chesapeake Sailing Yacht Symposia Papers

• Development and Initial Review of the Mark II Navy 44 Sail Training Craft. Paul Miller, David Pedrick and Gram Schweikert

• Tacking in the Wind Tunnel. Frederik C. Gerhardt, David Le Pelley, Richard G. J. Flay and Peter Richards

• Full Scale Measurements on a Hydrofoil International Moth. Bill Beaver and John Zseleczky

• *Alpha* and *Rocker* - Two Design Approaches that led to the Successful Challenge for the 2007 International C-Class Catamaran Championship. Steve Killing

The Twentieth CSYS – March 2011

• A Multidisciplinary Computational Framework for Sailing Yacht Rig Design & Optimization through Viscous, Vincent G. Chapin, Nolwenn de Carlan, Peter Heppel

• Experimental Full Scale Study on Yacht Sails and Rig under Unsteady Sailing Conditions and Comparison to Fluid Structure Interaction Unsteady Model, Benoît Augier, Patrick Bot, Frederic Hauville

• Photogrammetry Based Flying Shape Investigation of Downwind Sails in the Wind Tunnel and at Full Scale on a Sailing Yacht, Johannes Mausolf, Julien Deparday, Kai Graf, Hannes Renzsch, Christoph Böhm

• Pressure Distributions on Sails Investigated Using Three Methods: On-Water Measurements, Wind-Tunnel Measurements, and Computational Fluid Dynamics, Ignazio Maria Viola, Richard GJ Flay

• Advancement in the Application of Finite Element Analysis to the Optimization of Composite Yacht Structures, David Fornaro

• Free-Surface Effects of Variations in Appendage Vertical Volume Distribution: Where does a Bulb not See the Free-Surface?, Jonathan R. Binns, Robert Thompson, Paul A. Brandner, Leonard Imas

• RVPP: Sailing Yacht Performance Prediction fully integrated into a RANSE based flow code, Christoph Böhm, Kai Graf

• Effective Wind Tunnel Testing of Yacht Sails Using a Real-Time Velocity Prediction Program, David Le Pelley, Peter Richards

• To Twist, or Not to Twist? – A Scientific Attempt to Understand What We Think We Already Know About Sail Trim, William C. Lasher, Robert Ranzenbach

• The Effects of Streamlined Rigging on Sailboat Performance, Bruce J. Martin, Grzegorz P. Filip, Kevin J. Maki, Robert F. Beck, Eric Hall

• An Investigation into the Efficiency of Gybing Daggerboards, Douglas Slocum

• An Investigation of Transom Lift Devices on High Performance Open-Class Sailing Yachts, Seth Cooley

• The Use of Sailing Simulation to Increase Participation, Jonathan R. Binns, Mark Habgood, Norman R. Saunders, Paula Cunningham, John Mooney

• Sizing and Analysis of Folding Propellers for Auxiliary-Powered Sailboats, Donald MacPherson, Benjamin Segil, Sune Ehrenskjold

The Twenty First CSYS – March 2013

• Investigation Of Scale Effects In Sailing Yacht Performance Prediction By Numerical And Experimental Methods, Mustafa Insel And Ziya Saydam

• Least Squares Estimation of Sailing Yacht Dynamics from Full-Scale Sailing Data, Katrina Legursky

• Delft Systematic Yacht Hull Series Presentation, Lex Keuning

• Keelboat Yaw Gyradius Measurement, Peter Hinrichsen

• A New Real-time Method for Sailboat Performance estimation based on Leeway Modeling, Ronan Douguet, Jean-Philippe Diguet, Johann Laurent, Yann Riou

• Mainsail Planform Optimization For IRC 52 Using Fluid Structure Interaction, Robert Ranzenbach, Dave Armitage and Adolfo Carrau

• An Experimental Validation Case For Fluid-Structure-Interaction Simulations Of Downwind Sails, H. Renzsch, K. Graf

• A Refinement of the Method Used to Determine the Balance of a Sailing Vessel During the Design Phase, with Application to Sail Design and Subsequent Sail Selection and Sailing Operations, Capt. Iver Franzen

• Uncertainties In The Wind-Heel Analysis For Traditional Sailing Vessels: The Challenges It Presents For Forensic Analysis Of Sailing Vessel Incidents, Bruce Johnson, William Lasher, Matt Erdman, and Jan Miles

• The Evolution of Design: SALTS New Sail Training Schooner Project, Stephen Duff, Fabio

Fossati, Andy Claughton, Will Krzymowski, Tony Anderson

- On The Hydrodynamics Of A Skiff At Different Crew Positions, Ignazio Maria Viola And Joshua Enlander
- A Measurement System for Performance Monitoring on Small Sailing Dinghies, Christoph Boehm, Robert Brehm, Janek Meyer,Lars Duggen, Kai Graf
- A Wind Tunnel Study Of The Interaction Between Two Sailing Yachts, P.J. Richards, D.J. Le Pelley, D. Jowett, J. Little, O. Detlefsen
- The Development of the New Volvo Class, Britton Ward, Chris Cochran and Farr Yacht Design, Ltd.

www.ingramcontent.com/pod-product-compliance
Lightning Source LLC
Chambersburg PA
CBHW080616190526
45169CB00009B/3203